Applied Risk Management in Agriculture

Applied Risk Management in Agriculture

Dana L. Hoag

CRC Press is an imprint of the
Taylor & Francis Group, an **informa** business

CRC Press
Taylor & Francis Group
6000 Broken Sound Parkway NW, Suite 300
Boca Raton, FL 33487-2742

Printed in the United States of America on acid-free paper
10 9 8 7 6 5 4 3 2 1

International Standard Book Number: 978-1-4398-0973-0 (Hardback)

Library of Congress Cataloging-in-Publication Data

Hoag, Dana L.
Applied risk management in agriculture / Dana L. Hoag.
p. cm.
Includes bibliographical references and index.
ISBN 978-1-4398-0973-0
1. Agriculture--Economic aspects--United States. 2. Risk management--United States. I. Title.

HD1761.H57 2010
630.68'1--dc22 2009015508

Visit the Taylor & Francis Web site at
http://www.taylorandfrancis.com

and the CRC Press Web site at
http://www.crcpress.com

Contents

Preface

Over the last few years I have witnessed a great deal of interest in risk management education. However, based on my participation in more than fifty Ag Survivor educational and extension programs, I could not understand why more agricultural producers were not using risk management tools when so many were readily available. My suspicion was that the topic required the integration of many disparate and difficult parts, which is time consuming and overwhelming for producers who are already giving it their all to maintain a viable business. So, I set out to write a book about risk management and the fundamentals of building a risk management plan.

I was encouraged when I conceived (or stumbled on) the idea of using strategic management to organize the process of risk management. Strategic management has been around for a long time and has been widely accepted inside and outside the business community. This approach provided an almost ideal platform for organizing risk, since risk management is a business decision. As I developed the idea, the program grew into ten steps that fell neatly into the three traditional stages of strategic risk management. Fortunately, these steps could still accommodate a breadth of information in a practical, formulaic framework.

Over time, other colleagues joined me in order to bolster the quality of information I could offer. Their skills proved complementary and they expanded the risk management resources we could make available to readers. We also added Aaron Sprague, who was working on his master's degree with me, as a case study. Aaron and his family were gracious enough to allow us to use their actual farm, EWS Farms, and family information to test and demonstrate the SRM process. Although using a single case study about a single corn-wheat farm makes it hard for people to relate to if they don't grow the same crops, using one case study adds consistency and clarity to make the process flow from one step to the next. That is, we can connect the Sprague's financial health from Step 1 to their choices on risk management in Step 7, rather than using a corn example in Step 1 and tomato example in Step 7. The concepts are the same, whether we are talking about farming corn or catfish. As you read on, think *catfish*, or whatever enterprise you are into, when you see the word *corn*, and concentrate on the bigger strategic risk management process. It is our experience that producers have a lot in common, even when they don't grow the same crop.

As we commenced writing, we measured EWS Farms from every angle so as to make our example meaningful. In so doing, we estimated that the mean corn price would be about $2.00 per bushel based on ten years of real and local data. Despite our diligence, the very next year prices skyrocketed, earning corn producers over $6 per bushel. In this case, the past did not do very well on predicting the future. We considered updating the book with new prices, but the high prices did not last. After two years, prices returned to the $2 to $3 range. This lesson confirms that a process such as the one demonstrated in this book can be useful, but the process needs to provide flexibility and to look at risk and risk management from multiple viewpoints, which we have tried to do here.

On behalf of all the authors, I hope you find the book to be a well-organized, complete source of information about risk management. We have made the book accessible and meaningful for agricultural producers, college students, and people outside of agriculture who can adapt what they find here to their own conditions. Section 2 of the book provides a complete description of all ten steps in the SRM process, an example of how it was applied to EWS Farms, and a user's guide for the free tools available at www.RiskNavigatorSRM.com. Go to this Web site, or our companion Web sites, www.RightRisk.org and www.AgSurvivor.com, for a comprehensive collection of risk management information related to agriculture. Section 3 provides a discussion of a more traditional view of risk management and has two chapters that will start you on your way to customizing your own risk management programs. With this information, a reader can learn about risk and build risk management plans with Risk Navigator SRM, supplement this framework with traditional tools, and then move on to customizing complicated problems with advanced tools when ready.

Dana L. Hoag
Professor, Colorado State University

Acknowledgments

We wish to thank some of the many people who helped us make this book and the companion Risk Navigator SRM Web site possible. Specifically, we acknowledge and thank:

The United States Department of Agriculture Risk Management Agency
Collaborating authors for their expertise and diligence
Jim Bradney, Eihab Fathelrahman, and Jay Parsons for programming the SRM Web site and tools
James Richardson, Texas A&M University, for helping us with Simetar©
The Spragues for opening their personal and business life to make the EWS Farms case study possible

Editor

Dana L. Hoag is professor of agricultural and resource economics at Colorado State University. Dr. Hoag grew up on a small acreage in Colorado. He studied at Washington State University and started his career at North Carolina State University. He has worked with many different types of producers, from swine to wine, specializing in agricultural production, risk management, natural resources, and policy. Dr. Hoag helped found RightRisk, Ag Survivor, and Risk Navigator SRM.

Contributors

James C. Ascough II, PhD
USDA-ARS-NPA, Agricultural Systems Research Unit
Fort Collins, Colorado

Eihab Fathelrahman
Consultant
Fort Collins, Colorado

Duane Griffith
Department of Agricultural Economics and Economics
Montana State University
Bozeman, Montana

John P. Hewlett
Department of Agricultural and Applied Economics
University of Wyoming
Laramie, Wyoming

Catherine Keske
Department of Soil and Crop Science
Colorado State University
Fort Collins, Colorado

Jay Parsons
Department of Agricultural and Resource Economics
Colorado State University
Fort Collins, Colorado

James Pritchett
Department of Agricultural and Resource Economics
Colorado State University
Fort Collins, Colorado

Aaron Sprague
Farmer
Holyoke, Colorado

Section I

Introduction

We begin this book with an overview of risk management that lays the groundwork for understanding the educational information provided in the following Sections II and III. After discussing risk concepts in Chapters 1 and 2, we describe the case study farm in Chapter 3. We take the time to describe this farm to show that it is a real farm with real people, and to provide the context for the risk management situations we apply to this farm in later chapters. We describe the Risk Navigator SRM© process in Chapter 4 to set the stage for Section 2, which describes the ten-step SRM process in full detail.

1 Introduction

Dana Hoag

CONTENTS

News flash: Agriculture is risky! If you don't believe that, don't waste your time reading this book. If you do, then get comfortable—you're in the right place. *Applied Risk Management in Agriculture* is designed to help decision makers better recognize risks, understand their personal attitude toward risk, and manage risks specific to their own operations. The information provided within these pages does not just inform and educate, but also enables managers to develop workable risk management plans and put them into action; it also can equip college students with the tools to do the same when they move into their careers.

About two thirds of the book will focus on our newly developed collaborative Internet program called Risk Navigator SRM© (www.RiskNavigatorSRM.com). SRM stands for Strategic Risk Management, which borrows from the strategic management literature in business. SRM is a ten-step risk management process that provides a meaningful and accessible method for serious agricultural producers and students alike to assess and manage the risks inherent in agriculture.

An organized program like SRM makes the complex world of risk management smaller and more manageable. However, as a reader's experience grows, so will the need to break beyond the SRM boundaries and adapt the program to more difficult and customized problems. Therefore, the other third of this book presents risk management concepts without reference to the SRM framework. The book concludes with two chapters that show readers how to move past the limits of SRM by customizing their own management planning tools.

1.1 RISK AND AGRICULTURE

The business of agriculture is probably more risky than other forms of employment because almost everything a farmer or rancher does is at the mercy of Mother Nature or fickle input and output markets. The agro-environment, which includes weather, pest infestations, and other natural phenomena, makes agriculture more risky by affecting how much is produced, how much it costs to produce, the quality of the product, and how much is paid for the product at market. When the stars align just

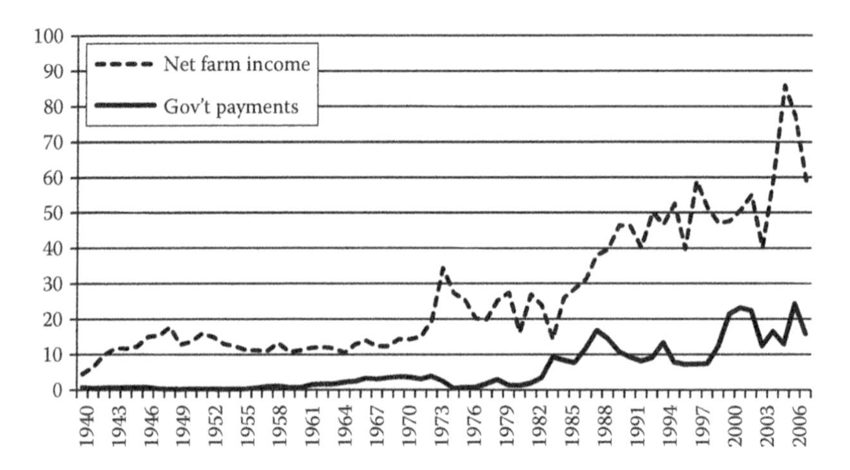

FIGURE 1.1 United States farm income, 1950–2006. Source: USDA Economic Research Service.

right, producers make money. But if a producer fails to produce a crop or faces low output or higher-than-expected input prices, he could find himself in financial ruin.

Consider net farm income for the United States as shown in Figure 1.1. Though it varies, in recent years it appears to hover around $50 billion per year; a lot of risk is hidden in these figures. In the 1996 Farm Bill, for example, the government made efforts to curb expensive commodity income and price support programs. At the time, farmers were happy to take the cuts in farm program payments because that meant they could grow more crops (since they would not be restricted on how much they could grow by government programs anymore) and take advantage of high prices. It did not take long, however, for farmers to overwhelm markets with surplus crops and drive the market prices so low that the government had to intervene with emergency price supports. In 1999, government support jumped from about $12 billion to $21 billion in an attempt to provide emergency support. Government payments comprised about half of the net farm income until 2002, when a new farm bill made appropriate adjustments. Without the government's support in these lean years, agricultural producers would have been devastated. Then, in a complete reversal, there were huge price spikes in 2006 and 2007. Who could have predicted that wheat could hit a record high of $25 per bushel on the Minneapolis Exchange and corn prices would hold at over $5 per bushel? Net farm income rose from $58 billion in 2006 to almost $87 billion in 2008. Would these prices hold or would prices return to historic levels? Corn prices had fallen back to the $2 per bushel range by the end of 2008. Predictably, farm income is forecast to fall to $71 billion in 2009 (Morehart and Johnson, 2009). That's a lot of variation.

More evidence of how variable farm income is can be found in Figure 1.2. The coefficient of variation is plotted for various components of farm income. The coefficient of variation is the standard deviation (a measure of how much something varies) divided by the mean. This is a measure of how much income varies in proportion to its mean. A higher number means more volatility. Notice that the coefficient of variation is much higher for farm income than off-farm income. Even more striking

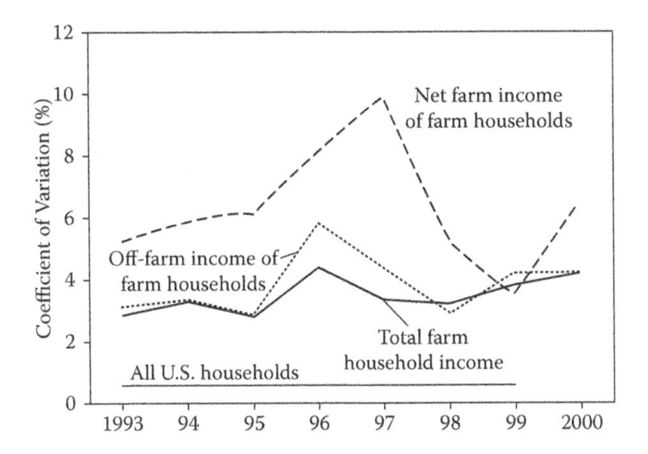

FIGURE 1.2 Variability of farm and off-farm income, United States, 1950–2006. Source: Mishra, Ashok, Hisham S. El-Osta, Mitchell J. Morehart, James D. Johnson, and Jeffrey W. Hopkins. 2002. *Income, wealth, and the economic well-being of farm households.* Agricultural Economic Report AER812, pp. 77, July.

is that the coefficient of variation is 4 to 6 times higher for farm income or total farm income (which includes on- and off-farm income) than for all U.S. households, seen at the bottom of the graph. Farming appears much more risky than earning a living somewhere else.

When we talk about risk, we mean dealing with uncertainty in business, whether by avoidance, elimination, or acceptance. The most common risks producers think about are production risks to their yields and price risks related to inputs and outputs. Water is probably the biggest production risk for many farmers, but yields can be affected by pests, weeds, hail, fire, operating funds, fuel, or a manager's ability (or inability) to produce a certain crop in a specific environment. Many production risks can be reduced or eliminated altogether by economic or agronomic practices. Irrigation, for example, can alleviate water limitations; crop insurance can offset losses due to hail. However, many shocks cannot be completely avoided. Sometimes a producer simply has to face a lower yield.

An unstable agro-environment not only affects producers but makes input and output prices more volatile for the entire agricultural sector. When there is a drought, the price of corn goes up, increasing profits for corn growers and decreasing profits for cattle feeders. When it floods, prices rise. When the weather is perfect, prices go down. And often, one region enjoys good weather while another suffers through bad conditions. The United States is not isolated from the rest of the world's agro-environmental swings either. Because the United States has become dependent on trade, the prices of beef, dairy, grains, and other commodities are heavily influenced by what happens in other countries. We live in a global society.

Many decisions in agriculture have to be made before the outcomes are certain. A farmer has to plant corn in April but won't know her price until August. A cattle backgrounder has to buy calves before he knows how much grass is available for grazing in the spring and summer. If too many calves are purchased, profits will fall because there is not enough feed per head. Purchasing too few stockers will mean profits lost to grass uneaten. Compounding the situation is the fact that producers have to make decisions for large tracts of land and expensive equipment complements or to plant crops that fruit for many years. Blueberry and wine grape producers, for example, make huge investments in their stock for tricky weather and fickle markets. Once a decision is made, it is difficult or impossible to reverse because resources have been committed.

1.2 MANAGING RISKS WITH RISK NAVIGATOR SRM

Some farmers seem to virtually disregard risk management (Harwood et al., 1999), despite noting that they think risk is important. A survey of 960 *Farm Futures* readers found that only 5% of producers use available tools to manage production and financial risk (Risk Management Agency [RMA], 1998), and these individuals seemed to be self-selected, high-end risk managers. As found in the survey, most producers do not manage risk for the following reasons:

- Resistance to change
- Too many decisions, not enough time
- Old fears die hard (remembering a bad experience)

Everyone knows that you don't have to manage risk formally to survive. There is a saying that it's better to be lucky than smart. But do you really want to live by this motto? Some folks will make poor decisions and survive, or even thrive in the face of risks, while others who make great decisions will be devastated by bad luck. Risk management is subjective and offers no guarantees. Nevertheless, accounting for risk exposes subjectivity, while ignoring it does not. Risk analysis does not overcome the frailties of human judgment but makes you think more deeply about how to manage risk (Hardaker and Lien, 2005). The fields of agricultural economics, economics, business, and finance offer farm and ranch managers many tools and techniques to manage risk. What is still lacking is a framework for pulling these various pieces together in a practical and accessible way that includes both the details and the breadth of experiences that farm and ranch managers face daily. The Strategic Risk Management (SRM) process provides such a framework, and it is currently available for use on farms, ranches, and in other businesses.

Traditionally, risk management identifies the sources of risk and provides the tools to manage it, given a manager's personal preferences and willingness to tolerate risk. SRM integrates long-standing risk management concepts into a more holistic approach based on principles found in strategic planning. Strategic planning is described on the Wikipedia Web site as:

> the art and science of formulating, implementing and evaluating cross-functional decisions that will enable an organization to achieve its objectives. It is the process of specifying the organization's objectives, developing policies and plans to achieve these objectives and allocating resources to implement the policies and plans to achieve the organization's objectives. Strategic management, therefore, combines the activities

of the various functional areas of a business to achieve organizational objectives. (Wikipedia.org)

The SRM process describes how to strategically manage risk. This process differs from other types of planning for the future because the user not only considers what the future may look like and what risks may be present, but he or she also figures out the best operational niche or position to occupy given risk preferences, goals, and anticipated threats or opportunities.

With SRM, we have organized risk management into ten practical and easy-to-understand steps. The strategic steps determine financial health, determine risk preference, and establish goals. Tactical steps determine risk sources, identify and rank management alternatives, and estimate likelihoods. The operational steps involve implementing plans, monitoring and adjusting those plans, and replanning for the future.

1.3 HOW TO USE THIS BOOK

Many books and articles have been written about risk management and decision analysis but most are not practical for farmers and ranchers or do not offer real-world applications for college students that study risk management. While it helps, no previous knowledge about risk management, economics, or finance is necessary here. Our goal is to provide a guide for farm and ranch managers, producers, and specialists, enabling them to evaluate their own risk environments and management options. This book is also useful to upper-division undergraduate students or graduate students in economics and related fields.

This book is organized into three sections. Section I includes Chapters 1 through 4, which are introductory. Chapter 2 introduces the reader to risk management. In Chapter 3 we introduce the reader to our case farm, EWS Farms. We apply all ten SRM steps directly to this real case farm in order to show that it can be applied directly from this book and our Web site. Chapter 3 describes the farm and the family that runs it. An overview of the SRM process is provided in Chapter 4.

Each of the ten steps of the SRM process is described in Chapters 5 through 12 in Section II. Each of these chapters is further broken down into three parts: (1) application of the SRM process to the EWS Farms case study; (2) a detailed description of the SRM fundamentals, which includes information on how to personally complete each step; and (3) a user's manual for the Risk Navigator SRM tools. These three parts are meant to stand independently so that the reader can choose the level in which they are interested. The first part shows how we applied the process to EWS Farms. Readers are given the objective of the step, the strategy taken by EWS Farms to accomplish the step, and how it was implemented.

The details, rationale, and theories behind how the step is accomplished are presented in Part 2 of each chapter. This section is meant to read something like a college textbook, while the first section is intended to read like an Extension bulletin. The last section, Part 3, is a user's guide to the free tools available at www.RiskNavigatorSRM.com. These tools were developed specifically to help with each step. A person could choose to read only Part 3 if he just wants to work through the tools for each step, or include a reading of Part 1 for an example of each step. It is highly recommended that you read Part 2 to gain a better understanding of each step so that you can get more out of the lessons within.

Chapters 13 through 17 in Section III provide more detailed information about risk management, outside the Risk Navigator SRM framework. In these chapters we discuss traditional ways to manage the five main types of risk: production, market and price, financial, human, and institutional. These chapters are meant to further inform the reader about managing risks, some of which won't fit into the SRM framework.

In each chapter you will find illustrations and practical examples designed to help you better comprehend the material. While the text provides a review of risk concepts in each chapter, you'll also find practical advice that cuts through the myriad choices to suggest specific strategies and working tools that we would use ourselves. Furthermore, you are invited to visit www.RiskNavigatorSRM.com, the Web site that matches our text and provides working tools designed especially for each chapter in Section II. The Web site, which is updated continuously, has information about where to find other resources as well.

REFERENCES

Hardaker, B., and G. Lien. 2005. Towards some principles of good practice for decision analysis in agriculture. Working paper 2005-1, Norwegian Agricultural Economics Research Institute, Oslo, Norway.

Harwood, J., R. Heifner, K. Coble, J. Perry, and A. Somwaru. 1999. *Managing risk in farming: Concepts, research and analysis.* Washington, DC: U.S. Department of Agriculture, Economic Research Service, Market and Trade Economics Division and Resource Economics Division, Report 774.

Mishra, A., H. S. El-Osta, M. J. Morehart, J. D. Johnson, and J. W. Hopkins. 2002. *Income, wealth, and the economic well-being of farm households.* Agricultural Economic Report AER812, pp. 77, July.

Morehart, M., and J. Johnson. Farm income expected to decline in 2009. Amber Waves, USDA Economic Research Service. Accessed April 5, 2008 from www.ers.usda.gov/amberwaves.

Risk Management Agency (RMA), U.S. Department of Agriculture. 1998. *Building a risk management plan: Risk-reducing ideas that work.* Washington, DC.

Wikipedia. April 2008. Accessed April 9, 2008 from http://en.wikipedia.org/wiki/Strategic_management.

2 Managing Risks and Risky Decisions

Dana Hoag

CONTENTS

2.1 WHAT IS RISK?

All businesses encounter risk and uncertainty, but few people know how to define it. Taking a risk involves making a decision to take an action (or to take no action) where there is some uncertainty.

Many situations encountered by a farmer or rancher involve uncertainty. Some decisions are simple or routine, while others are difficult and could make or break an operation. Common decisions include deciding how much grass will be available for spring grazing, what the price of corn will be, and whether or not it will rain tomorrow. More difficult decisions include whether to change the crop traditionally produced or whether to buy a new, expensive harvester. Sometimes the chance or probability of a certain outcome can be determined. For example, you may know the chance of rain tomorrow is 20%. But sometimes a person doesn't even see a risk coming, let alone know what its probability is.

The difference between risk and uncertainty is that risk involves making a choice that exposes the business to an uncertain future. For example, whether or not it will hail this summer is an uncertainty, not a risk; yet if a producer decides to plant a crop, she is making a choice that exposes her to a risk based on that uncertainty. Of course, people take risks because there is some chance at a reward that makes it all worthwhile. Managing risk means balancing the trade-off between taking risks and getting returns.

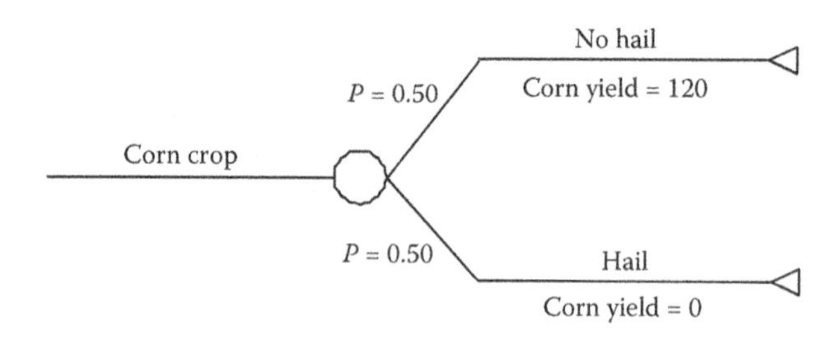

FIGURE 2.1 A decision tree for corn crop yields with and without hail.

Risk has two components: a payoff and a likelihood that the payoff will occur. Some payoffs are quantitative—the outcome is measured in numbers (a dollar value or a yield). Other payoffs may be more qualitative—the outcome is simply considered important or significant to the operation. Regardless of whether the payoff is quantitative or qualitative, the outcome is the end result of an action, an exposure to risk. The likelihood of an event occurring is measured the same way; it can be quantitative—a 50% chance that corn prices will be above the loan rate—or qualitative.

Risk management is the process of taking actions to shape the likelihood and the outcome. For example, hail is a risk faced by many producers who plant corn. This risk can be represented in a decision tree as shown in Figure 2.1. (Decision trees are described in detail in Chapter 9.) The likelihood of hail is 50%. If there is no hail, the crop should yield 120 bushels per acre. If it hails, there will be nothing to harvest. In this case, taking a risk means exposing oneself to the chance of an undesired outcome; that is, losing half the crop to hail damage. The likelihood of the event is 50%, and the two potential outcomes are 120 bushels per acre and 0 bushels per acre. The expected value of planting the corn crop is a weighted average of the two outcomes, and the weights are the probabilities. In Figure 2.1, the expected payoff is 60 bushels per acre because $50\% \times 120 + 50\% \times 0 = 60$. Hail insurance will alter the expected payoff of planting the crop by reducing the probability of a 0 bushel yield, but will also reduce the payoff by the cost of the hail insurance premium.

People take risks when the expected reward for doing so outweighs the cost. Whether the producer chooses a certain crop over another will depend upon whether the profit is high enough to make it worth managing the risks. But a fine line exists between taking a risk and gambling. If a person cannot identify the possible outcomes or probabilities in a situation, the decision maker is gambling on the hope that everything will work out. Or if a person takes on a risk with an expected reward worth less than the cost, he is also gambling on the outcome. A lottery ticket that costs $1 has an expected payoff in most American states of less than $0.50. Spending $1 for an average payoff of $0.50 is gambling. Managers do "gamble" from time to time. For example, a manager might want to try a new technology that could pay off big if it works out, like a new seed variety, breed, or tillage system. In such cases, many people try the new technology on a limited basis so the loss isn't devastating if the attempt doesn't work out. This is considered a form of risk management.

It is also important to draw out the difference between variability and uncertainty. Variability is easier to live with than uncertainty because a producer can plan for it.

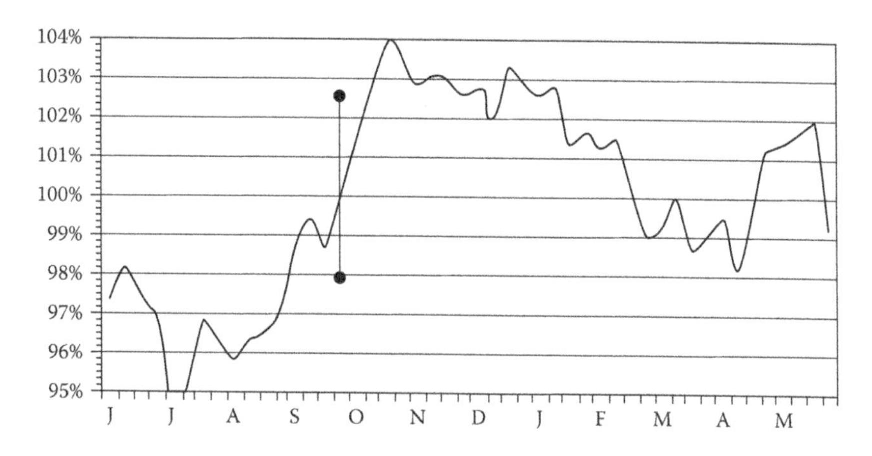

FIGURE 2.2 Mkt$ense, weekly seasonal wheat prices plotted for six Kansas locations by the Kansas Farm Bureau. Source: Kansas Farm Bureau. 2008. *Kansas Wheat Prices and Wheat Marketing*, http://www.kfb.org/commodities/commoditiesimages/Kansas_20Wheat_20Prices.PDF. (Accessed February 2009)

Not all variability carries a risk; some variability is predictable. Consider the price pattern for wheat in Kansas, as shown in Figure 2.2. The Kansas Farm Bureau collects prices every Wednesday from six Kansas locations (Goodland, Dodge City, Beloit, Walton, Topeka, and Fredonia) for their MKT$ense program. They then index the wheat prices by dividing each weekly or monthly average price by the season's average price, converting each to a percentage above or below the average (www.kfb.org). The index is typically at its lowest in July when local supplies are large because of harvest and the local demand is relatively small. As grain is shipped abroad, prices gradually rise through October, coinciding with peak seasonal export demand. They then fall slowly until harvest. Local wheat prices take on a rather predictable, seasonal pattern over the year. While the week-to-week price changes, the seasonal price pattern is stable because it follows a harvest and wheat shipment cycle. The point is that the seasonal pattern is variable, not uncertain. Producers can even benefit from this price pattern because variability itself is predictable—the same way casinos benefit from predictable patterns in Blackjack Poker. She could choose to store wheat or to sell at harvest based on the known pattern of variability. Likewise, wheat processors can plan for the known price pattern. Each actually makes more money by planning for the variation.

Even though a seasonal wheat price pattern exists, the actual week-to-week price won't follow the pattern exactly and will deviate from the seasonal price path. These deviations represent uncertainty. The actual prices for the beginning of October, for example, probably won't fall exactly on the expected price line. You can represent a range of potential prices around the seasonal pattern by drawing a vertical line, as shown in Figure 2.2. In this example, the predicted price in early October is 100% of normal, but the actual prices range historically between 98% and 103%. The range represents the uncertainty portion of the price.

Calculating risk is quite a bit more complicated than the examples discussed above. After all, while watching the weather forecast for hail, you are also checking the change in prices at the Chicago Board of Trade and scouting your corn crop for pests. Managing risk is complex, and as your understanding of risk improves, you can read more advanced sources and utilize the more advanced risk management tools discussed in later chapters.

2.2 RISKS IN AGRICULTURE

According to economists Robison and Barry (1987), producers face about 10,800 uncertain situations. If this is the number of risks they could count, imagine how many more there really must be. The 1996 Farm Bill created the Risk Management Agency (RMA) to help producers manage these risks, and the RMA grouped the many types of risk into five primary sources (RMA, 1997; Harwood et al., 1999):

- Price and market
- Production
- Institutional (legal, regulatory, and government)
- Human resource
- Financial

Most producers think about market and production risks more than they think about financial risk, and they don't think much at all about human or institutional risk. On the face of it, this makes sense, since at the heart of every farmer's financial plan is the following simple equation:

$$\text{Profit} = \text{Price} \times \text{Quantity} - \text{Costs}$$

Producers want a high price and quantity and low costs. While institutional, human resource, and financial risks do not directly appear in the profit equation, they can bankrupt any operation that dares to ignore them. Growing sunflowers might be a great money-making endeavor, but you might not be able to pursue it for any number of limitations brought on by financial, institutional, or human risks. Your banker might deny your operating loan. You might have to plant corn to maintain base acreage for a government program. Perhaps your landlord won't let you plant sunflowers because he thinks it's not that profitable. These financial, institutional, and human risks, respectively, represent just a few of the many things that producers face every day.

Consider the example of a farmer in Longmont, Colorado. He arranged to purchase a consistent supply of chicken manure from a neighboring farm to stabilize his nutrient costs and eliminate as much commercial fertilizer as he could. He thought his environmentally conscious neighbors would support him, but instead found himself facing numerous threats of a lawsuit over odor problems. He had to take a loss on his long-term contract by cancelling it when he failed to plan for this type of human or institutional risk. In another case, a rancher in North Park, CO, uses draft horses in winter because they are more reliable than machinery in the cold, harsh winter climate (Figure 2.3).

FIGURE 2.3 Ranchers in Colorado high country rely on draft horses to pull their hay sled in winter to avoid risks of equipment breakdown. Photo by Catherine Keske, 2009.

For the reader's benefit, each of the five types of risk identified by the U.S. Department of Agriculture Risk Management Agency is described briefly below. Note that risks in one category can affect another. For example, high crop prices in 2007 actually created financial risks for farmers and storage facilities that had to meet margin calls for hedged grains. One risk is not isolated from the others.

2.2.1 Price and Market

Price and market risk are probably the risks that producers spend the most time worrying about. Price or market risk simply means the risk of getting a low net price. Prices can be low due to management practices or business risk, such as poor quality, transportation, storage and handling, or created by forces outside the farm gate like crop yields in international markets or government policies within or between countries. Price risk is often on the minds of farmers and ranchers because of its impact on profitability and because it is largely out of their control.

2.2.2 Production

Production risk is tied to the amount a producer can sell or harvest and how well the manager can influence costs. This is surely the risk producers spend the most time controlling because production risks must be faced on a daily basis and are in direct control of producers. Decisions have to be made about which task will be done today and which one will have to wait for tomorrow. Managers have to decide which seed variety to use, which pesticide, which vaccines, and which feeds to use. They have to decide where to invest their labor, when to water, and when to harvest. The list is endless, but the goal is the same: produce the most you can at the lowest cost.

2.2.3 Institutional

Institutional risk refers to the way rules affect profits; this includes both laws and policies. In the early part of this new millennium, more than half of net farm income emanated from government price and income support programs (Environmental Working Group, 2009). These programs have many rules that, if changed, can have a big impact on the bottom line. To make matters worse, the programs are overhauled every five years. These five-year farm bills set up the rate of support, as well as the rules to qualify for support. In the past, this has meant idling land, limiting production base, and meeting soil conservation standards. These programs have created huge swings in the amount of acreage planted to wheat, corn, and other major crops. Farm bills, however, represent only a tiny piece of the institutional pie.

Farmers and ranchers also have to deal with animal stewardship requirements, waste requirements, licenses, transportation laws, labor laws, and the potential for lawsuits over everything from environment to inheritance. One of the biggest examples of institutional risk a producer faces today is what will happen to the estate if he or she passes away. An estate or succession plan can be the difference between whether or not a farm or ranch survives the owner's passing.

Most people forget to recognize that institutional risks have a positive impact too. The government invests in research and education, makes trade deals, and creates safety guidelines that improve quality of life.

2.2.4 Human and Personal

Managing human and personal risk means paying attention to the people around you. How do you treat your labor, your landlord, your family, and your business associates? And how does that affect your risk? Immigration is a hot issue in the United States. It may be cheaper to employ immigrant laborers, but it's risky because some people may not be legal residents.

Labor is an important resource to be allocated on the farm, but the bottom line doesn't always dictate its use. One producer I interviewed in Idaho has not been willing to adopt conservation tillage because it would mean that he would need one less worker and that employee would be his brother. He won't adopt conservation tillage if it means having to fire his own brother. A farm in Mississippi failed when the owner was hospitalized and no one at home knew what to do. Human risks are more difficult to define and understand than other types of risk, but they should be managed just as diligently as the others.

2.2.5 Financial

Producers are exposed to financial risk when they have debt. Since most people borrow operating money from a bank or other financial institution, they are at a lender's mercy. What if your equity falls low enough that banks won't lend you money? What if your loan payments on fixed assets become too great to cover and the farm is seized? What if your bank fails? New Frontier Bank in Greeley, CO, left hundreds of producers in trouble when it went under in 2008 because local banks simply could not finance the volume of operating loans needed. The Federal government eventually had to step in to resolve the crisis by making $253 million available for loans. It is important to determine the operation's financial health, as it affects the ability to take on projects that earn profits. Consideration of the impact of financial risk when making decisions about any of the other risk categories is crucial to the success of an operation.

2.3 PROFILING RISK

Frank Knight was one of the first authors to formally take up risk and uncertainty when he wrote about it in 1921. Knight thought of risk as an event; he thought a probability could be described objectively like a coin toss, and he believed that uncertainty was a different situation where probabilities could not be assigned except subjectively. Knight's ideas evolved as others began to research risk and, in particular, examine the notion of probability. Probability is a *subjective* interpretation of the degree of belief that an outcome will occur. Probably all estimates of probability are therefore subjective. Even something as sophisticated and scientific as a weather forecast is still subjective in the sense that the *Farmer's Almanac* may out-predict the weather service on any given day. Therefore, it would be very difficult to meet Knight's criteria for risk; that is, how would you know whether the probability is really known, except in very simple cases like rolling dice?

One way to describe risk is to create a risk profile (described fully in Chapter 10), which contains information that helps decision makers discriminate one management action from another. A risk profile may include descriptive information, such as the name and location of the operation, technical details, and financial information. In addition to these details, a risk profile may focus on uncertainty, describing a project in terms of its statistical properties, including central tendencies, range, and potential worst-case outcomes. When there is uncertainty, the decision maker then has to compare a profile of possibilities for each project.

A probability density function (PDF) is one type of risk profile to describe outcomes and likelihoods. Most people know a PDF by the term bell-shaped curve. Let's look at the risk associated with hail and planting corn. We could graphically show the outcome (hail or no hail) and the probability (50% or 0.5) with a histogram, or bar chart, as shown in Figure 2.4. A histogram is simply a graph with a bar for each possible outcome that is as tall as its probability. In this case, each bar is 0.5 high. The probabilities for all of the outcomes will always add up to 1.0.

Figure 2.4 is a simple risk scenario with just two outcomes. More bars can be added when there are more outcomes. Let's consider rolling a pair of dice. The possible outcomes are 2, 3, 4, 5, 6, 7, 8, 9, 10, 11, and 12. Only one combination, a pair of ones, yields a 2 and only one combination would deliver a 12, a pair of sixes. But there are six combinations that yield a roll of 7: 1 + 6, 6 + 1, 2 + 5, 5 + 2, 3 + 4, and 4 + 3. We can plot all of these into the histogram shown in Figure 2.5. In this histogram, the outcomes are the numbers between 2 and 12. The vertical axis measures the number of rolls that will give rise to those numbers. If we connect the tops of the bars with a solid line, we get the familiar bell-shaped PDF.

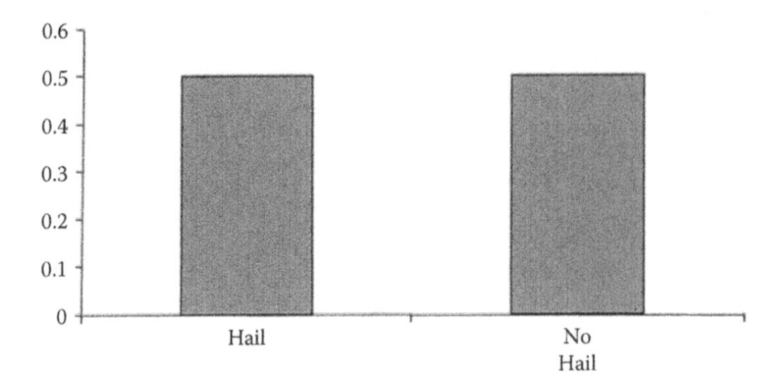

FIGURE 2.4 Probability of hail or no hail.

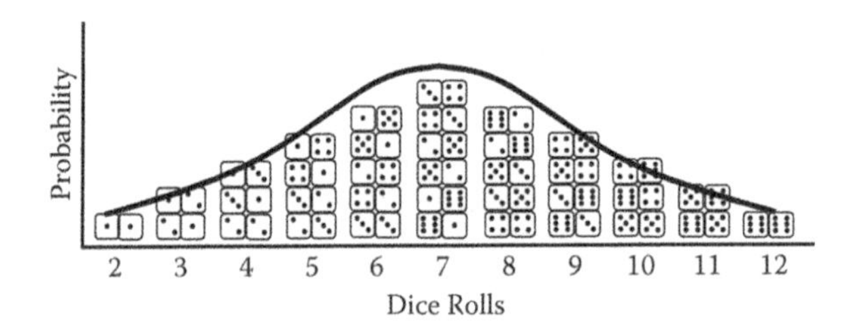

FIGURE 2.5 Probability density function for rolling two dice.

Let's interpret the PDF drawn in Figure 2.5. The tallest (or fattest) part in the middle shows the outcome that is most likely to happen; rolling two dice has the greatest probability of yielding a sum that equals 7. But it is possible to land in the skinny part on either side by rolling a sum other than 7. The statistical properties of the PDF can be calculated, too. The mean, median, and mode, for example, are 7 for the roll of two dice. Measures of dispersion, like the variance, indicate the range of the outcomes, or the spread; in this case, the outcomes range from 2 to 12.

For the risk manager, a PDF can be created that shows the potential outcomes and their likelihoods for two decisions—like buying hail insurance or not. The manager can use the PDF both to better understand just how risky the decision is and to decide whether the benefits of managing risk are really worth the costs. A PDF with higher measures of central tendency is more desirable because there is a better chance of experiencing a good outcome. Skinnier PDFs are less risky than fatter ones since they have a smaller range of outcomes—good or bad. A PDF with a longer tail on the left side is less preferred because when a rare event happens, its outcome will be worse. (We've just scratched the surface on risk profiles. Chapters 9 and 11 describe how to use risk profiles, and Chapter 10 provides a detailed description of how to build them.)

2.4 RISK MANAGEMENT AND DECISION ANALYSIS

Managing risk is important because people are usually risk averse. Risk management is the "systematic application of management practices, procedures and practices to the tasks of identifying, analyzing, assessing, treating and monitoring risk" (Hardaker et al., 1997, 12). Important risks are worthy of systematic analysis to determine a best-bet course of action. Decisions are difficult because they are complex and uncertain, and because a person almost always has more than one objective, which means trade-offs are inevitable. Admittedly, risk management is subjective and full of faults. Nevertheless, accounting for risk exposes subjectivity, while ignoring it does not. Risk analysis does not overcome the frailties of human judgment, but it makes a person think more deeply about how to manage risk.

The Risk Navigator SRM process, described in Chapters 4 through 12, takes advantage of methods and tools developed in several disciplinary sciences, such as economics, finance, and decision sciences, where researchers have designed methods to manage risk. After figuring out your ability to tolerate risk and setting your goals in the strategic portion of the SRM, the tactical section organizes information into a handy format, called a risk payoff matrix, which will enable you to make sound risk decisions in the third and final stage of SRM.

Let's dig into the case study we will use as an example throughout this book. For the Sprague family, owners of EWS Farms, the risk management problem is summarized in Figure 2.6. For now, think of this payoff matrix as a map of what we will be doing as we move through the SRM process. The Spragues use the techniques shown in Chapter 8 to decide which risks to prioritize first. The likelihoods, developed in Chapter 10, are reported in the column marked "probability" for each potential state of nature in the column headed "states." The two states of nature are a short crop and a normal crop. Three managerial responses are considered based on the information presented in Chapter 9—cash market sale, forward contract, and hedge. Each of these decisions has an outcome dependent upon whether the United States has a short crop or a normal crop.

Type of Risk to be Managed: Price of Corn (Chapter 8)				
Probability: U.S. Production (Chapter 10)		**Management Actions: Cash, Contract or Hedge** (Chapter 9)		
States	**Probability**	**Cash Market**	**Forward Contract**	**Hedge**
		Outcomes		
Short U.S. crop	0.65	**342500**	332700	336200
Normal U.S. crop	0.35	287700	**305900**	295400
Expected value		**323320**	**323320**	321920
Minimum		287700	**305900**	295400
Maximum		**342500**	332700	336200
Maximum regret		18200	**9800**	10500
Regret 1		0	9800	6300
Regret 2		18200	0	10500

Rank best outcome (Chapter 11)

FIGURE 2.6 A risk payoff matrix of EWS Farm's corn marketing problem.

Finally, we have to choose which action we want to employ. Based on what you see in Figure 2.6, which practice would you choose? With a gross revenue of $342,500, the cash market is the best option if there is a short crop, but forward pricing is the way to go if there is a normal crop. Since there is uncertainty, you don't know which is better. The risk profile proves that it is difficult to choose because each action has its own pros and cons. There is often no clear winning strategy.

We describe many methods to rank risk in Chapter 11. Essentially, we try to match your risk personality to a tool that ranks management actions based on the criteria you think are important. For example, if you are very concerned about avoiding bad outcomes, we would pick a tool that focuses on picking strategies that don't have bad outcomes. If you want to earn the most money possible, we focus on strategies that maximize your expected value.

2.5 SUMMARY REMARKS

This chapter presented a flyover view of what risk management is all about. The remainder of the book will be more detailed since our intention is to help readers be able to manage real risks. The next chapter describes a farm where the producer agreed to apply the process as he moved back to his family's operation. The rest of the chapters are all devoted to explaining risk in such a way as to make the reader able to start a complete, comprehensive, and capable risk management program on their farm, ranch, or other agricultural operation.

REFERENCES

Environmental Working Group. 2009. Available at http://farm.ewg.org/sites/farmbill2007/.

Hardaker, J. B., R., B. M. Huirne, and J. R. Anderson. 1997. *Coping with risk in agriculture.* Oxfordshire, UK: CABI Publishing.

Harwood, J., R. Heifner, K. Coble, J. Perry, and A. Somwaru. 1999. *Managing risk in farming: Concepts, research, and analysis.* Washington, DC: U.S. Department of Agriculture, Economic Research Service, Market and Trade Economics Division and Resource Economics Division, Report 774.

Kansas Farm Bureau. 2008. *Kansas wheat prices and wheat marketing.* Accessed February 2009 from http://www.kfb.org/commodities/commoditiesimages/Kansas_20Wheat_20Prices.PDF.

Knight, F. 1921. *Risk, uncertainty, and profit.* Houghton Mifflin, Co., New York.

Risk Management Agency. 1997. *Introduction to risk management: Understanding agricultural risks: Production, marketing, financial, legal and human resources.* U.S. Department of Agriculture, Washington, DC.

Robison, L., and P. Barry. 1987. *The competitive firm's response to risk.* Macmillan, New York.

3 A Case Study of EWS Farms

Aaron Sprague and Catherine Keske

CONTENTS

The authors of this book realize that most practitioners have very busy schedules that don't allow enough time to do endless financial analyses, especially if that involves "book" work. With this in mind, we have asked Aaron Sprague, who is working on his master's degree at Colorado State University and returned to his farm in northeastern Colorado (Figure 3.1), to complete the entire Navigator SRM process. This information is presented to show that Navigator can be used in a real-world setting and to demonstrate its simplicity. It provides a detailed description of Aaron's farm for those readers who find the context of his farm helpful in understanding how to use Navigator.

FIGURE 3.1 Cattle and cropland on EWS Farms in Phillips County, Colorado. Photo by Jay Parsons, 2008.

3.1 BACKGROUND

EWS Farms is an agricultural production partnership owned by the Sprague family in Holyoke, Colorado. The Spragues have graciously agreed to provide production, marketing, and financial information to apply risk management lessons as a case study. The majority of the acreage in the operation lies five miles north of Holyoke, the county seat of Phillips County (Figure 3.2), the second most northeastern county in the state. Although relatively small compared to some of the other rural counties in Colorado, Phillips County is extremely productive on a per-acre basis. With irrigated corn yields of about 190 bushels per acre and nonirrigated yields hovering between 50 and 60 bushels per acre, the county consistently ranks in the top three producing counties in the state. Much of the acreage in the area is under center pivot irrigation that uses the Ogallala Aquifer as the sole source of irrigation water. Like many farms in northeastern Colorado, although irrigation has increased rapidly over the past few decades, a substantial amount of farmland is still used for dryland production.

EWS Farms has been in the agricultural production business since the early 1940s. Having weathered the various storms (like droughts, inflated and crashing land values, and low prices) the industry has encountered over the last 60 to 70 years, EWS Farms is no stranger to risks inherent in every facet of production agriculture. The four Sprague brothers who founded the business were born and raised on the family's original homestead not far from the current site of operation. Due to pressures affecting production agriculture during the late 1920s to late 1930s, each of the four brothers decided to pursue other interests deemed more stable than production agriculture. Despite the fact that they had all gone their separate ways, they decided to make their lifelong dream of operating a family farm come true in the early 1940s. All four brothers returned to the Holyoke area and began the Sprague Brothers Cooperative Farm (later named the Sprague Brothers Partnership). Over the years, the partnership, still active today and known as EWS Farms, has endured times of great success, as well as times of seemingly insurmountable failure.

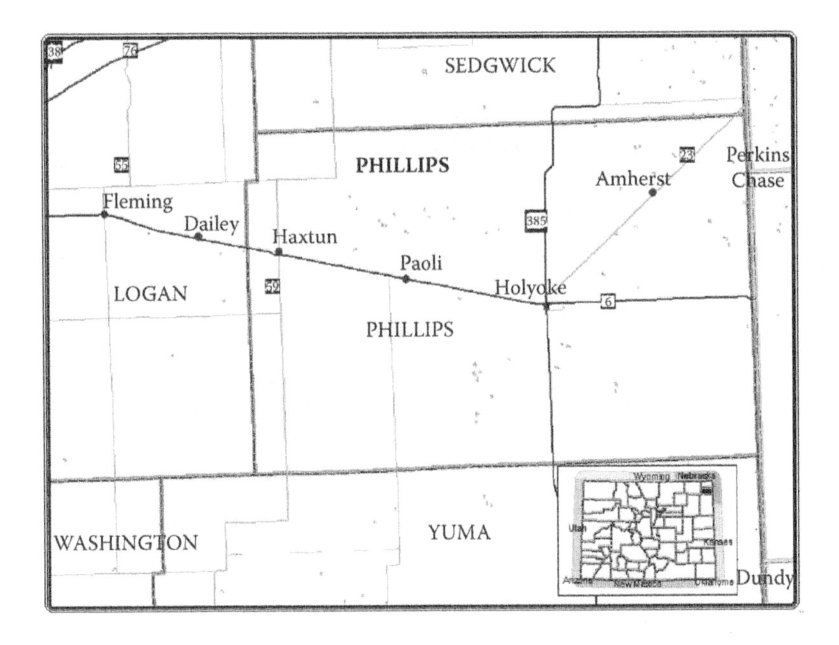

FIGURE 3.2 Phillips, County, Colorado, with inset showing location of the county. Data sources: U.S. Census Bureau, Colorado Division of Local Government, and Colorado Department of Transportation.

3.2 THE SPRAGUE FAMILY

As mentioned previously, the entire family is an integral part in the operation of the business. Although the farm is and has been family owned from the time of its inception in the early 1940s, the second and third generations of the Sprague family are now charged with management of the operation. Russell (Figure 3.3) and his wife Kimberlee are the primary decision makers for the operation. Because he moved into this role directly out of high school, Russell's more than 30 years of experience operating the business with his father and his family make him a valuable resource. Kimberlee is also a critical component in that she does most of the record keeping for the operation. In addition to all of the time spent operating the business, they have raised five children ranging in age from 19 to 28. The family consists of two daughters, Desiree and Brianne, who are the oldest and youngest children, as well as three sons, Aaron, Russell, and Dustin. The farm is now being co-managed with son Aaron and son-in-law Aaron.

While it is important for the continuation of the family business for family members to return to the area, all of Russell and Kimberlee's children also understand how important their education is to the success of farm. The three oldest children, as well as their spouses, have received undergraduate degrees. At this writing, Desiree is pursuing a master's degree in occupational therapy and Aaron is working on his master's degree in agricultural and resource economics. Russell Jr., the third child,

FIGURE 3.3 Russell, left, pictured with son Aaron and son-in-law Aaron. Photo by Jay Parsons, 2008.

has completed his master's degree in speech communications and is now working on his law degree. These three oldest children are married with their own families. At this writing, the youngest siblings, Dustin and Brianne, are pursuing their undergraduate college degrees in business and nursing, respectively.

In addition to academics, athletics plays a very large role in the lives of all of the siblings and their spouses. Aaron and Russell both played football at Colorado State University. Desiree's husband Aaron played football at Doane College, and Dustin plays football at the University of Colorado. Desiree, Amber (Aaron's wife), and Billie (Russell Jr.'s wife) were all cheerleaders in high school and college. Brianne is currently on the volleyball team at Regis University.

The Sprague family is well known for their community involvement, particularly with the local school district (Holyoke RE-1J). Of the three families involved in the business (Russell and Kimberlee, Desiree and Aaron, Aaron and Amber), all work in the local school district. Unlike their parents, Desiree and Aaron, and Aaron and Amber (Figure 3.4), decided that additional off-farm income is necessary. They have supplemented farm income with full-time work in the school district. Desiree teaches English as a second language (ESL) and Amber is a speech therapist. Russell, Aaron, and Aaron still have active roles in the local high school sports programs. On the business side, all of the family members who are directly involved in the operation are also part of many local cooperatives and community groups. Clearly, EWS Farms supports the rural economy and the social fabric of the community in every possible way.

Consistent with the history of the operation, it is the family's vision to support both family members who choose to return to the area and directly participate in the operation and those who choose to pursue other professional interests. Although the family realizes that this is a lofty goal that presents numerous challenges, the family business has recently taken steps toward fulfilling this mission by including the two

FIGURE 3.4 Aaron and family: wife Amber, Miles (left), Reid (center), and Wyatt (right). Photo by Jay Parsons, 2008.

oldest siblings and their families in direct roles. The farm also benefits from this decision because family involvement has added both stability and the assurance that the business will have sufficient manpower to operate in the future. The addition of off-farm income also provides less income uncertainty.

3.3 THE FARM OPERATION

At the time of this writing, EWS Farms intensively crops about 2,500 acres of productive land in northern Phillips County, Colorado. The acreage is split between irrigated and dryland acres. The main irrigated crop is corn raised for grain, while the dryland production includes a two crop, three-year rotation of wheat, dryland corn, and fallow. As will be discussed later, the operation has also made considerable attempts to increase profitability by reducing costs, with the goal of becoming one of the region's lowest-cost producers. With the recent addition of two families to the operation, EWS Farms' historic and viable business is facing critical issues in the near future. Like many agricultural production firms today, EWS Farms will inevitably be exposed to all types of agricultural risks, including production, marketing, financial, legal, and human.

We proceed in the following sections with a description of various aspects of the farm, from the physical and capital resources, to management and financial performance indicators. A summary of this information is found in Table 3.1.

3.4 PHYSICAL AND CAPITAL RESOURCES

3.4.1 Climate

Northeastern Colorado lies in the western part of the midsection of the United States. Although technically not in the nation's traditional corn or wheat belts, the region's marginal production is often very high. An important fact to remember when

TABLE 3.1
Summary of Enterprise Data for EWS Farms

Category	Description
Farm size	2500 acres, 500 irrigated and 2000 dryland
Labor and Management	This operation is a family-oriented production agriculture business. Traditionally, the father has provided all of the labor and management while drawing on the wife or children when extra help is needed. Recently, however, the two oldest of the five children and their families have returned to the area to be directly involved with the operation. The addition of the son and son-in-law has taken pressure off of the mother and father, but has also added increased financial demands on the operation. It is the intent of the family to continue the family operation with plans to adequately manage the new operational demands.
Crop Rotations	The irrigated acreage of the operation uses a continuous corn for grain rotation. The dryland acreage in the operation has recently shifted to two crops in a three-year rotation. This rotation includes production of hard red winter wheat, dryland corn, and a year of summer fallow.
Tillage Practices	Historically, all of the acreage in the operation has been tilled traditionally. However, the operation is in the process of shifting to the sole use of reduced-till methods in all enterprises. Currently, the irrigated corn is produced using reduced-till methods while the dryland corn is produced using no-till methods. The winter wheat enterprise is the last of the operation's enterprises to be converted to the reduced tillage methods.
Soil Fertility and Weed/Pest Management	With the conversion of the tillage practices to reduced-till methods, soil fertility and weed/pest management has become very important to the success of the operation. The management typically works in conjunction with crop production specialists at the local cooperative to develop herbicide, insecticide, and fertilizer programs for each individual field in the operation. Therefore, the management depends greatly on the expertise of the specialist in the planning and recommendation of the particular program used.
Irrigation Practices	Despite the fact that most of the acreage in the area has historically been used for dryland production, irrigation has increased greatly over the past few decades. Irrigation in the area is achieved through the extraction of groundwater contained in the Ogallala Aquifer. The majority of the region's irrigation systems are electrically powered pumps that provide water to electrically powered center pivot systems. Although irrigation has enabled acreage to reach extremely productive levels at the margin, recent regional drought conditions have brought aquifer overdraft issues to the forefront and caused some uncertainty as to the future of the irrigation practices in the area.
Marketing Strategies	The current marketing scheme of the operation is not formalized. Although management has a solid grasp of costs of production and breakeven levels, there is no formal decision as to how and when to market the commodities. Most often, cash sale at or after harvest occurs. The operation also takes advantage of positive pricing opportunities with the use of forward contracting through the local cooperative.

TABLE 3.1 (*Continued*)
Summary of Enterprise Data for EWS Farms

Category	Description
Production Performance	Historical production data indicate that the operation is capable of producing high levels of commodities on a per-acre basis. Expected yields for the operation are approximately 200 bushels per acre for irrigated corn, 60 bushels per acre for dryland corn, and 40 bushels per acre for dryland winter wheat. All of these levels compare positively to the historical county- and state-level yields.
Financial Performance	Financially, the operation seems to be stable. A concerted effort to obtain low cost of production as well as a decision to limit hired labor and machinery ownership costs has enabled the operation to achieve this stability. Also, a management decision to evaluate and execute an expansion in the near future will enable the operation to more adequately service the recently increased financial demands on the operation.

evaluating the climate of the area is that northeastern Colorado is west of the 100th prime meridian, often referred to as the "20-inch line." This designation is critical when evaluating the amount of annual precipitation received. Regions to the east of this line generally receive greater than 20 inches of precipitation per year, while regions to the west of this line usually encounter less than 20 inches of precipitation per year. As a result, production agricultural practices west of the 100th prime meridian are drastically different from the nation's Corn Belt east of the meridian. This is chiefly due to increased exposure to greater climatic risk, an important consideration in the EWS Farms risk management plan.

One of the main climatic sources of risk to the farm is regional weather. Although the crop conditions are favorable for almost the entire growing season, the possibility of severe summer thunderstorms always looms in the background. It is not uncommon for production in the county to be drastically reduced (often by nearly half) from the effects of one or two severe thunderstorms throughout the growing season that pound local crops. Another substantial production risk in recent years has been the severe drought that has taken a toll on a large area of the nation. With an average rainfall of only 17 inches per year in the area (Table 3.2), slight deviations from normal precipitation levels often result in dramatic drought effects being realized, especially in the dryland crops within the county. Winters in the area are relatively mild with the occasional severe blizzard, which can affect the production sector as well as the regional economy.

3.4.2 Soil and Soil Conservation

The composition of the soils in the study area varies dramatically. Within Phillips County, the soils are generally split into two different types from the north to the south by U.S. Highway 6. To the north of the highway, the soils are very heavy and dark, with high levels of organic matter. Most of the land has very little slope and much of

TABLE 3.2
Normal Monthly Temperatures and Precipitation in Holyoke, Colorado

Month	Normal Temperature (°F)	Normal Precipitation (inches)
January	27.3	0.40
February	32.1	0.41
March	38.2	1.10
April	48.6	1.60
May	58.8	3.30
June	68.7	3.15
July	74.5	2.71
August	72.8	2.02
September	63.3	1.27
October	51.5	0.94
November	37.9	0.59
December	29.8	0.33
Average Annual	**50.3**	**17.82**

the tillable acreage in the area continues to be managed with conventional tillage practices, which will be discussed later in the chapter. In the southern half of the county, the soils are sandy. It is not uncommon to find large areas or even entire fields of *sugar sand*, a very fine type of sand susceptible to many different erosion factors such as wind and water. The tillage practices are mostly in line with conventional practices; however, the acreage in the southern sandy areas is treated carefully with respect to winter and early spring cover. The extreme susceptibility of the soils to wind erosion in particular is the primary reason for this difference. Where much of the traditional tillage is performed in the fall in the northern part of the county, the majority of the tillage is performed in the early spring in the southern half of the county.

The case farm has acreage of both soil types. The base of the farm is in the northern part of Phillips County. These acres are all very heavy, dark soils that are relatively resistant to erosion factors. By adding additional acreage, the operation has moved into the southern part of the county, directly in the heart of the sandy soils of the region. Until recently, the heavy soils of the operation were managed with traditional tillage practices; however, with the development of a soil and water conservation plan through the Soil Conservation Service of the U.S. Department of Agriculture (USDA) in the late 1980s, the farm has gradually shifted to more use of reduced tillage. The addition of the sandy acreage has also prompted the use of reduced tillage practices over all acreages of the farm.

The current EWS Farms conservation plan contains three primary areas of emphasis: conservation cropping sequence, crop residue use, and surface roughening. The actual crop rotations and tillage practices used on the farm will be discussed later in this chapter. It is pertinent to note that three of the conservation compliance goals can be achieved through managing tillage practices. The first goal is to implement cropping systems that will hold soil loss to conservation compliance levels, improve the physical condition of the soil, and help to control insects, weeds, and disease. The second goal is to maintain sufficient crop residue on the soil surface to reduce soil erosion in the critical wind erosion period (typically from late fall to early spring). The third goal of the plan is to use surface roughening in the event that less-than-adequate soil surface cover exists. Surface roughening prevents wind-induced soil erosion by roughening the soil surface with ridge- or clod-forming tillage practices. All of these goals are more successfully met with increased use of reduced tillage practices.

3.4.3 Irrigation

Throughout history, irrigation (Figure 3.5) has had paramount importance in many areas of the world. From China to the United States and nearly everywhere in between, often irrigation has been one of the most important aspects and contributors to both social and economic development. In Western states, there is a saying that goes: "Whiskey is for drinking, and water is for fighting." Irrigation also plays an important role in Phillips County and to the case farm. Irrigation has enabled the region to enjoy an extremely rich agricultural heritage and productivity, and nearly all of the local economy is tied in some way or another to the irrigation and its agricultural infrastructures. Despite significant geographic diversity, including flat plains in the north and the rugged sand hills in the south, much of the acreage in the county is used to produce irrigated crops. Corn for grain is the dominant irrigated crop.

Much of the world's irrigation is achieved through the diversion and application of surface water. In this region of Colorado, however, irrigation is achieved almost solely through the extraction of groundwater from the underlying aquifer, otherwise known as groundwater irrigation. In the northern half of the county, center

FIGURE 3.5 Sprinkler irrigation system for corn at EWS Farms. Photo by Jay Parsons, 2008.

pivot irrigation systems with accompanying wells dot the horizon for as far as the eye can see. In the southern half of the county, the same technology is used over nearly every crest of the sand hills. This has enabled the productivity of the area to reach extremely high levels; for example, 190 bushels per acre as compared to nonirrigated yields of 35 bushels per acre. In the 2005 to 2006 growing season, the Spragues observed irrigated corn yields of 300 bushels per acre. Although 300 bushels per acre is not common to most of the acreage in the area, it is not uncommon to find irrigated yield potential in the area to consistently reach well over 200 bushels per acre.

The irrigation systems in the area are operated primarily through the use of electric-powered pumps and center pivots. The amount of electrical resources required to operate the irrigation infrastructure is massive. Keeping in mind the expansive electrical network that crosses diverse terrains, it is logical that a type of natural monopoly structure exists for all of the electrical power distribution in the area. Highline Electric Association (HEA), a member-owned cooperative, is the institution in charge of providing customers with electric power.

Since its incorporation on December 19, 1938, the cooperative has grown dramatically (HEA Web site, www.hea.coop, accessed June 9, 2009). When the first section of the cooperative was energized in early 1940, it had 189 meters and sold 288,000 kilowatt hours (kWh). In 2002, those figures increased to 9,737 and 460,000,000, respectively. The members of HEA total 9,473 and range over two states and 11 different counties. The cooperative has nearly 5,000 miles of transmission and distribution lines, including 24 different substations. HEA has sold over 423 billion kWh of electricity and now expects annual revenues of nearly $25 million. HEA headquarters are located in Holyoke, Colorado, where it employs 50 full-time employees, and HEA has branch offices that are located in Sterling and Ovid, Colorado, as well. One of the most interesting facts about the cooperative is that it now supplies over 2,900 irrigation wells with the electricity for operation. A conservative estimate of the total acreage that these wells are able to irrigate is 377,000 acres. Considering the high production capabilities of these acres, the value of the electricity-enabled irrigation to the area seems to be great.

3.4.4 Ogallala Aquifer

To many of the early farmers and ranchers in the 1940s, the HEA cooperative could do no wrong. How could something that brought the convenience of electricity to the area provide anything but benefit? Nevertheless, with all of the new information about environmental impacts and long-term sustainability of the agricultural infrastructure in the area, some concerns about the irrigation practices have arisen and seem to conflict with the traditional viewpoints. One of the main concerns includes overuse of the Ogallala Aquifer, the source of all groundwater in the area. Replenishing aquifers takes thousands of years, and there are now concerns about drawing the aquifer down to levels that are uneconomic for agricultural production, or even worse, losing the aquifer completely.

The Ogallala Aquifer, which originated from the Ogallala Formation in Texas, is the southernmost extension of the major water-bearing unit underlying North

America. It is a huge underground reservoir that lies west of the Mississippi River and east of the Rocky Mountains. The Ogallala Aquifer region includes six states from South Dakota to Texas and New Mexico to Kansas. It ranges about 800 miles from the north to the south and 400 miles from the east to the west.

Most overdraft concerns arrive as a result of the increased awareness of water scarcity at the global level in recent history. Increased growth rates, in addition to increased competition between the different water demands (industrial, municipal, environmental, and agricultural) has put much stress on water resources across the world. It is estimated that 26 countries are now in a situation of water shortage (less than 1000 cubic meters per person). Estimates also indicate that within the next 15 years, nearly one-third of Africa's forecasted population will have water shortage problems. Although municipal water shortages in northeastern Colorado are not at these proportions, viewpoints and concerns on managing the Ogallala Aquifer have evolved in the past decade. An increasing number of people in the area are shifting their focus from seeing the aquifer as a reliable water source for irrigation and thus increased productivity, to viewing it as a depleting, finite water source that needs to be conserved for the sustainability of the entire area.

Most of the stakeholders involved realize the importance of irrigation in ensuring food security and improving rural welfare economically and socially. However, many are starting to recognize the downfalls of irrigation that relate to environmental issues like water scarcity from groundwater overdraft. In Phillips County, agricultural producers and residents are realizing there must be a balance between extraction and retention. Although the producers have been able to achieve high levels of productivity, there are concerns that the practices associated with this type of productivity are not consistent with actions that would ensure the long-term sustainability of the aquifer.

Historically, concerns of water scarcity in this area have been initiated by prolonged and devastating droughts. As previously mentioned, the effect of droughts can be devastating due to the climate and geography of the area. Recalling that the area in which the case farm lies only receives approximately 17 inches of precipitation per year, the effects of equal percentage decreases in annual precipitation will be far more dramatic in the study region than those that exist in a region with plentiful precipitation.

The lack of surface water is also an area of concern. The surface water present is found primarily in a few rivers and lakes or reservoirs in the area. Some of the rivers in the area actually serve as "drains" to the Ogallala Aquifer when the water table is below that of the actual aquifer. Other rivers do not recharge the groundwater. Frequently, most of the rivers in the area are dry, making the rivers a poor recharge source for the aquifer and compounding the scarcity issue.

The area lakes are the only depressions in the region that catch surface water. They are not self-fed and therefore are not a source of recharge to the aquifer. Soil types also may decrease the recharge of the aquifer. The combination of all these factors and recent legal disputes over water rights have increased awareness of the region's dependence on the Ogallala Aquifer as the economic lifeblood of the area. It has become important to many of the area residents to balance aquifer use in order to achieve long-term agricultural sustainability.

3.4.5 Machinery and Facilities

The EWS Farms operation maintains a machinery complement in its agricultural production. Maintaining a high-quality used-machinery complement is one of the main ways that the operation maintains low operating costs. But the operation must balance the risk of employing used machinery and potentially lower productivity due to inefficient equipment with the opportunity for lower costs and lower capital equipment debt. One of the keys to achieving this balance involves maintaining a skilled labor force.

As for facilities, EWS Farms also has one machinery storage building and one 8,000-bushel grain bin with a 1,000-bushel batch dryer in the top. In addition to these facilities, a set of cone-bottom bins that will house 2,800 bushels is located at Russell and Kimberlee's homestead. These facilities are used primarily to store winter wheat retained for seed. Private storage is not needed at this time since two large cooperatives are very near the operation, serving to minimize operating costs.

3.4.6 Labor and Management

Currently, all of the labor and management is provided by the three families involved in the operation (Russell and Kimberlee, Desiree and Aaron, and Aaron and Amber). Historically, Russell has been in charge of both the business management and production aspects of the operation, with either Kimberlee or the children providing supplemental labor as necessary. Although this has worked well in the past, the operation is in a time of change with the addition of Desiree and Aaron's newly expanded families. This is a good opportunity to increase the efficiency of the operation by using the various talents and skills that Aaron and Aaron possess. More specifically, Russell's son Aaron has had an intensive, high-quality education in the field of agricultural business and economics, which gives him a solid understanding of cost management and maximizing production efficiency. Desiree's husband Aaron also has a college education and has a special charisma and business sense that works well in business negotiations. Both are intelligent young farmers who are excited about the business and who are willing to accept most of the physical requirements for agricultural production. Meanwhile, Russell continues to provide an extensive amount of business experience and skills in working with tight production margins and budgets. The goal of the business is to increase the efficiency of operations with the infusion of two new operators who present a diverse set of skills. This will also reduce the business risk of relying on a single operator.

On the other hand, the addition of the two new families to the operation has increased the financial demands on the operation, as the revenues must now support three families. In order to accommodate the recent changes, the operation needs to expand. This decision to expand has tightened operating margins even more than before, and the management is concerned that without increased resources to offset these demands, the business will lack sufficient revenues and profits. With this in mind, EWS Farms will continue to monitor business performance and evaluate new opportunities for expansion.

3.5 MANAGEMENT FEATURES

3.5.1 Crop Rotations

Currently, EWS Farms is solely a commodity production business. Managing crop rotations is key to production output, sustainability, and the long-term success of the operation. In the early years of the business, nearly all the acreage was in dryland production; however, as previously mentioned, some of the acreage is now being used in irrigated production. Currently, the operation consists of approximately 2,500 acres of tillable land, of which 500 acres are irrigated using groundwater from the underlying Ogallala Aquifer. Since the introduction of irrigated acreage, the crop rotation on these acres has been relatively stable. The dryland rotations, however, have recently undergone a transformation from the traditional one crop in two years rotation of winter wheat followed by summer fallow to now using a two-crop program with a three-year rotation of winter wheat, dryland corn, and summer fallow.

As evidenced by the rotations, the primary commodities produced by the operation are corn for grain and hard red winter wheat. The dryland acres, with the two-crop, three-year rotation, are used to produce hard red winter wheat, as well as some corn. The farm's winter wheat is typically planted in the middle of September and harvested the following July. The irrigated acres are primarily in a continuous corn crop rotation, with planting generally in the middle of April and harvest at the end of October. The dryland corn is typically planted in the middle of May and harvested in the beginning of October. This rotation has served the business well in recent years, despite the terrible drought. Improved weed management, better soil erosion management, and increased soil fertility have been the major benefits of the rotation.

From a long-run perspective, crop rotations play an important role in the operation's strategic goals. In order to attain the maximum efficiency of the current resource base and future expansion opportunities, the management aspires to incorporate new and complimentary enterprises into the operation. For example, as previously mentioned, water scarcity issues are forcing the operation to investigate the addition of new irrigated cropping technology to the business. Adding a livestock enterprise to capture the grazing benefits of pasture acres also is seen as a good business expansion opportunity.

3.5.2 Tillage, Soil Fertility, and Pest and Weed Management Methods

Historically, nearly all of the acreage in the area has been conventionally tilled. Conventional fall tillage is described by the USDA Soil Conservation Service as tillage practices mostly completed in the fall immediately following harvest. After harvest, the residue left is disked and plowed into the soil before the winter freeze. The acreage is left with a rough surface after fall operations to aid in control of wind and water erosion throughout the winter months. The seedbed is then prepared in the spring of the following year just prior to planting. In addition to the tillage, herbicides and pesticides are often used throughout the growing season to manage weed and insect pressure. Most of the acreage managed by the case farm followed these well-established principles for years. However, with the business's expansion, the additional acreage is not suited for these conventional practices. Voluntary changes in management and farming practices have resulted in the decision to implement more conservation-oriented tillage practices called *minimum tillage practices*. These minimum tillage practices provide several economic and conservation benefits, which will be described momentarily. Although the operation had been moving toward this for many years, 2005 was the first year that widespread minimum tillage practices were used.

With the conversion to new tillage practices, it is the operation's goal to use only reduced-till methods with existing crop rotations. Referencing USDA Soil Conservation Service descriptions of tillage methods in the area, reduced-till methods are employed in the spring, rather than the fall. In this process, crops are still harvested in the fall, but the residue is left standing throughout the winter period. The acreage is then tilled in the spring, leaving approximately 25% or more of the previous year's crop residue on the soil surface at planting time. Herbicides and pesticides are still used throughout the growing season to control the weed and insect pressure. In some instances, the decision to use reduced tillage on the acreage is accompanied by the decision to use increased amounts of herbicides to control the weed pressure on the acreage.

EWS Farms has been successful in reaching the goals of reduced tillage in most of the crop enterprises. In fact, all of the operation's irrigated corn in 2005 was produced using the new-to-the-area strip-till method. The strip-till method embodies the principle that only a small area of the field needs to be tilled—an area just wide enough to plant the seed. Instead of tilling the land horizontally, the strip-till method focuses on tilling the land vertically, consistent with how plants naturally grow in the field. This enables nearly the entire crop residue from the previous year to remain on the soil surface as protection against wind and water erosion. Another major benefit to these tillage practices is the significant gains in water conservation. Water conservation alone may induce widespread shifts to strip-tillage in the area. Strip-tillage can also better direct fertility products to where the plant will be throughout the growing season, resulting in large fertility efficiency gains, as well as reduced fertility pollution. All of the dryland corn in the operation is now produced using either strip-till or no-till methods. No-till indicates that the only pass of tillage over the acres is to plant the crop.

Like the rest of the agricultural sector, recent technological improvements and increased demand for this type of tillage have made one-pass tillage and planting equipment more efficient and easy to operate. Benefits of reduced tillage are decreased costs of production, decreased manpower requirements, decreased soil compaction, improved soil composition, and improved wind and water erosion control. Despite the benefits of one-pass tillage, it is also important to note that EWS Farms is willing to revert to conventional tillage in infrequent applications if gains are expected from such action, as may be the case with winter wheat production.

The winter wheat acreage on the farm is conventionally tilled during a summer fallow period. Although much of the dryland acreage in the area has converted to a rotation involving two crops in three years, many people still believe the climate in the area will not support continuous cropping on the dryland acreage. Some producers in the area have begun to use chemical fallow for the idle period; however, uncertainties and concerns have influenced EWS Farms in its decision to continue using the conventionally tilled winter wheat production practices. Areas of concern include administration of the proper dosage, application costs, general use of chemicals, and asset fixity/high initial capital requirements of converting dryland equipment to reduced-tillage capabilities.

An important factor in the transformation of the tillage practices is the management of weeds and soil fertility. In addition to purchasing a large amount of inputs at the cooperative, the operation also takes advantage of the weed management and soil fertility services provided by the cooperative's agronomy service. With this program, an in-house crop production specialist aids in planning and recommending herbicide, pesticide, and fertilization programs, based primarily on the operation's yield expectations. In this case, the operation depends highly on the services of the local cooperative.

3.5.3 Cost of Production

As previously stated, one of the case farm's comparative advantages is that it has been managed for low-cost production. As can be seen in Table 3.3, EWS Farms is somewhat successful at reducing costs; its production costs are lower than standard crop budgets prepared by Colorado State University Extension. Wheat costs, in particular, are much lower. These reductions have not come at the expense of production either, since the yields at EWS Farms are comparable to the county average (see Table 3.4). Although management is careful not to decrease costs of production to the point of negative consequences, some of the input costs have been identified as opportunities to decrease overall production costs.

One of the main areas identified is the cost of hired labor. To achieve low labor costs, EWS Farms should only utilize family labor, as it has done in the past, so no hired labor costs are incurred by the operation unless absolutely necessary. The family is responsible for all farm operations, including capital equipment maintenance and repairs. Another strategy for keeping low production costs involves running a used machinery compliment with low levels of machinery debt. (This is evidenced from the accompanying set of financial information that will be seen later.) Although there is a balance of machinery cost and efficiency loss, the operation has made a substantial effort to remain on the low end of the machinery ownership and operating costs without losing efficiency, compared to other agricultural producers its size. Site-specific calculations on seeding rates, fertilizer, and chemical application rates and irrigation requirements are also computed to achieve low levels of production costs. The operation uses economic and biologic information at the field level to make the best decision for production. This differs from other operations that base managerial decisions on the performance of the entire farm. As a result of field-level management practices, in many instances EWS Farms is able to reduce production costs.

3.5.4 Marketing

The operation's marketing scheme is not formalized. Although management has a solid grasp on production costs and breakeven prices as a reference, there is no

TABLE 3.3
Cost of Production Comparison of EWS Farms and Colorado State University Extension Enterprise Budgets

Category	EWS Irrigated Corn ($)	CSU Irrigated Corn ($)	EWS Dry Corn ($)	CSU Dry Corn ($)	EWS Winter Wheat ($)	CSU Winter Wheat ($)
			Operating Costs			
Seed	40.00	35.70	20.00	14.90	4.00	6.39
Fertilizer	48.00	56.71	18.00	15.27	14.00	17.28
Herbicides	30.00	27.35	25.00	19.50	0.00	4.00
Insecticides	30.00	9.33	0.00	0.00	0.00	0.00
Irrigation Energy	60.00	63.81	0.00	0.00	0.00	0.00
Irrigation Repair	10.00	10.00	0.00	0.00	0.00	0.00
Crop Insurance	20.00	11.79	15.00	6.45	8.00	7.85
Sprinkler Lease	0.00	60.00	0.00	0.00	0.00	0.00
Custom Spray	0.00	4.50	0.00	0.00	0.00	4.00
Crop Consultant	7.00	6.00	0.00	0.00	0.00	0.00
Machinery Operating Costs	40.00	46.27	16.00	28.80	13.12	24.88
Custom Hauling	0.00	28.21	0.00	2.12	0.00	6.24
Total Operating Costs	**285.00**	**359.67**	**94.00**	**87.04**	**39.12**	**70.64**
			Ownership Costs			
Machinery Ownership Costs	8.14	56.29	3.56	29.55	3.45	60.04
General Farm Overhead	2.00	10.00	2.00	10.00	2.00	10.00
Real Estate Taxes	6.00	5.00	2.00	1.86	2.00	1.80
Total Ownership Costs	**16.14**	**71.29**	**7.56**	**41.41**	**7.45**	**71.84**
Total Direct Costs	**301.14**	**430.96**	**101.56**	**128.45**	**46.57**	**142.48**

Source: CSU Extension, 2006. (www.colostate.edu)

TABLE 3.4
Comparison of EWS Farms Historical Yields to County Averages (bu/acre)

Crop	1990	1991	1992	1993	1994	1995	1996	1997	1998	1999	2000
Irrigated Corn Grain											
County	163.50	166.00	148.00	112.00	178.50	124.50	136.50	158.50	165.50	165.50	181.50
EWS Farms	170.33	191.62	180.00	118.00	194.98	184.32	195.05	189.46	177.85	121.46	190.00
Dryland Corn Grain											
County	65.00	53.50	73.00	52.00	49.50	39.50	63.50	64.00	67.50	73.00	34.50
EWS Farms	60.73	53.24	51.57	60.83	53.28	63.94	65.31	59.39	8.96	64.47	59.00
Winter Wheat											
County	39.00	33.00	31.50	44.50	26.50	41.50	33.50	34.50	45.50	44.00	28.00
EWS Farms	44.95	22.31	38.82	53.55	24.82	46.43	18.83	32.36	53.74	46.34	35.00

Source: US Department of Agriculture, National Agricultural Statistics Service (www.nass.usda.gov; accessed June 9, 2009).

formal decision or plan as to how and when to market commodities. EWS Farms recognizes that this is an area of future opportunity because the operation has much to gain from the use of an organized marketing plan. At the moment, nearly all of the grain produced by the operation is stored at and marketed through the local cooperative. In some instances, cash forward contracts from the local cooperative are used in an attempt to capture positive pricing opportunities. However, most of the commodities are marketed through a cash sale at harvest or at some point after harvest. Government loans on sealed commodities are also utilized.

3.6 PERFORMANCE INDICATORS

3.6.1 Production Performance

Although the area is diversified with respect to soil type, as well as terrain from the north to the south, all areas in the county produce at relatively high corn and wheat yields compared the rest of the state (Table 3.4). Irrigated corn yields for the farm typically range near the 200-bushel-per-acre mark while the dryland corn yields usually fall in the 50- to 60-bushels-per-acre range. Winter wheat yields are usually between 35 and 45 bushels per acre for the farm. When compared to the Phillips County average yield for irrigated corn of 155 bushels per acre (as reported by the National Agricultural Statistics Service of the USDA), so the expected yield of 200 bushels per acre seems to be strong. This is also the case for the winter wheat, with the county yield being approximately 36 bushels per acre versus the operation's average, which is reported as 41 bushels per acre. The dryland corn in the area is a relatively new rotation and, therefore, the reported averages should be evaluated with some caution. However, the county average for dryland corn is reported as approximately 58 bushels per acre, much in line with the expectation of the operation for 60 bushels per acre.

There are several important facts that pertain to the evaluation of the farm-level yield data. First, recent dryland yields in the area have been skewed from normal due to the presence of the severe drought in the region. Also, when comparing the farm-level expectations to the county-level expectations, it is important to remember the effects of weather risk in the area often cause production yields to vary greatly. As stated earlier in the chapter, it is not uncommon for farm yields in the area to be cut in half by even a single storm during the growing season. For example, EWS Farms has experienced some hail over the period of yield evaluation, but only a couple severe losses have been due to weather. As the data indicate, overall production capabilities on the farm seem to be adequate for the area and are consistent over time.

3.6.2 Financial Performance

A detailed evaluation of the case farm's financial performance follows in Chapter 5; however, it is worthwhile to mention some of the financial measurements that will be evaluated in subsequent chapters. The first three measurements are taken from the set of coordinated financial statements prepared for the case farm. Financial data include net business income, annual net cash flow, and change in equity from beginning to end of year. The other two measurements are ratio calculations, such as the debt-to-asset ratio and the rate of return on business assets, which are also found on the financial statement. The accompanying coordinated financial statements also reveal a rudimentary look at the overall debt structure of the case farm.

Net business income can be interpreted as the return to the farmer for unpaid labor, management, and owner equity. In other words, it is a measure of profitability. Net income is calculated by matching revenues with the expenses incurred to create those revenues plus the gain or loss from the sale of capital assets before taxation occurs. Net income for EWS Farms was $29,151 for the year analyzed, as reported at the bottom of the accrual adjusted income statement.

Annual net cash flow is a measure of the operation's ability to meet cash requirements throughout the year. Cash flow is calculated as the difference in the cash inflows and cash outflows of the operation over a certain time period. Annual net cash flow is reported in the financial statements at the bottom right-hand side of the cash flow statement as $7,591. The distinction between cash inflows and outflows versus incomes and expenses is important and will be further detailed in subsequent chapters.

The last measurement from the set of coordinated financial statements is the change in equity from the beginning to the end of the year. This measurement is analyzed on the last line in the right-hand column of the balance sheet. It is calculated by finding the difference in the business net worth from the beginning to the end of the year. This measurement indicates the amount of growth or loss that occurred within the business over the time period studied. For the case farm, the change in equity from the beginning to the end of the year is a negative $15,849.

The first of the two ratios analyzed for the case farm is the debt-to-asset ratio. The debt-to-asset ratio is a solvency measure that expresses the proportion of total farm assets owed to creditors. More plainly said, it is a measure of the claims against the assets of the operation. In traditional agricultural operations, a smaller percentage of debt to percentage of asset owned is considered more favorable. For example, it is common for lenders in the agricultural lending sector to prefer an operations debt-to-asset ratio not exceed 50%. The debt-to-asset ratio calculated for the case farm is 17.77%, clearly under the 50% mark. This is a solid ratio for the case farm. A driving factor behind this low ratio is the management decision to run a used machinery line that incurs very little ownership cost.

The second ratio evaluated is the rate of return on business assets. This ratio measures the rate of return on the operation's assets and can be used as an overall measure of profitability. The rate of return calculated for the case farm is 1.32%, which seems to be in line with the agricultural production sector.

A complete discussion about what these financial indicators mean is found in Chapter 5 on financial health. A summary of selected indicators is provided in Table 3.5.

3.7 CONCLUDING REMARKS

EWS Farms is a perfect case study because it is a traditional farm that is in an expansion mode. Aaron needs to evaluate his risks as he contemplates how to include his family in the operation. To demonstrate that the Risk Navigator tools and techniques

TABLE 3.5
Select Financial Indicator Ratios for EWS Farms

Measure/Ratio	Value	Ratio
Liquidity	**($)**	**(%)**
Current Ratio		
Current Assets	240,991	
Divided by Current Liabilities	36,448	6.61
Solvency		
Debt/Asset Ratio		
Total Business Liabilities	279,288	
Divided by Total Business Assets	1,571,491	17.77

can be used in real-life settings, we apply each SRM step to EWS Farms in the first section of each chapter. We hope this will make it easier for agricultural managers to relate to the tools used and to apply them to their own businesses.

4 The Strategic Risk Management Process©

Dana Hoag and John P. Hewlett

CONTENTS

Most agricultural producers are not involved in production agriculture just to make money. When asked, many respond that they also enjoy the lifestyle and being close to the land. Many also comment that it is a great way to raise a family in a small-town, rural environment. Some inherited their operations from family members. But risk can threaten any of these values. One of the best ways to avoid unwanted change is to develop a strategic plan. This chapter shows you how to do that specifically for risk.

Agricultural economics, economics, business, and finance offer farm and ranch managers many tools and techniques for decision making when it comes to risk. The concepts of profit, marginal analysis, cash and noncash costs of production, enterprise analysis, and financial analysis all provide decision makers with valuable information. What is still lacking is a framework for pulling these various pieces together in a practical and accessible way to include both the detail and the breadth of experience that farm and ranch managers face on a daily basis. For this purpose we developed the Strategic Risk Management process (SRMP or SRM process) and called it Risk Navigator SRM©. The SRM process provides such a framework and is designed for use on farms, ranches, and other agribusinesses. You will manage

FIGURE 4.1 Strategic Risk Management Process©. Source: Dana Hoag, The Strategic Risk Management Process. Unpublished proposal, Department of Agricultural and Resource Economics, Colorado State University, 2005.

risks better if you follow the steps, read about each step, and use the tools on our Risk Navigator SRM Web site that are specifically designed for each step (www.risknavigatorsrm.com).

The idea for SRMP came from Hoag (2005), as he searched for a way to make risk management more accessible. He found that risk management concepts could be portrayed through the well-established strategic management literature from business, as described in Figure 4.1. The process will assist agricultural businesses by developing a risk management plan that takes into consideration strategic objectives, like resources, risk preferences and the long-term goals of the operation, tactical tools to address risk problems, and operational steps to see that the plan is implemented, monitored, adjusted, and reapplied.

The process is divided into three main parts—strategic, tactical, and operational—containing ten specific steps. The process is cyclic with feedback and reevaluation as conditions change. That is, risk management requires continuous evaluation and reevaluation. Management decisions are based on the operation's goals, actual performance, and consider current and forecast conditions that affect all types of risk.

The SRMP differs from other types of planning for the future. Users not only consider what the future may look like and what risks may be present in their operations, but they also learn how to determine the best niche or position for their business to occupy given risk preferences, goals, and anticipated threats and opportunities. We discuss each stage below.

4.1 THE SRMP STRATEGIC STAGE

The SRMP is an *integrated* process, meaning the strategies identified in the initial steps are linked to the objectives and plans developed in subsequent steps. These interconnections of the various steps help ensure that all planning levels are interrelated and that all players are aware of the desired outcomes and plans for achieving them. When all of a business's resources can be aligned toward its identified goals, success, while not guaranteed, is much more likely. In addition, the process must consider stage-of-life issues for the people involved. For example, young families are more vulnerable than those who are more established. Transition of business ownership and management from one generation to the next must be carefully executed to maintain viability.

This stage of the process is labeled *strategic* because it involves setting strategy. In a military sense, strategy describes the overall plan designed to achieve a particular outcome. In a competitive situation, a winning risk management strategy is one that results in sustainable positive outcomes over the long run with acceptable levels of risk. In this way, strategic risk management is focused on resources and goals that better position the business for the future; the tactical stage that follows focuses on the details to attain those goals. At this point in the process, discussions and analysis are solution neutral (Robinson and Robinson, 2004).

The strategic portion of the SRMP includes the following three steps: (1) determine financial health, (2) determine risk preference, and (3) establish risk goals. These steps define the endowment and attitude of the decision maker, which set some boundaries on which management strategies are most realistic. Someone with a lot of money can take on different risks than someone without any money, for example. It is important to determine how able you are to take on risk before proceeding to making decisions about how risks will be managed. Subsequent steps of the SRMP describe the specifics of how to move the business toward its desired future, and we will look closely at those as we move along.

4.1.1 Step 1: Determine Financial Health

Determining financial health refers to assessing the well-being of the business's financial resources. This process will identify areas of financial strength and weakness within the business. Doing so helps management better understand vulnerabilities, and allows for the development of plans to reduce them to acceptable levels. In addition, the practice may help identify areas of underutilized capacity, perhaps offering the option to capitalize on developing opportunities.

Just as athletes who train for a particular sport must develop strength in the specific areas the sport demands, they also must be aware of their overall health. Developing only the muscles necessary for lifting heavy weights may make an athlete well able to lift weights beyond the strength of others, but may make him ill-suited to win a 100-yard dash. Financial resources also may be overdeveloped in some areas and less so in others. In general, a balance of performance in each area of interest leads to a healthier business better able to withstand the shifting winds of change in the general economy. Put another way, the chain of financial health is only as strong as its weakest link. For this reason, it is important to assess where the weaknesses lie.

Financial health is multifaceted. As such, several measures are used to determine the health of this important resource. In general there are five areas of financial health that any business should be concerned about: (1) liquidity, (2) solvency, (3) repayment capacity, (4) profitability, and (5) financial efficiency. Each area is unique and important to the overall performance of the financial resource. A series of 16 ratios or indexes are utilized to evaluate financial health. Data for the ratios is derived from the four basic financial statements: the balance sheet, a cash flow statement, an accrual adjusted income statement, and a statement of owner equity. Finally, you can develop your own personal credit score as an indicator of financial health using the Risk Navigator SRM tools developed for this step.

4.1.2 Step 2: Determine Risk Preference

Step 2 in the SRM process is unique to risk. Most people prefer more money to less, but every person is different when it comes to how much risk they will tolerate to get more money. For example, would you be willing to use your savings to become a day trader in the stock market or to take a job in a war-torn country in order to earn big bucks? Many people would not, but some people would.

People have three basic types of risk preferences or tolerance. Individuals who exhibit risk-neutral preferences seek to maximize income while ignoring the presence of risk. Risk-loving people intentionally seek risk, just as people who have an addiction to gambling do. Most people exhibit risk-averse preferences, meaning they are willing to give up income to avoid risk. For example, suppose you were willing to pay $1,100 per year for automobile insurance. Insurance companies are in business to make money, so they have to charge more than they expect to pay out. If it costs only $900 per year to provide the car insurance, you would be paying a $200 premium above and beyond the cost of insurance, which you are willing to do in order to avoid the risk of paying a large settlement in case of an accident. The insurance company would be earning a $200 premium to take on that risk for you. Likewise, a farmer or rancher might be willing to accept lower profits from a marketing contract that reduced his or her price risk or to accept lower profits by vaccinating livestock to avoid disease outbreaks.

How risk averse a person can be is highly variable, which is why economists and other financial professionals have difficulty helping individuals determine the best course of action to follow. Recommendations cannot be made until the advisor knows how risk averse the individual is under different circumstances. The tools presented in Chapter 6 assist in assessing what a given individual's tolerance for risk might be under various conditions. Once you know your financial health and risk tolerance, you can think about the next step: setting specific risk goals.

4.1.3 Step 3: Establish Risk Goals

The next step in the SRM process is to set your risk goals. This is an extremely important step because goals guide the rest of the planning process. Goals should identify both family desires and where the business should be in 5 to 20 years.

To aid in this process, draft a mission statement for the operation to capture the operation's focus and purpose. This statement describes the direction the operation takes in the future.

Next, craft your risk goals. These goals should be SMART: Specific, Measurable, Attainable, Related to other goals and constraints, and Traceable over time. Prioritize your goals based on a list of the resources required. Since resources are limited, address higher-priority goals before moving on to lower-priority goals. The next step is to identify the necessary resources required for each goal to be attained. For example, how much credit is required? Are there needs for additional training? Are my soils the right type for the crops I want to produce? It is important to be honest in the assessment and determine if the goals are realistic given current resources, soil conditions, or livestock herd, etc.

Next, the process moves to setting tactical objectives. These are the methods used to accomplish the goals. Improving yield would be a tactical goal. The final step is to describe the operational plans. These are the action steps taken to achieve the tactical objectives and strategic goals. Chapter 8 provides a more detailed presentation of the specifics of the SRM process for setting risk goals.

4.2 THE SRMP TACTICAL STAGE

In the tactical stage of the SRMP, agriculturalists must evaluate various alternatives for reaching their vision for the future outlined in the strategic stage. That is, you make a solid plan. Individuals address resource constraints, consider alternative methods of risk management, and outline specific steps to follow. Creativity and a willingness to work through the details of various plans are needed to successfully complete this segment. Careful, diligent planning will be rewarded by fewer unexpected bumps in the road and, quite possibly, a more favorable outcome in the end.

Again following the military example, *tactical* refers to a maneuver or action taken to achieve a particular goal. Two management actions that result in the same profit may not be equal in either their resource costs or associated risk levels. For this reason, the tactical stage is acutely focused on evaluating situations for their impact on the resource base, implications for costs and returns, and more importantly, for the levels of risk.

Specific steps in the tactical level of the SRM process include: (1) determine risk sources, (2) identify management alternatives, (3) estimate likelihoods, and (4) rank management alternatives.

4.2.1 Step 4: Determine Risk Sources

The first step in the tactical phase is to determine when risks will come and where they come from, and to prioritize where risk management efforts will pay off most. No one has the time and money to address every risk. Navigator helps you identify the risks that you face and determine which ones need to be prioritized for best management.

We start with identifying risks. To that end, we list the five major types of risk in agriculture: production risk, market or price risk, financial risk, institutional risk, and human resource risk. We then provide a comprehensive list of agricultural risks commonly faced in each of these areas. (A number of tools and techniques are presented in Chapter 8 to help list and describe these risks.) Finally, we offer a variety of tools on the Web site to help organize and prioritize these risks. For example, a contributing factor diagram helps you describe the various risks that contribute to a desired outcome, such as making a profit or being able to pay off a loan. Often just identifying the risks can be a bigger challenge than developing a method for managing them. This flow-charting technique helps the user to think through the various factors needing attention.

When you are done charting your risks, you should have more information than you can possibly pay attention to. Since all risks do not have the same financial impact, a method for prioritizing the sources of risk is also helpful. We offer risk-impact and risk-influence tools to help. These charts highlight each risk by how much impact it has on you and how much influence you have on it. Another helpful tool we discuss is SWOT analysis. A SWOT analysis identifies Strengths, Weaknesses, Opportunities, and Threats to accomplishing your goals.

4.2.2 Step 5: Identify Management Alternatives

Decision makers have to decide how to manage risks after identifying and prioritizing them. There are four basic ways to manage risk: assume it, avoid it, reduce it, or transfer it. The objective is to find the appropriate trade-off between the risk and achieving your personal goals. Some people will choose to assume risk in order to capture the returns that are often associated with it. Of course, even someone who assumes risk will also try to reduce it. At the other extreme, some people are so uncomfortable with uncertainty that they will avoid risk altogether. Risk can be transferred to people who are better prepared to handle it. Farmers, for example, can shift yield risk to insurance companies and the government by purchasing crop insurance.

The objective of any particular risk management strategy is to manipulate the risk profile into a more acceptable form. Management efforts are focused on narrowing risk by squeezing the probability density function (profile) or increasing the expected value of the outcome. For example, production risk can be managed through diversification or by installing irrigation. Marketing risk can be managed with storage or by using the futures market. Maintaining credit reserves will help with financial risk and having a backup management plan can reduce human risks.

4.2.3 Step 6: Estimate Likelihoods

The sixth step in the SRM process provides the tools for estimating the likelihood of various alternatives. This is the last step required to build the actual plan and choose your risk management strategy. Most people find statistics very difficult to understand. Remember that bell-shaped curve you learned about in high school or college? How well do you remember what it means? Could you estimate one of these probability density functions (PDFs) or risk profiles? They contain a wealth of information about how risks affect you. In Chapter 10, the SRMP will provide a detailed description about what these important curves are, how to interpret them, and even how to compute them. In short, you will learn how to create a risk profile.

Also in Chapter 10, several Navigator tools are provided to help develop sophisticated statistical descriptions about how important variables are distributed over time or space. A farmer or rancher, for example, could determine his or her average yield and the PDF will show how much chance there is of getting lower or higher yields, and much more. This information is used to determine which management strategy best suits your personal situation.

Chapter 10 also discusses how to find information and how to build a PDF using this information. You will even learn how to build a PDF using just your own knowledge or opinion when information is scarce.

4.2.4 Step 7: Rank Management Alternatives

The final step of the Tactical Stage of the SRM process is to rank the various alternatives considered to this point and select those with the most desirable outcomes. The SRM process uses the concept of a payoff matrix along with powerful tools in Excel™ to analyze and compare different risks. These are the same methods that highly paid consultants use. Two or more risks can be compared by looking at their returns, the probability of good and bad outcomes, and by factoring in the personal risk preferences of a decision maker.

The Risk Navigator tool called the Risk Ranker organizes your information into a payoff matrix about risks, probabilities, and management alternatives, and creates five different types of analyses that help you compare which risks are right for you. But before doing that, we start by looking at whether risk even needs to be considered. Sometimes, the risk is not important enough to make one management alternative preferred over another. For example, if using irrigation reduces risk and improves profits, risk is not a factor. If risk is important, we provide risk profiles of each management alternative for comparison purposes. Likewise, you can use graphical profiles, such as fan graphs or stoplight graphs, to compare risks. We discuss several preformed risk management strategies such as Stop Light diagrams and Maximin for a more formal comparison.

Finally, we consider a person's personal risk tolerance scores to rank one management system over another. In all, there are over ten different ways to compare and rank management alternatives based on risk and returns.

4.3 THE SRMP OPERATIONAL STAGE

The third stage of the SRM process is the operational stage. It is within this segment that the action plans are implemented by actually taking the planned risks. The focus here is on the day-to-day duties of management. Watching how the plans unfold and adjusting to the inevitable bumps along the way are necessary. It has been said that when implementing a strategic planning process, the user is off-course from the original plan most of the time. Reaching the destination then depends entirely on making course corrections as needed to ensure the business moves in the desired direction. These activities are the focal point of the operational level.

Specific steps in the operational portion of the SRM process include implement plans, monitor and adjust, and replan. Strategic plans are often left in a drawer and never fully carried out, usually due to a lack of diligence in developing all levels of the plan—the vision for the future, consideration of alternative methods for reaching that future, selection of the preferred method and the implementation, and the monitoring and readjustment necessary to see the plan through to completion. This stage reduces that risk. Finally, the process does not end, rather it continues into the future as the business matures, is transferred to the next generation, or evolves to offer a new array of products. The SRM process is depicted as a circle to illustrate this cycle and the need for continuing the process.

The best-drafted strategic goals, tactical objectives, and operational plans will result in nothing but frustration if the planning process does not contain action on the part of the people involved. The operational level is focused on the activities of day-to-day work. At this stage, the planning process should influence and affect what takes place week by week, month to month, and season to season. For best results, there should be a structured approach used as the activities of labor and management are carried out.

Operational-level activities include making sure the operational plans are carried out, that resources are available when and where they are needed, that those responsible for various stages of the production process are providing the needed oversight, and that systems are reacting appropriately when contingencies come to light. In addition, successful shifting toward the future requires simultaneous work on a number of strategic goals and tactical objectives with differing time frames. Operational-stage management provides the needed oversight and coordination to ensure smooth functioning of the business and resource use.

4.3.1 Step 8: Implement Plans

Ultimately, the management team must put whatever plans have been made into action. Implementation of the plan involves acquiring the necessary resources, scheduling the tasks to be completed, and overseeing all aspects of the plan. This is the area where management teams are usually most comfortable. These are the normal day-to-day, get-it-done steps in the process. With proper planning, however, even this step can be accomplished more effectively.

An ordered approach to organizing this phase of management can be of great assistance, especially during busy portions of the year. Implementation includes detailing who is responsible for which activities, where the resources will come from, when the resources are needed, what risks and contingencies are involved, and what priorities are placed on the choices available. Implementing the plans from earlier steps of the SRM process involves addressing all these issues and more.

4.3.2 Step 9: Monitor and Adjust

As plans are implemented, resource use must be monitored and adjustments made as needed. Rarely are plans implemented exactly as outlined on paper, particularly where uncontrollable factors, such as weather and markets, are involved. Implementation and execution must be monitored and midcourse adjustments made if goals and objectives are to be realized. The importance of this step is often overlooked by managers; however, evaluation of how closely plans are being followed is critical to reaching goals. Midcourse adjustments are necessary in nearly every activity people pursue.

Good records are essential for monitoring, evaluating, and making necessary adjustments. Implementation of even basic record keeping when resources have not been previously monitored is essential to successfully completing this step of the process. Where risks are higher or consequences greater, more complete or detailed record keeping may be justified. Yet if management does not spend the time to review the information and remain open to using that information to make periodic adjustments, even the best, most detailed record keeping becomes time and effort wasted.

Management of resource supply and demand is a critical function in the implementation step. However, just as critical are monitoring and adjusting to associated increases in threats or opportunities presented by the changing environment in which the business operates. Coupled with these environmental changes are changes in the risks and contingencies that accompany them. Having spent time considering alternative management strategies and ranking those alternatives in the tactical level of the SRM process provides management with the background necessary to make sound decisions in the face of adverse circumstances. Contingency planning covered in the implementation step also provides the ammunition for quick decision making when adjustments are needed.

4.3.3 Step 10: Replan

Replanning is often ignored, probably because it seems to highlight what was not achieved. However, as most individuals will readily acknowledge, people tend to learn more from failures than from successes. Recognizing what was not accomplished is the first step toward addressing the deficiencies responsible. Identifying reasons that goals are not achieved is crucial in continuing the strategic planning cycle. Although replanning occurs throughout the year as resource use is monitored, it also should occur at year end.

A year-end analysis evaluates the performance of all ranch and farm resources. This analysis includes determining how closely actual performance matches the budget, with consideration of any benchmarks set for resource performance as part of transition planning. After completing this analysis, plans should be drafted for the coming year, completing the SRM process cycle.

As plans are implemented, the assessments should be made with full knowledge of how the implementation came to pass. The positive and negative influence of risk should be accounted for, in addition to identifying areas where portions of the planning process broke down. Only then can replanning begin with confidence. The replanning process is the first step in beginning the cycle anew as the manager moves toward revisiting his or her risk preferences and strategic goals.

Finally, it should be noted that the SRM process is ongoing into the foreseeable future. While some, maybe most, strategic goals and objectives will be met, others will prove too difficult or conditions will change to make them unimportant or too costly to achieve. That outcome is reasonable and acceptable. Accomplishment of all goals and objectives is not the sole measurement of success. Rather, progress toward those goals that are most important will provide many positive returns, particularly if systems have been implemented to allow management the capability of measuring that progress over time.

4.4 CONCLUDING REMARKS

Risk Navigator SRM packages the ten-step SRM process for you. A complete description of each step can be found in Chapters 5 through 12. Each of these chapters describes what the objective of each step is and multiple ways to accomplish it. The www.RiskNavigatorSRM.com Web site contains easy-to-use Internet tools that are the methods we prefer for each step. For more clarity, we have provided an example of how EWS Farms conducted each step at the beginning of each chapter.

REFERENCES

Bastian, C., and J. Hewlett. 2004. *Safety first: A RightRisk lesson guide*. RightRisk Education Team (RR-L-1 January 04). Accessed June 2, 2009 from www.RightRisk.org

Hewlett, J. *Strategic planning for risk management*. Risk and Resilience in Agriculture series. Accessed June 2, 2009 from http://agecon.uwyo.edu/RnRinAg

Hewlett, J. P., C. Bastian., D. Kaan, and J. Trane. 2004. *Goal setting for strategic RightRisk management*. RightRisk Education Team. Accessed June 2, 2009 from http://www.RightRisk.org

Hoag, D. 2005. The strategic risk management process. Unpublished proposal, Department of Agricultural and Resource Economics, Colorado State University.

Robinson, D. G., and J. C. Robinson. 2004. *Strategic business partner: Aligning people strategies with business goals*. San Francisco, CA: Berret-Koehler Publishers, Inc.

Section II

The Strategic Risk Management Process

In this section we discuss all ten steps of the Strategic Risk Management process in full detail. We have organized each of these chapters in three parts in order to make them more effective. You choose the level that best meets your interest. Part 1 explains how we applied the step to our case farm, Part 2 discusses the fundamental theories and concepts for the step, and Part 3 is a user's guide for the tools developed for the step.

PART 1: EWS FARMS

The first part of each chapter is concerned with how the step was applied to EWS farms. This makes it easier to understand what the step is about by relating it to a real farm problem and threads together a full example of the Strategic Risk Management process from start to finish. Some of the information in this section may be a little difficult to fully understand for those people not well versed in the concepts used. However, the concepts are carefully explained in Part 2 for anyone with interest in reading further. In each case, we discuss the EWS farm objectives, strategy, and implementation.

PART 2: FUNDAMENTALS

The fundamentals section is designed to give the details, background, theory, and examples that back up each step. This section is written to provide the information needed to fully understand each step. Many people won't need this level of detail and you can selectively read only the parts you are most interested in if you don't want to wade through it all. We recommend that you read it all whenever possible.

PART 3: RISK NAVIGATOR MANAGEMENT TOOLS

Think of Part 3 as a user's manual for the tools that we develop and demonstrate in Parts 1 and 2. In Part 3 our emphasis shifts from teaching concepts to simply explaining how to use the tools. We demonstrate the tools using the actual EWS farms example. Therefore, you might notice some of the results have already been used in Parts 1 and 2, when we used the tools to help EWS farms complete the steps. In addition, when you download the tools, you can choose this same EWS example or a blank sheet with nothing prefilled.

Many people will be able to pick up the tools and use them without reading the book. More people will be able to use them after reading the example in Part 1. A richer understanding comes from including a thorough reading of the fundamentals section.

5 Step 1
Get a Financial Health Checkup

Duane Griffith

CONTENTS

Producers often enjoy a short break after harvest and a long summer of hard work. Machinery may be stored for the winter, but producers would not think of pulling their equipment into the fields for the next season without thoroughly reviewing its readiness and reconditioning it to ensure good performance. The financial risks of downtime during critical production periods can be substantial.

Similarly, risk managers need to pause and review their financial health, the well-being of a business as measured by adequate financial analysis. Just as machinery and equipment are maintained in good condition, the financial health of a business must be monitored and maintained. Financial resources may be in better shape in some areas and less so in others. An operation may have a strong net worth, but weak net cash flows. In general, healthy performance across all financial measures leads to a healthy business that is better able to withstand the shifting winds of change in the general economy and business environment. Analyzing a business's financial health and the potential damage of weak links is important for long-term survival.

Weak or failing financial health on farms and ranches has contributed to a reduction in farm numbers from 6.4 million in 1910 to 2.1 million in 2003. From 1929 to 2008, the total value of farm production, measured in nominal dollars, rose from $13.8 billion to a forecasted level of approximately $350 billion. Total production expense and net farm income followed a similar pattern (Figure 5.1); however, adjusting income and expense for inflation paints a different picture. In 1929, inflation-adjusted net farm income was $51.5 billion, rising to a high of $108.1 billion in 1946.

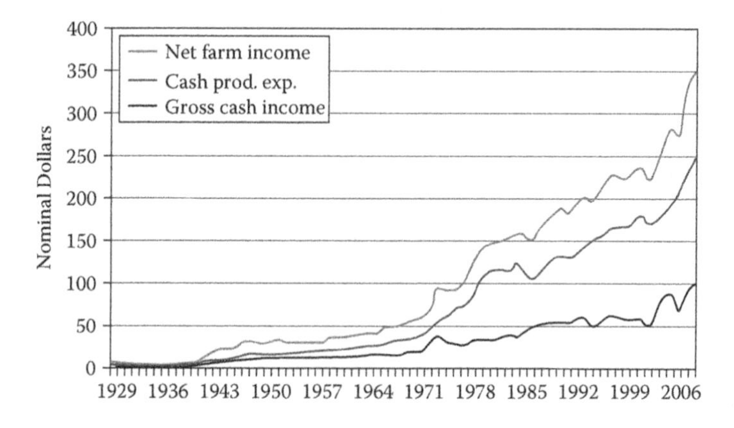

FIGURE 5.1 Farm income and expense stated in billions of nominal dollars, 1929 to 2008-F. (From Bureau of Economic Analysis, accessed June 10, 2009 from http://www.bea.gov)

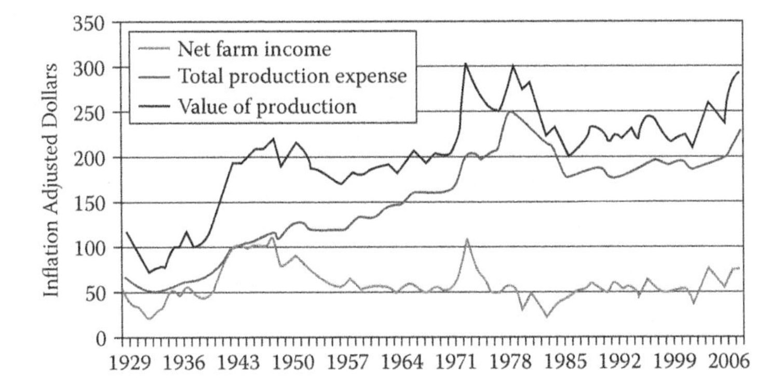

FIGURE 5.2 Inflation adjusted farm income and expense in billions of dollars, 1929 to 2008-F. (From US Department of Agriculture, Economic Research Service, accessed June 10, 2009 from http://www.ers.usda.gov)

By 2002, inflation-adjusted net farm income had fallen to $38.5 billion, but at the time of this writing, it is forecasted to be $75 billion in 2008 (Figure 5.2). Inflation-adjusted net farm income has remained relatively constant for the last three decades and indicates a fundamental characteristic of the agricultural sector. Agriculture is a competitive industry and that competitiveness keeps a lid on net income (Bureau of Economic Analysis).

5.1 PART 1: EWS FARMS FINANCIAL HEALTH CHECKUP

5.1.1 Objectives

It some respects, setting goals and objectives can be relatively easy. We do it for each new year: lose weight, exercise more, don't work so hard so we can spend more time with the family, and the list goes on. Actually achieving these goals can be more difficult. With a new baby on the way, it might be hard to justify the new speed boat. And so it is with agricultural operations. The first step to reaching goals and objectives is to get an accurate picture of your business's financial *position* and past *performance.* This financial information provides the necessary framework to plan for the future.

The objective of Step 1 is to develop a set of financial statements that accurately measure the business's financial position and performance and can be used to evaluate the financial implications of tactical plans used to meet EWS Farms' objectives.

5.1.2 Strategy

Our strategy is to develop the four types of financial statements required to accurately measure business position and performance: Beginning and Ending Balance Sheets, Cash Flow, Accrual Adjusted Income Statements, and a Statement of Owner Equity (SOE) (Frey and Klinefelter, 1980; Farm Financial Standards Council, 1997). Some people may not be familiar with the SOE statement, which uses information from the Accrual Adjusted Income Statement to reconcile the net worth listed on the Beginning and Ending Balance Sheets. Including the SOE is a necessary step in ensuring accurate financial analysis. This can be accomplished without regard to any particular goals or objectives or without regard to any of the other steps in the SRM process, as the financial statements measure current business position and past financial performance for the most recent accounting period.

Since we are looking backwards, historical financial income and expenses and cash inflows and outflows for the business are relatively easy to retrieve from an operation's record-keeping system. But not everything is as easy. We must still make some estimates, such as the market value of land or crops held for sale and what assets and liabilities should be included, to prepare a Balance Sheet. The methods we use can have a significant impact on the resulting measures of financial health. To collect the necessary documents, we start by meeting with EWS Farms' management to gather records and an experienced estimate of the current market value of their assets.

EWS Farms is substantially ahead of the vast majority of producers with the completeness and accuracy of its financial health monitoring. Management already prepares historic cash flow and income statements, which provide true measures of financial *performance*, in addition to a Statement of Owner Equity to reconcile the other financial statements. EWS Farms can use this information to start tactical financial planning to meet business and family goals and to ensure there is no pending future financial disaster.

5.1.3 Implementation

Six-step Strategic Implementation:

1. Develop a Beginning Balance Sheet
2. Develop a Historic Cash Flow Statement
3. Develop a Historic Income Statement
4. Reconcile these financial statements with the SOE
5. Develop financial ratios and measures for the business's financial position and performance from the financial statements.
6. Identify initial strengths or weaknesses based on the financial information developed and share this information with the EWS Farms management team.

We will not bother to go through each of these steps in detail, but we will provide a brief review for awareness of the detail that should be considered. Our focus will be on the big-picture health results for EWS Farms. In general there are 5 areas of financial health a business should monitor: liquidity, solvency, repayment capacity, profitability, and financial efficiency. Data derived from the 4 basic financial statements are used to calculate a series of 16 ratios and measures that help evaluate each of these 5 areas and, taken as a whole, the overall financial health of the business. These statements are described for EWS Farms in this part of the chapter, with more information about how to build and use them provided in the fundamentals section in Part 2.

EWS Farms has provided all of the required financial statements. These statements are in a condensed format but constitute the complete set of financial statements needed for a financial health exam. The condensed form of the statements is provided by an Excel spreadsheet called RDFinancial. The EWS Farms example of RDFinancial is provided as a Risk Navigator SRM tool called RDFinancial EWS Farms in both Excel and as a Flash file. This condensed format is designed to help explore the financial analysis process, while minimizing the efforts necessary to prepare detailed financial statements, maximizing the ability to learn financial management concepts and providing a relatively accurate assessment of financial health. The condensed form is used to focus on learning financial analysis concepts, but detailed statements should also be prepared for the management team.

These statements, seen in Figure 5.3, indicate that EWS Farms is relatively financially healthy. Total assets on the Ending Balance Sheet are listed as $1,571,491 and total liabilities are listed at $279,288, resulting in a Business Net Worth of $1,292,204. The debt-to-asset ratio is a healthy 17.77% on the Ending Balance Sheet, which is solidly in the green. There is $7,591 cash on hand, Ending Net Cash Flow and Ending Balance Sheet cash on hand, with no operating loan carryover. The Cash Flow Statement also indicates there is enough cash available for a family living withdrawal of $45,000. Change in Equity from the beginning to the end of the year, however, is a negative $15,849.

The Accrual Adjusted Income Statement shows gross revenue of $358,424 and total operating expenses of $312,520. Cash interest for Term Debt and capital leases of $12,540 and $4,865 of cash operating expenses are added, along with a negative $651 noncash interest to get a total expense of $329,274. Net Business Income from Operations and Net Business Income are reported as $29,151. These are the same number due to the process used to condense the financial statements. In this instance, Net Income is also reported as $29,151 because the tax estimator built into RDFinancial was turned off. The SOE indicates that the beginning and ending net worth on the Balance Sheets are reconciled, the discrepancy is equal to $0, and we can be confident the information in these financial statements is an accurate representation of the financial position and past performance of EWS Farms at this point in time.

Information provided by this condensed form of financial statements can be used in two ways. The first is to help understand the flow of information between financial statements and using the dollar amounts shown on the financial statements to evaluate financial position and performance. The second is using information on the financial statements to calculate ratios or other measures of financial position and performance. Our initial focus is on using RDFinancial and EWS Farms to help understand the flow of information between the financial statements and basic financial management concepts. Ratios and other financial business performance measures are discussed later.

Note that Net Business Income, Net Cash Flow, and Change in Equity are not the same number (see Figure 5.3). While it may seem obvious that these numbers should not be the same, producers usually cannot explain why they differ. As we start to explore these relationships, follow the dollar values in the condensed financial statements to see how information flows between the statements. Figure 5.3 will be used as a reference point to show how information contained on the financial statements interacts to provide an accurate picture of EWS Farms' financial position and performance as we explore the implementation of family goals listed in Chapter 3.

Percent Crop Revenue	100%	Percent Livestock Revenue	100%
Percent Cost of Production - Crops	100%	Percent cost of Production - Livestock	100%
Percent Gov. Payments	100%		

Balance Sheet

	Assets Beginning	Assets Ending		Liabilities Beginning	Liabilities Ending
Cash on Hand	1,500	7,591	Accounts Payable (Exp)	2,000	2,000
Crops Held for Feed (Exp)	8,400	8,400	Accrued Interest (Exp)	12,540	11,889
Crops Held for Sale (Inc)	200,000	200,000	Current Principal	11,908	12,559
Market Livestock (Inc)	0	0	Other Current Liability (Exp)	10,000	10,000
Other Current Assets (Inc)	15,000	15,000	Short Term Notes Payable (Exp)	0	0
Cash Invt Growing Crops (Exp)	0	0	Other Current Liab. (Not Adj.)	0	0
Supplies&Prepaid Exp. (Exp)	10,000	10,000	Def. Tax on Current Assets	0	0
Total Current Assets	234,900	240,991	Operating Loan Carryover	0	0
Non-Current Assets			**Total Current Liab.**	36,448	36,448
Mach. & Equipment	325,000	292,500	**Non-Current Liabilities**		
Breeding Livestock	0	0	Prin. on T.D. & C.L.	255,399	242,840
Real Estate (Land, Bldgs, Impr)	1,040,000	1,038,000	Total Business Liab.	291,847	279,288
Total Business Assets	1,599,900	1,571,491	**Business Net Worth**	1,308,053	1,292,204
			Change in Equity From Beginning to End of Year		(15,849)

Cash Flow Statement

		Inflows			OutFlows
Crop Sales & Net Insurance Payments		319,920	Operating Expenses	No Interest >	278,020
Mrkt & Cull Livestock Sales		0	Other Cash Business Expense		0
Government Payments		38,505	Cash Int. Exp. - T.D. & C.L.*		12,540
Other Cash Business Income		0	Cash Int. Exp. - Operating	7%	4,865
Operating Loan Proceeds	50.0%	139,010	Loan Prin. Payments - T.D. & C.L.		11,908
Loan Proceeds Capital Assets		0	Breeding Livestock Asset Purchases		0
Non-Business Inflows/Revenue			Mach & Equip & Real Estate Purchase		0
Other Nonfarm Inflows, net of taxes		0	Owner withdrawals		45,000
Other Nonfarm Inflows, net of taxes		0	Cash Taxes Paid (Income & SS)		0
Total Cash Inflows		$497,434	Other Cash Outflows (Not Expenses)		0
				Subtotal	$352,333
			Operating Loan Prin. Payments		$139,010
			Total Cash Outflows		$491,343
			Annual Net Cash Flow (never < zero)		7,591

T.D. = Term Debt, C.L. = Capital Lease

Income Statement - Accrual Adj.

	Income
Cash Income (adj. for cull lvstk sales)	$358,424
Non-Cash Income Adjustments	0
Non-Cash Income (Raised Brdg Lvstk)	0
Capital Gain/Loss on Breeding Lvstk (Net)	0
Gross Revenue	$358,424
	Expense
Cash Expense (Excluding Interest)	278,020
Non-Cash Feed Inventory Adjustment	0
Other Non-Cash Non-Interest Expense	0
Depreciation (Land, Bldgs, Equip.)	34,500
Total Operating Expense	312,520
Cash Int. Exp. - T.D. & C.L.	12,540
Cash Int. Exp. - Operating	4,865
Non-Cash Interest Expense	(651)
Total Expense	$329,274
Net Business Income From Operations	29,151
Net Business Income	29,151
Income & SS Taxes (Cash & Non-Cash)	0
Net Income	$29,151

Statement of Owner Equity

Beginning Net Worth (Cost/Mrkt)		1,308,053
Net Income	+	29,151
Non-Business Cash Inflows	+	0
Owner Withdrawals (Cash)	-	45,000
Asset Valuation Change or Cont./Distrib.	+/-	$0
Calculated Ending Net Worth	=	1,292,204
Reported Ending Net Worth (Cost/Mrkt)		1,292,204
Discrepancy		0

FIGURE 5.3 Condensed financial statements for EWS Farms.

5.2 PART 2: FUNDAMENTALS FOR MONITORING FARM FINANCIAL HEALTH

Financial management can be a daunting task for most producers. A lot of data and information must be tracked, organized, and analyzed. Our intention is to provide a good description of the concepts that will help you become a better financial manager. We explain how to use farm data and information to build the financial tools that you will use. This includes a description of financial statements and how to use this information to determine financial health.

5.2.1 Fundamentals of Preparing Financial Information

Farmers and ranchers typically prepare a Balance Sheet at year end that indicates current financial position, but Balance Sheets do not say anything about how the business arrived at its current position. Producers will typically prepare a cash flow *projection* for the coming year. A projected cash flow estimates cash inflows and outflows for the coming year, ignoring the past year's cash flow performance. Since the Balance Sheet is primarily a measure of financial position, past business performance that resulted in the current financial position is largely lost in a lack of detailed information. Management decisions are based on poor financial information and producers may not react quickly enough to past management decisions that fundamentally threaten the business's ability to survive.

Preparing and *reconciling* all four financial statements is necessary, but understanding the flow of information between the financial statements and how that flow of information affects the reconciliation process is a critical first step in knowing how to use the information to analyze financial position and performance. Reconciliation is the process of verifying how two values at different points in time are related. Think about the process of reconciling the checkbook after receiving the bank statement at the end of every month. This process verifies the ending cash balance on successive bank statements, adjusting for deposits, checks, and other transactions that are not on the bank statement or charges and other transactions made by the bank that are not in the check register. The process accounts for all transactions by both the bank and the checking account owner to verify the amount of cash in the checking account. Reconciliation allows the checking account owner to see exactly how and why the account balance has changed.

From a financial health perspective, the circle in Figure 5.4 can be viewed as a reconciliation process that tells the business owner exactly how the beginning and ending net worth values are related, both measures of financial position at a point in time. Business and family transactions during the year link the beginning and ending net worth values and allow the business owner to understand how and why net worth increased or decreased during the year due to business performance, family related activity, or other outside influence.

A recent survey reported that, on average, only 12% of participating producers felt comfortable with the process of preparing a financial analysis (Coble, Knight, Patrick, Baquet, 1999). Because of their discomfort, only a small number of producers

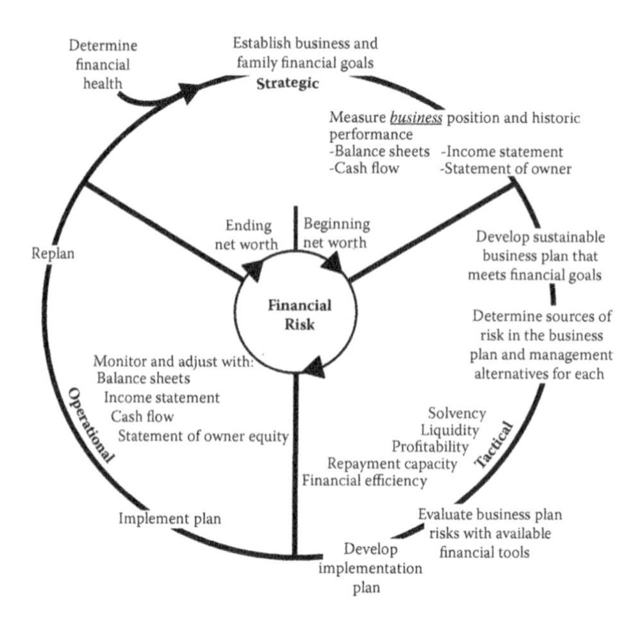

FIGURE 5.4 Application of the financial management process to the SRM process.

actually complete an adequate financial analysis, and this can be the first step toward financial disaster.

Strategic financial analysis is useful in setting a future direction for the operation. Once the direction has been decided, the next step is tactical financial planning, the process of evaluating the financial implications of the selected strategic plan. Tactical planning will help ensure a desired result can be achieved before a plan is implemented. The tactical plan must contain enough information to evaluate impacts on both the financial position and business performance. An example might be a strategy to use off-farm income to support the cash flow needs of the business and families. If that strategy is selected, the entire management team must recognize the contributions of those generating the off-farm income toward meeting the goals of the family business. An important element of success in the financial planning process is that this information must be shared with all parties involved to ensure all family members and business partners (lenders) understand their role in achieving the combined family and business goals.

Procedures to evaluate risks associated with selected tactical plans and risk management strategies are covered in Chapters 6 through 12. Chapter 15 covers evaluation of financial risks related to the goals and objectives of EWS Farms and the tactical plans to help meet these goals and manage risks.

5.2.2 Exploring Financial Statements

While preparing the financial statements, EWS Farms' management team must understand the relevant issues related to each individual financial statement. When completing their financial analysis, they must be aware of the potential pitfalls created by using particular methods and/or including or excluding information from one or more of the statements.

To illustrate the use of the condensed form of the financial statements and start the tactical planning process, let's explore the impact of the family structure on the operation. EWS Farms' family concerns fall into two categories—the ability of the operation to provide an adequate family living for the existing families and additional families who wish to join the operation, and providing support for family members who may not return to the operation. The number of families involved with EWS Farms indicates it may have a difficult time providing support for all possible families. EWS Farms has already indicated that off-farm jobs are on the list for tactical financial analysis. Let's start by measuring the potential impact of bringing two of the children back to the operation without considering supplemental off-farm income. For example purposes, each of the two children and their families are paid $20,000 for family living.

Figure 5.5 shows the results of adding $40,000 to "Owner Withdrawal" on the outflow side of the Cash Flow Statement, bringing total owner withdrawals to $85,000. Previously, the annual net cash flow on the Cash Flow Statement was $7,591, so we can expect that an additional $40,000 cash withdrawal would cause cash flow problems. The historic annual net cash flow is now $0, a result of the way RDFinancial is written, and a $32,409 operating loan carryover is listed on the Ending Balance Sheet. In addition, the value of cash on hand for the Ending Balance Sheet is listed at $0 and the change in equity from the beginning to the end of the year is now a negative $55,849. Net business income, however, did not change and the discrepancy amount on the SOE is still $0. Examination of the SOE before and after this transaction shows the calculated and reported ending net worth (cost/market) have changed, but the new values are identical and still result in a $0 discrepancy.

Why is there no change in net business income? Owner withdrawals in a sole proprietorship, while an *expenditure* (cash outflow), are not an *expense* to the business. Owner withdrawals are assets (cash) withdrawn from the business. This transaction affected the cash flow, Balance Sheets, and the SOE, but the financial statements remain reconciled. Appropriate information on all financial statements is automatically changed by RDFinancial.

The fact that net business income did not change is a key in analyzing the financial implications of this transaction. The income statement is rarely prepared for agricultural operations, but without an income statement, there is no opportunity to compare net cash flow and net business income. The added family structure is a support requirement that the current resource base did not need and cannot provide.

Let's explore the interaction of information on the financial statements a little further. When comparing financial statement values in Figure 5.3 to Figure 5.5, we get before and after results of the added family living withdrawal. It is fairly intuitive to follow the additional cash outflow trail created. The zero cash flow and cash on hand at the end of the year, along with the operating loan carryover of $32,409 are a direct result of having a positive cash flow of $7,591 before adding an additional $40,000 of owner withdrawal ($40,000 additional outflows minus $7,591 previous available cash = $32,409 operating loan carryover). Changes on the Balance Sheet are captured by the SOE under "Reported Net Worth."

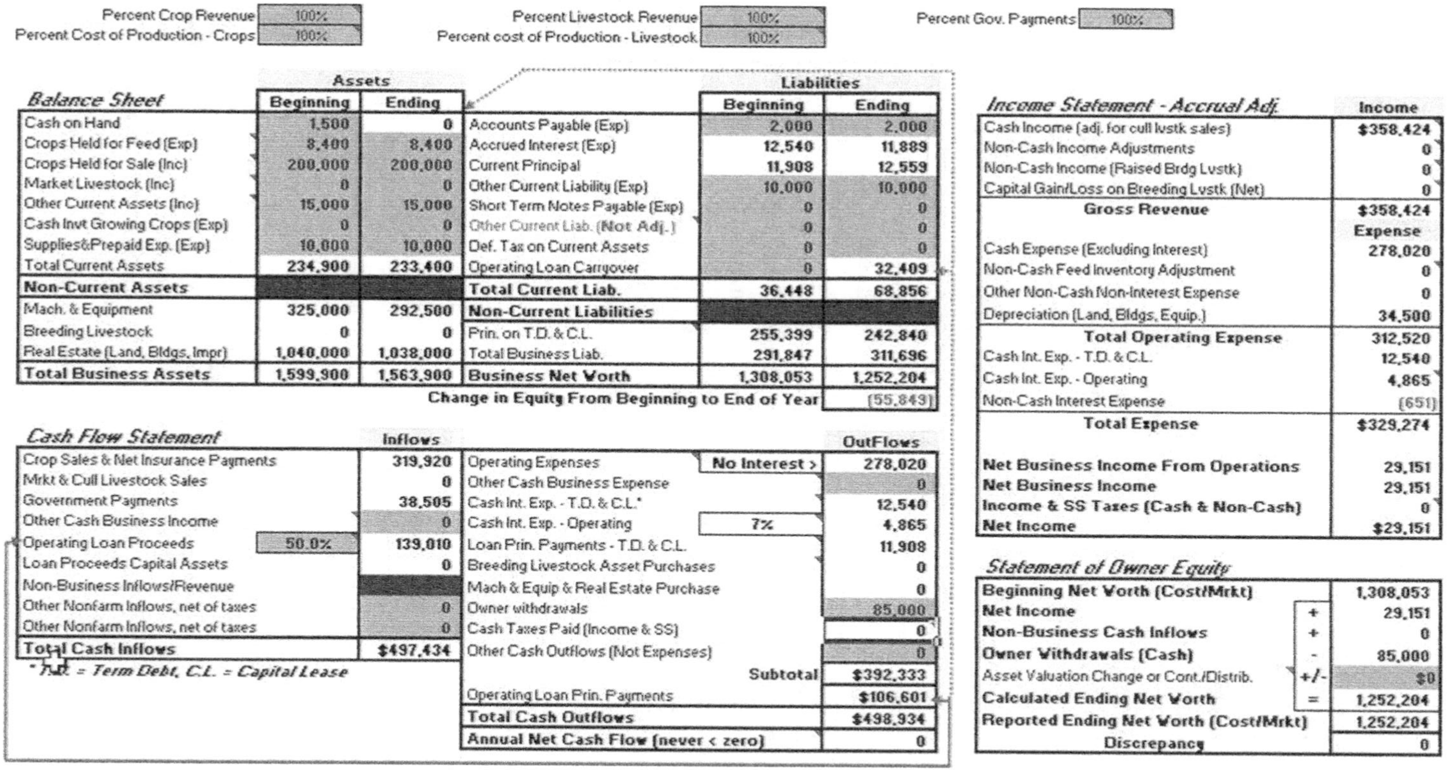

Percent Crop Revenue	100%
Percent Cost of Production - Crops	100%
Percent Livestock Revenue	100%
Percent cost of Production - Livestock	100%
Percent Gov. Payments	100%

Balance Sheet

	Assets Beginning	Assets Ending		Liabilities Beginning	Liabilities Ending
Cash on Hand	1,500	0	Accounts Payable (Exp)	2,000	2,000
Crops Held for Feed (Exp)	8,400	8,400	Accrued Interest (Exp)	12,540	11,889
Crops Held for Sale (Inc)	200,000	200,000	Current Principal	11,908	12,559
Market Livestock (Inc)	0	0	Other Current Liability (Exp)	10,000	10,000
Other Current Assets (Inc)	15,000	15,000	Short Term Notes Payable (Exp)	0	0
Cash Invt Growing Crops (Exp)	0	0	Other Current Liab. (Not Adj.)	0	0
Supplies&Prepaid Exp. (Exp)	10,000	10,000	Def. Tax on Current Assets	0	0
Total Current Assets	234,900	233,400	Operating Loan Carryover	0	32,409
Non-Current Assets			**Total Current Liab.**	36,448	68,856
Mach. & Equipment	325,000	292,500	**Non-Current Liabilities**		
Breeding Livestock	0	0	Prin. on T.D. & C.L.	255,399	242,840
Real Estate (Land, Bldgs, Impr)	1,040,000	1,038,000	Total Business Liab.	291,847	311,696
Total Business Assets	1,599,900	1,563,900	**Business Net Worth**	1,308,053	1,252,204
			Change in Equity From Beginning to End of Year		(55,849)

Cash Flow Statement

		Inflows			OutFlows
Crop Sales & Net Insurance Payments		319,920	Operating Expenses	No Interest >	278,020
Mrkt & Cull Livestock Sales		0	Other Cash Business Expense		0
Government Payments		38,505	Cash Int. Exp. - T.D. & C.L.*		12,540
Other Cash Business Income		0	Cash Int. Exp. - Operating	7%	4,865
Operating Loan Proceeds	50.0%	139,010	Loan Prin. Payments - T.D. & C.L.		11,908
Loan Proceeds Capital Assets		0	Breeding Livestock Asset Purchases		0
Non-Business Inflows/Revenue			Mach & Equip & Real Estate Purchase		0
Other Nonfarm Inflows, net of taxes		0	Owner withdrawals		85,000
Other Nonfarm Inflows, net of taxes		0	Cash Taxes Paid (Income & SS)		0
Total Cash Inflows		$497,434	Other Cash Outflows (Not Expenses)		0
				Subtotal	$392,333
			Operating Loan Prin. Payments		$106,601
			Total Cash Outflows		$498,934
			Annual Net Cash Flow (never < zero)		0

*T.D. = Term Debt, C.L. = Capital Lease

Income Statement - Accrual Adj.

	Income
Cash Income (adj. for cull lvstk sales)	$358,424
Non-Cash Income Adjustments	0
Non-Cash Income (Raised Brdg Lvstk)	0
Capital Gain/Loss on Breeding Lvstk (Net)	0
Gross Revenue	$358,424
	Expense
Cash Expense (Excluding Interest)	278,020
Non-Cash Feed Inventory Adjustment	0
Other Non-Cash Non-Interest Expense	0
Depreciation (Land, Bldgs, Equip.)	34,500
Total Operating Expense	312,520
Cash Int. Exp. - T.D. & C.L.	12,540
Cash Int. Exp. - Operating	4,865
Non-Cash Interest Expense	(651)
Total Expense	$329,274
Net Business Income From Operations	29,151
Net Business Income	29,151
Income & SS Taxes (Cash & Non-Cash)	0
Net Income	$29,151

Statement of Owner Equity

Beginning Net Worth (Cost/Mrkt)		1,308,053
Net Income	+	29,151
Non-Business Cash Inflows	+	0
Owner Withdrawals (Cash)	-	85,000
Asset Valuation Change or Cont./Distrib.	+/-	$0
Calculated Ending Net Worth	=	1,252,204
Reported Ending Net Worth (Cost/Mrkt)		1,252,204
Discrepancy		0

FIGURE 5.5 Financial impacts of increasing owner withdrawals to $85,000.

Reasons for the negative equity change, –$55,849, are not so obvious. If an Accrual Adjusted Income Statement is not prepared, the reason for this change cannot be determined and Balance Sheet values may never be calculated correctly. Successive Balance Sheets prepared independently and unreconciled cannot explain how net worth changed. Examining the SOE, we find why the change in net worth is negative. Net income, $29,151, from the Accrual Adjusted Income Statement, is less than owner withdrawals, $85,000, by the exact amount that net worth changed during the year, $55,849. Asset revaluation and contributions or distributions were not allowed in this example, so there is only one way equity can increase during the year. Net income must be greater than owner withdrawal, i.e., you must earn growth in equity by operation of the business and that earned growth must be greater than owner withdrawals. The impact of two additional family members has increased the size of the negative earned growth in equity shown in Figure 5.3 and created an operating loan carryover.

Is this operation now in financial danger? In the short term, no. Continuing without recognizing the relationship between family living and the resource base's ability to earn a profit (net business income), in addition to the resulting financial implications, will eventually create serious financial and family problems. Since the vast majority of producers do not prepare an Accrual Adjusted Income Statement, they cannot reconcile the financial statements they do prepare and never see the relationships between the recommended financial statements. Without a good understanding of these relationships, it is easier to make decisions with negative long-term consequences. It is even easier to allow delayed reaction to these consequences.

We are purposely ignoring a tremendous number of details that may accompany this type of change in the real world. If, however, we assume the change in owner withdrawals is representative in aggregate and captures most of the details, there are serious long-run financial and family consequences for this operation. These transactions help us explore the condensed form of the financial statements and learn basic financial management concepts. We will continue to use transactions throughout this text that are representative of the real world, while ignoring nitpicky details.

5.2.3 Individual Financial Statements

The four basic financial statements have been introduced to indicate their significance as a complete set of financial statements for analysis and to pose questions for the reader. If one or more of these statements is not prepared, what measures of business position and performance are being missed and how does this hinder understanding of the firm's financial health? How would additional measures obtained from a full set of financial statements help the firm assess the financial health of the business? What is the result of using financial ratios and measures prepared using the wrong information? Lack of adequate financial analysis may doom any strategic planning process. Producers who do not know how they arrived at their current financial health *position* may have difficulty implementing a plan for where they want to go. Is the current financial position the result of good business management, efficient operation, asset appreciation, contributions from off-farm income, gifts and inheritances, or other factors? What portion did business performance versus asset appreciation play in the current financial position of the business?

Format, contents, and procedural issues for each of the recommended financial statements are reviewed briefly below with the intent to provide some background for future discussion and interpretation of the results from each financial statement. It is not our intent to cover detailed preparation of these statements. References to material for detailed preparation have already been provided. The intent is to raise awareness about how recording detailed financial information while preparing financial statements can affect the results and interpretation of the financial information.

5.2.4 Business Only versus Business and Personal

Before we review particular issues with the Balance Sheet, cash flow, and income statements, one issue can be addressed that affects the content and format of all three. The agricultural industry typically includes personal assets and liabilities on financial statements. For an analysis of *business position and performance*, it would seem logical to eliminate all personal activity. While easily said, this is difficult to implement, and the distinction between business and personal assets is often fuzzy (Langemeier, 2004a; Farm Financial Standards Council [FFSC], 1997). Lenders may require information about off-farm activity that helps support the family and business. Record-keeping procedures used by producers also make it difficult to separate family and business activity.

Before preparing financial statements, decide what information will be used. If off-farm wages and personal assets must be included to satisfy a lender, include all relevant information on each financial statement. Even though personal information may be included, accurate measures of *business* position and performance can be produced if the nonbusiness assets, liabilities, income and expense, and cash inflows and outflows are separated from their business counterparts. The format of the financial statements must include nonbusiness categories to allow this separation to occur. RDFinancial's condensed form allows separation of family cash inflows and outflows but ignores family assets and liabilities.

5.2.5 Balance Sheet Content

The Balance Sheet lists all assets, liabilities, and the business net worth. Liabilities are the dollar value of claims by others on total assets. Equity is calculated as total assets minus total liabilities and estimates the dollar amount of total assets the owner would receive if business assets were liquidated in an orderly fashion. Figure 5.6 shows a condensed form of the information contained in a Balance Sheet.

The Balance Sheet is a major source of financial information for producers and lenders in agriculture as it provides a snapshot of the financial *position* at a point in time. Trends developed from successive Balance Sheets provide key measures and indications of the financial *performance* over time. A key indicator, if not *the* key indicator, of financial performance for a farm or ranch business is its ability to earn growth in equity through operation of the business. If a business is not able to generate *earned* growth in equity, it must rely on infusions of cash and other resources from outside the business in order to survive.

In general terms, every dollar of business income increases net worth and every dollar of business expense decreases net worth. Nonbusiness inflows (off-farm inflows) included on the business/family financial statements increase net worth and nonbusiness outflows, increased family living withdrawals, affect cash on hand listed on the Balance Sheet, and decrease net worth. Information on the income statement and cash flow affect asset and liability accounts on the Balance Sheet, and asset and liability accounts on the Balance Sheet can affect information on the income statement. Accurate measures of business position and performance over time must include all four financial statements.

5.2.5.1 Purpose of the Balance Sheet

Balance Sheets are used for two primary purposes. The first is business analysis for an ongoing operation. This type of Balance Sheet plays a large role in a lender's decision to advance credit based on collateral available for loans. The second purpose of the Balance Sheet is an inventory of assets and liabilities to estimate the liquidated value of a business and the outcome of a voluntary or forced sale on all parties concerned. A Balance Sheet prepared to estimate the results of a forced liquidation often use values for assets and/or includes items for liabilities that may be substantially different than values used to analyze an ongoing business. The focus of our discussion is business position and performance measures for an ongoing business entity and we will not discuss Balance Sheets prepared for forced sales. For more detail about this topic, visit the Financial Accounting Standards Board (FASB) Web site (www.fasb.org).

5.2.5.2 Balance Sheet Issues

Issues that must be addressed when preparing a Balance Sheet include whether to use cost or market values for assets, the possible use of one or two columns to record both cost and market values, the number of asset and liability categories to be included on the Balance Sheet, and including or excluding personal assets. The next section of this chapter briefly reviews a Balance Sheet's format, organization, and content to provide a frame of reference for future analysis. The following discussion provides a taste of Balance Sheet issues without becoming mired in details.

5.2.5.3 Format and Organization of the Balance Sheet

All Balance Sheets, regardless of their particular format, have major categories of assets and liabilities. The majority of Balance Sheets prepared for agricultural operations are divided into current, intermediate, and long-term asset and liability categories. Agriculture is the only industry using an intermediate category on Balance Sheets. The Farm Financial Standards Council (FFSC, 1997) recommends only current and noncurrent asset and liability categories as a move toward standardizing financial statement preparation that conforms to generally accepted accounting principles (GAAP). This recommendation is easily implemented by combining the Intermediate and Long-Term Assets into a Non-Current Assets category and combining Intermediate and Long-Term Liabilities into one Non-Current Liability category. Net worth is calculated as total assets minus total liabilities.

Column 1	Column 2	Column 3	Column 4
Balance Sheet	*Cash Flow Statement*	Cash Basis Income Statement	**Accrual Adj. Income Statement**
Assets:			
Current Assets: (CA)	*Beginning Cash Balance*		
Checking & Savings			
Supplies & Prepaid Expense	+ *Cash Inflows:*	Income:	Income:
Livestock Held For Sale	Cash Business Income	Cash Business Income	Cash Business Income
Crops Held For Sale			
Feed Inventory			**Current Asset Adjustments**
Investment in Growing Crops	*Other Inflows:*		**Crops Held for Sale**
Accounts Receivable	*Gifts*		**Livestock Held for Sale**
Gov. Payments Receivable	*Inheritance*		**Hedging Accounts**
Marketable Securities	*Off-farm Inflows*		**Accounts Receivable**
Other Current Assets	*New Loan Proceeds*		**Government Payments**
Total Current Assets (CA)			
	+ *Capital Asset Sales*		
Long-term Assets: (LTA)			
Machinery & Equipment			
Breeding Livestock			
Real Estate	= *Total Cash Inflows*	Total Business Income	**Total Business Income**
Buildings & Improvements			
Other Long Term Asset			
Total Long-Term Assets (LTA)			
Total Assets = (CA + LTA)	- *Cash Outflows:*	Expenses:	Expenses:
	Cash Business Expenses	Cash Business Expenses	Cash Business Expenses

Liabilities:			
Current Liabilities: (CL)	*Other Cash Outflows*		**Current Asset Adjustments**
Accounts Payable	*Principle Payments*		**Supplies & Prepaid Exp.**
Accrued Interest	*Family Living Draw*		**Investment in Growing Crops**
Principal Due in 12 Mo.	*Savings and Retirement*		**Crops Held for Sale**
Personal Property Tax Due	*Other Owner Draw*		**Current Liability Adjustments**
Real Estate Tax Due			**Accounts Payable**
Accrued Lease Payments			**Short Term Notes**
Accrued Payroll Tax			**Accrued Interest**
Short Term Notes Payable			**Accrued Taxes**
Other Current Liabilities			**Accrued Lease Payments**
Total Current Liabilities (CL)			
Long-term Liabilities: (LTL)			
Machinery		Depreciation Expense	Depreciation Expense
Equipment			
Breeding Livestock			
Real Estate			
Buildings & Improvements	*- Capital Asset Purchases*	± Gain or Loss on Capital	± Gain or Loss on Capital
Other Long Term Loans	*and Down Payments*	Asset Sales	Asset Sales
Other Long Term Loans			
Total Long Term Liabilities (LTL)			
Total Liabilities = (CL + LTL)	*= Total Cash Outflow*	Total Business Expense	**Total Business Expense**
Net Worth			
	= Cash Surplus or Deficit	Cash Net Business Income	**Accrual Net Business Income**
= (Total Assets - Total Liabilities)			
	= (Total Cash Inflows	= (Total Business Income	**= [Total Business Income**
	- Total Cash Outflows)	- Total Business Expense)	**- Total Business Expense]**

FIGURE 5.6 Balance Sheet, Cash Flow, Cash Basis Income, and Accrual Adjusted Income Statements compared.

5.2.5.4 Asset Valuation

Almost all Balance Sheets prepared for agricultural operations use market values for assets. GAAP recommends valuing assets at their cost or book value. The cost or book value is the asset's original purchase price plus capital improvements to the asset made since its original purchase minus depreciation taken on the original purchase price and the cost of improvements. For raised products, such as grain or market livestock held for sale, investment in growing crops, or raised breeding livestock, capitalized cost of production can be used as a cost basis asset value, but this is difficult to implement.

Market valuation is much easier to implement and has the benefit of satisfying one of the primary purposes for which the Balance Sheet is used—loan collateral analysis. Unfortunately, using market valuation introduces the influence that changing asset values have on financial business position and performance.

As mentioned, a key indicator of business performance is *earned* growth in equity through time. This requires separating the ability of the business resource base to create positive or negative change in equity from changes in equity due to influences like asset appreciation. For this reason, it is often suggested that assets should be valued at cost to evaluate financial performance and to eliminate introducing the influence of changing an asset's market value somewhat randomly. For example, let's say a producer has 20,000 bushels of wheat in storage at the beginning of the year and values it at $3 per bushel for a total of $60,000 current market value. During the year, the producer sells all of the current year's wheat production plus 3,000 bushels of wheat in storage at the beginning of the year. The producer uses a value of $3.75 for the 17,000 bushels in storage at year end, for a total value of $63,750. A price increase (estimated market value) and a quantity decrease during the year result in a higher value of wheat on hand at year's end. Should the increased value be attributed to business performance? The additional $3,750 in the value of stored wheat contributes toward a potential increase in equity during the year, but was it earned by the business, i.e., was the owner smart or just lucky? Is the estimated market price of $3.75 high or low?

GAAP's recommended cost or book value approach helps eliminate confusion about the business performance introduced by using market valuation. In our example, assume the cost or book value of the wheat in storage is $2 per bushel. Only the quantity change is considered if the cost or book value is held constant on successive Balance Sheets. This results in a decline of $6,000 in the value of stored wheat rather than an increase of $3,750. While this eliminates confusion about business performance introduced using estimated market values, it also generates some of its own confusion by ignoring real-world market valuation changes of current and non-current assets. Confusion about what is being measured is why it is important to understand issues related to each financial statement and the basic concepts of financial analysis.

Both financial *position* and *performance* measures are necessary for a business operation. The previous discussion hints that financial position should be measured with financial statements that have been prepared using market values. Financial performance, on the other hand, should be measured using Balance Sheets that have

TABLE 5.1
Asset Valuation Procedures Recommended by the FFSC

	Required Disclosure		
	Cost	Market	Other
Marketable Securities	Yes	Yes	
Inventories[a]	No	Yes	
PIK Certificates	Yes	Yes	
Accounts Receivable[b]	Yes	No	
Prepaid Expenses	Yes	No	
Cash Investment in Growing Crops	Yes	No	
Purchased Breeding Livestock[c]	Yes	Yes	
Raised Breeding Livestock[c]	Recommended	Yes	Base value if cost not included
Machinery and Equipment[c]	Yes	No	
Investments in Capital Leases	Yes	No	
Investments in Cooperatives	No	No	Net equity
Investments in Other Entities	Yes	No	Net equity if enough ownership exists to exert control
Cash Value of Life Insurance	No	Yes	
Retirement Accounts	Yes	Yes	
Other Personal Assets	No	Yes	
Real Estate[c]	Yes	Yes	
Buildings and Improvements[c]	Yes	Yes	

[a] GAAP requires lower cost or market value except for certain inventories ready for sale.
[b] Less an allowance for doubtful accounts
[c] Market value information permitted only as supplementary data for GAAP statements.

been prepared using cost or book values. Strict classification of a Balance Sheet as either a cost or market value Balance Sheet is misleading. Balance Sheets are almost always prepared using some combination of cost and market values. Table 5.1 lists asset valuation procedures recommended by the FFSC and can be used as a guide when preparing a Balance Sheet. As you can see, a mix of cost and market values are recommended. While these involve more work to prepare, the two-column Balance Sheets display both cost and market valuation methods.

5.2.5.5 Liabilities

Listing liabilities on the Balance Sheet is relatively easy. Liabilities are measured in dollar amounts owed at a particular point in time. Unlike asset values, dollar amounts for most liabilities can be calculated with relative ease.

The single biggest consideration about liabilities listed is whether to include a category called deferred taxes, or contingent tax liabilities. A deferred tax liability is a liability triggered by an event that has not yet happened. Since the liability is contingent on an uncertain event and must be estimated, these liabilities are seldom included by lenders or producers, especially for ongoing concerns.

TABLE 5.2
Impact of Contingent/Deferred Income Tax on Net Worth

A. Total Current Assets	$50,000
B. Total Non-Current Assets	1,500,000
C. Total Assets (A + B)	1,550,000
D. Total Current Liabilities (Excludes Contingent Taxes)	30,000
E. Total Non-Current Liabilities (Excluding C.T.)	1,200,000
F. Total Liabilities (Excluding C.T.) (D + E)	1,230,000
G. Total Net Worth excluding C. T. (C – F)	320,000
H. Contingent Taxes on Current Assets	10,000
I. Contingent Taxes on Non-Current Assets	420,000
J. Total Liabilities Including Contingent Taxes (F + H + I)	1,660,000
K. Net Worth Including Contingent Taxes	–$ 110,000

Some deferred tax liabilities occur with regularity (e.g., accrued income taxes), while others may never be triggered. Listing accrued income tax liability from annual business operations (a deferred income tax) on the Balance Sheet requires estimating taxable income as a result of last year's taxable income and the marginal tax rate at which the income will be taxed. This estimate of a deferred tax is listed in the current liability section of the beginning and ending Balance Sheets. These estimates are used to make accrual adjustments to the income statement. This type of deferred income taxes has a relatively small impact on the overall financial picture for an operation over the long run, but may have substantial impacts in any given year. Due to the estimation procedures involved and the relatively small impact, this type of liability is almost never included on the Balance Sheet.

In contrast, contingent income taxes on the sale of current and non-current assets can have a tremendous impact on total liabilities and net worth. Income taxes triggered by the sale of capital assets (the capital gains tax) on long-term assets, such as land, are another example of deferred tax. Most Balance Sheets prepared today exclude these liabilities for an operation that is considered financially healthy with no imminent plans for a voluntary sale or a forced liquidation. Table 5.2 illustrates the possible impact of excluding an estimate of deferred taxes on capital assets.

Excluding deferred tax liabilities can prove disastrous for tactical financial planning. These contingent liabilities affect the dollar values calculated by the financial statements and the ratios and measures used to evaluate the financial position and performance (FFSC, 1997; Frey and Klinefelter, 1980). Eventually, they may be the cause of considerable financial distress. Owners of many operations that were forced to liquidate during the farm financial crisis of the mid-eighties found themselves owing tens if not hundreds of thousands of dollars of income taxes after liquidating. In these cases, Balance Sheets prepared prior to the farm sale ignored the tax consequences of the sale, and therefore overestimated the equity position for the operation. Producers who thought they would come away from the sale with some equity not only had none, but owed thousands of dollars in taxes.

Producers are often unaware of the need to list contingent liabilities, but lenders realize that excluding these tax liabilities from the Balance Sheet can overstate net worth. Lenders often use conservative market values for assets to compensate for excluding contingent tax liabilities. But using conservative market values as a substitute for estimating contingent tax liabilities may still overstate an operation's net worth, which can and does lead to incorrect management decisions for the business, retirement, and estate transfer planning. Every effort should be used to get a fair, objective value of all assets and all liabilities in order to provide accurate information for decision makers.

RDFinancial's condensed version of the Balance Sheet includes a deferred tax listing in the current liabilities section for illustrative purposes. EWS Farms has followed industry practice and has not listed any contingent liability. The condensed Balance Sheet does not have a place to enter contingent/deferred liabilities on the sale of non-current assets.

5.2.5.6 Net Worth (Equity)

Net worth is calculated as total assets minus total liabilities and indicates the dollar value of assets owned by the operator. It follows that net worth is a fundamental indicator of how well the business is doing. Increased net worth occurs because of asset appreciation, from retained earnings generated from operations, or contributions from outside the business. Assume for a minute that the market value of assets and family living withdrawals are held constant. Trends in net worth should then reflect the overall financial health of the business. If enough income is generated to cover expenses and expenditures, such as principal payments and owner withdrawals, and there is still money left over, the leftover amounts show up on the Balance Sheet in successive years as an increase in net worth. These are referred to as *retained earnings*. If the opposite is true, a steady decline in net worth is occurring. Positive retained earnings and growth in equity are desirable, but net negative values do not necessarily indicate a problem with the business operation. During certain time periods, larger than normal sums of money or other business assets might be taken out of the business for children's education, retirement funds, or other activities, thereby holding net worth relatively constant or in decline for short periods of time.

To help evaluate business performance as measured by trends in net worth, it is useful to separate changes in net worth into three components: (1) the original equity provided by the owner plus any additional contributions of equity after the business was started, (2) changes in net worth from retained earnings, and (3) increases or decreases in asset values due to changing market values over time. Each of these components can tell a story about the operation's financial health, so each should be carefully measured and monitored. For example, assume two operations start at the same time and have the same characteristics, but one is located in an urban area and the other in a rural area. These operations currently have the same net worth, but the components of net worth are very different (Table 5.3).

Using only this limited information, earned growth in equity indicates the Rural Operation has more opportunity to survive as a viable agricultural operation on its own rather than depending on outside influences for equity growth.

TABLE 5.3
Components of Net Worth

Components of Net Worth—Rural Operation	
Original Equity Contribution	$10,000
Retained Earnings (Earned Growth in Equity)	50,000
Asset Valuation Increase	30,000
Total Current Net Worth	$90,000

Components of Net Worth—Urban Operation	
Original Equity Contribution	$10,000
Retained Earnings (Earned Growth in Equity)	10,000
Asset Valuation Increase	70,000
Total Current Net Worth	$90,000

5.2.6 Cash Flow Statement Content

The Cash Flow Statement is the second financial statement typically prepared for agricultural operations. Producers prepare a monthly cash flow *projection* that accompanies the Balance Sheet when annual operating loans are renewed. Typical categories of *inflows* and *outflows* in a Cash Flow Statement are shown in Figure 5.6. The detail in these categories varies, depending on the needs of those using the information, but enough detail should be provided to use the Cash Flow Statement for management decisions. Categorizing cash inflows and outflows helps determine whether the business is capable of meeting cash flow requirements or if additional inflows must be provided from outside sources. The level of detail will depend upon the size of the operation and the need to communicate financial information to others involved in the operation such as lenders, partners, or stockholders. If the normal business operation does not meet cash flow needs, long-term cash infusions into the business are required. The inability of a business to provide for its cash needs does not necessarily mean it is poorly managed. In many instances, the business is simply not big enough to generate sufficient cash flow to support a particular family structure. Identifying this particular weakness allows for long-term cash flow planning.

5.2.6.1 Purpose of the Cash Flow Statement

Cash flow projections are a best guess as to how cash will flow in and out of the business during the accounting period. Projections are made with full knowledge of uncertainties in agricultural production, marketing, financing, and family living, and it is certain that initial projections are not a perfect forecast. The Cash Flow Statement is used to identify fund sources and to note how these funds flow between the farm and family operations. Flow identification helps lenders evaluate the risks of extending credit. Comparing cash flow projections to historical cash flows is also valuable in identifying how and why projections, inevitably, are never perfect.

The terms *positive* and *negative cash flow* are used to describe the results of cash flow analysis. A positive cash flow means there were enough cash inflows to cover all cash outflows for the business and family. If a historical cash flow is prepared rather than a projected cash flow, a positive cash flow amount should equal the ending cash balance on the Balance Sheet. Unlike a positive cash flow retained in the business as cash, a negative cash flow does not generate a pile of negative cash at the end of the year, but something close to it. A negative cash flow for the previous year typically means that a lending institution or vendors provided the cash to meet the needs of the business and family. A negative cash flow shows up as an operating loan carryover or another type of liability (accounts or notes payable) listed on the ending Balance Sheet for the producer.

5.2.6.2 Format and Organization of the Cash Flow Statement

The format of the Cash Flow Statement provides information for financial planning and should allow for separation of business and family activities. The Economic Research Service of the U.S. Department of Agriculture (USDA) reports that as much as 90% of total farm income comes from off-farm sources. This is by design, meaning that these households indicated it was a choice to make agriculture a secondary source of income (Mishra, El-Osta, Morehart, Johnson, and Hopkins, 2002). In addition to the separation of family and business, major categories of inflows and outflows should be identified. To accomplish this, cash flows can be grouped by crop and livestock enterprises with as much detail as necessary to satisfy members of the management team. The Cash Flow Statement can be accompanied by schedules that provide very detailed inflows and outflows for business operation, loan repayments, and family activities (Griffith, 1990). Family activities can also be grouped by type of activity to provide additional information to the management team. Family living withdrawals for retirement separated from day-to-day outflows for family living might be examples of more detailed family cash flow information.

5.2.7 Accrual Adjusted Income Statement Content

Like the Cash Flow Statement, an income statement is for a period of time, usually prepared for a calendar year. The income statement lists *income* earned and *expenses* incurred from business operations during the year. Income is defined as the gross increase in capital attributable to business activities. Expense is defined as costs that have been consumed in the process of producing revenue (Niswonger and Fess, 1969). These definitions do not include a description of whether the income was actually received in cash or the expense was actually paid in cash. An accurate measure of business income and expenses for a period must include all cash and noncash income and expenses received or incurred during that period. Figure 5.6 lists information typically contained on an Accrual Adjusted Income Statement. Net Business Income, the bottom line, is a measure of profitability for the operation. To provide an accurate measure of profitability, the income statement must contain accrual adjustments from the beginning and ending Balance Sheets (Farm Financial Standards Council, 1997; Langemeier, 2004c; Frey and Klinefelter, 1980). However, farmers and ranchers in the United States are typically cash-basis taxpayers, and the IRS allows them to use a modified definition of income and expense, those that are actually received or paid in cash. This modified definition has set a de facto standard for the type of financial records and financial analysis necessary. The result is that a Schedule F and its measure of Net Farm Profit or Loss is often used as a replacement for a true income statement.

Industry acceptance of substituting a Schedule F for an income statement was an easy out. Unfortunately, it also dooms producers to make management decisions that will determine the financial health of their operations based on inadequate or incorrect information. Lack of an Accrual Adjusted Income Statement prevents the process of reconciliation and gaining a true measure of business financial position and performance. This practice has established a mindset about the level of record keeping required and encourages poor financial statement preparation.

For cash-basis taxpayers, the only Internal Revenue Service (IRS) requirement is a record-keeping system that is adequate to support and verify the information used to prepare a tax return (IRS, 2007). This record-keeping system can be any combination of records, including a shoe box full of receipts and deposits, a hand-kept journal of income and expenses, a computerized system of recording income and expenses, or any combination of the above. Other records, such as a depreciation schedule prepared by an accountant, are typical.

Comparing cash and accrual adjusted net income from farm business management records indicates that cash net income is from 11 to 68% higher or lower than accrual adjusted net income for the same operations (Ellinger, 1999). With this said, the IRS *does recommend* that adequate financial statements be prepared to measure business financial performance rather than using taxable income as a proxy for a true business performance measure of net income.

5.2.7.1 Purpose of the Accrual Adjusted Income Statement

An Accrual Adjusted Income Statement provides an accurate measure of business *performance* for the most recent accounting period under the heading *net business income*. Net business income measures the ability of the resource base to generate a profit from the use of business resources during a specific period of time, usually a calendar year. It does not, however, measure whether the resource base is being used in a fashion that optimizes profitability. Net business income is *the key performance measure* used to determine the ability of the business resource base to generate earned growth in equity. Combined with other information on the Statement of Owner Equity, an accurate measure of net business income is critical in determining whether the resource base is capable of supporting itself and the family or if the family must supplement the business to meet combined business and family goals.

5.2.7.2 Format and Organization of the Accrual Adjusted Income Statement

Income statement formats vary by financial institution, software vendor, and the need for specific types of income and expense detail required by business owners. Figure 5.6 illustrates the basic format of both a cash income statement and an Accrual Adjusted Income Statement. A Schedule F tax form and a cash-basis income statement present the same basic information. However, as indicated, the labels on both of these forms are misleading. The Schedule F bottom line, row number 36 on the 2007 tax form, is labeled "Net Farm Profit or (Loss)" (see Figure 5.7). While it

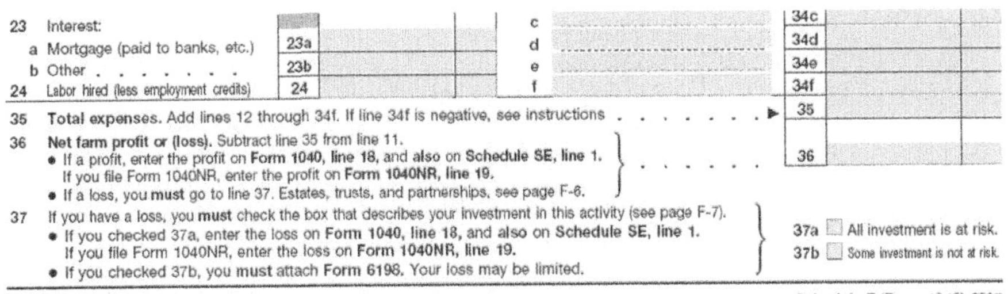

23 Interest:			c	34c
a Mortgage (paid to banks, etc.)	23a		d	34d
b Other	23b		e	34e
24 Labor hired (less employment credits)	24		f	34f

35 Total expenses. Add lines 12 through 34f. If line 34f is negative, see instructions . . . . ▶ 35

36 Net farm profit or (loss). Subtract line 35 from line 11.
- If a profit, enter the profit on Form 1040, line 18, and also on Schedule SE, line 1. If you file Form 1040NR, enter the profit on Form 1040NR, line 19.
- If a loss, you must go to line 37. Estates, trusts, and partnerships, see page F-6.

. . . . 36

37 If you have a loss, you must check the box that describes your investment in this activity (see page F-7).
- If you checked 37a, enter the loss on Form 1040, line 18, and also on Schedule SE, line 1. If you file Form 1040NR, enter the loss on Form 1040NR, line 19.
- If you checked 37b, you must attach Form 6198. Your loss may be limited.

37a ☐ All investment is at risk.
37b ☐ Some investment is not at risk.

For Paperwork Reduction Act Notice, see page F-7 of the instructions. Cat. No. 11346H Schedule F (Form 1040) 2007

FIGURE 5.7 Portion of Schedule F showing labeling for taxable income, 2005 tax year.

is labeled Net Farm Profit or (Loss), it is not a measure of net farm profit for any measure of business performance, except taxable income. A more appropriate label would be Net Farm Taxable Income. This labeling is also true of a cash-basis income statement (see Figure 5.6). While it measures cash-basis income, that measure is not a true measure of business performance; it is just a measure of taxable income. To reiterate, the cash income statement is included in Figure 5.6 as a comparison for the type of information it does *not* contain. The condensed form of the Accrual Adjusted Income Statement shown in Figure 5.6 illustrates the basic format and organization, but detailed examples are provided by Griffith or Frey and Klinefelter.

The vast majority of producers are cash-basis taxpayers, but we have indicated an Accrual Adjusted Income Statement is necessary for an accurate measure of net business income. Rather than recommend a complete accrual accounting system, the FFSG recommends an alternate approach, using accrual adjustments to a cash-basis income statement. The resulting Accrual Adjusted Income Statement yields the same answers as an accrual income statement produced by an *accounting system* and is easier to prepare using the information available from typical farm records.

An example may help illustrate the difference between cash and accrual. Farmer Jones has a grain and cow-calf operation. Gross sales from the cow-calf enterprise are about $200,000. Typically, calves are sold around weaning time in the fall. This year, Farmer Jones decides to keep the majority of the calf crop to feed through the winter. Part of the reason for this decision is that Farmer Jones has a tax problem. The previous year's grain crop was carried over and both the previous year's and current year's crop were sold this year. Delaying calf sales until next year will reduce total taxable income this year.

While Schedule F measures taxable income, there is still a question about the actual business performance during the year. Does the Schedule F with two years of grain crop sales and only the cull income from the cow-calf enterprise represent how the business performed during the year? The business did not produce the grain crop that was in storage at the beginning of the year nor does the Schedule F reflect the value of the calves produced during the year that are in inventory at the end of the year. Expenses on the Schedule F do not include the previous year's cash expense incurred to grow last year's crop sold this year, even though the revenue is included, but it includes all expenses for producing the calf crop. The use of Schedule F or a cash income statement masks true business performance because revenue and expenses are not matched in the appropriate accounting period.

Beginning Net Worth
+ Net Income
+ Non-Business Cash Inflows
– Owner Withdrawals (family living)
± Asset Valuation Change
± Contributed or Distributed Capital
= **Ending Net Worth**

FIGURE 5.8 Condensed format for the Statement of Owner Equity.

Measuring business performance for a given period of time using an Accrual Adjusted Income Statement is the process of measuring production from the resource base and the income and expenses associated with the production process during that period. The accounting profession calls this the *matching principle*. While this definition seems simple and straightforward, it is not easy to implement. The production process often carries over from one year to the next for crop and livestock enterprises and the record keeping for accurate income and expense by period can be difficult.

5.2.8 Statement of Owner Equity Content

The SOE contains the information necessary to reconcile the change in net worth *reported* on the beginning and ending Balance Sheets. Ending net worth is also calculated on the SOE. The reported ending net worth is then compared to the *calculated* ending net worth to determine if there is a discrepancy. The calculated ending net worth requires net income from the Accrual Adjusted Income Statement, owner withdrawals from the cash flow, and the value of contributions, distributions, and asset revaluation that may have taken place during the year.

5.2.8.1 Purpose of the Statement of Owner Equity

The SOE accomplishes two purposes. The first is to reconcile the change in equity and the second is to identify the source of the equity change. A positive change in equity may be good news, but the source of the change should be analyzed before too much celebration. If the reason equity increased was due to increases in the fair market value of land, machinery, and buildings and the net income from the business is actually negative, it is a cause for concern.

Figure 5.8 shows a condensed version of this statement. For a moment, let's assume away three categories on the SOE: non-business cash inflows, asset valuation change, and contributed or distributed capital. This assumption is realistic for many operations and simply indicates that per-unit asset values for crops held for sale or feed, land, machinery, breeding livestock, etc., are not changed during the year; that gifts of capital assets are not given or received; and that off-farm wages or income from off-farm investments are not used to support the operation. With these assumptions, there is only one way that net worth can increase during the year. Net income *must* be greater than family living withdrawals or ending net worth will be less than beginning net worth. In short, the resource base *must* generate growth in net worth.

5.2.9 Too Much Detail

While detailed financial statements are necessary, collecting information and preparing adequate financial statements is like swatting a beehive with a broom. Once you start the process of financial analysis, there is all this detail buzzing around your head and you're just waiting to be stung by the incorrect handling of one of the details. It is not our intent here to focus on detailed financial statement preparation. We can easily lose sight of the objective, why we should do a better job of measuring business financial position and performance, while worrying about whether or not we have used the right values for a particular asset, whether all the liabilities for each individual loan were listed correctly, and whether all income and expenses and cash inflows and outflows were recorded correctly. The objective of strategic and tactical financial analysis deteriorates into a tedious process of managing the blizzard of record-keeping details. What we really want is to get past the *how* related to details that must be managed and learn about *why* it is important to do a better job of financial analysis.

In general, farmers and ranchers are typically production oriented. They lack strong backgrounds in financial analysis, which means they do not feel comfortable preparing and interpreting financial statements (Coble, Knight, Patrick, Baquet, 1999); however, the issues we have just discussed for each statement, must be addressed. The previous discussion was intended to provide a review of the issues, not to discourage adequate financial analysis. The blizzard of details that must be managed is the primary reason agricultural financial analysis is not completed in a timely, accurate manner. While the details must be handled correctly, *forget about the details for now.* We will focus on what a good set of financial statements can tell us about an operation, and *why* we should worry about the type of financial analysis used for management decisions.

5.2.10 What Is Financial Health?

An individual producer's goals and objectives make the definition of financial health a moving target. Goals and objectives change over time. Someone just getting started in farming has different financial needs than someone making the transition to retirement years and turning management over to one or more of the children. This moving target poses other critical questions that are key to understanding financial health.

With a moving target as a background, financial health is evaluated in the broad categories previously identified: liquidity, solvency, repayment capacity, profitability, and financial efficiency (FFSC, 1997; Langemeier, 2004a,b,c; Ellinger, 1998; Miller, Boehlje, Dobbins, 2001). These categories can be evaluated using the financial statements directly, but these categories are further defined by a specific set of 16 ratios or dollar values calculated from a reconciled set of financial statements. These ratios and measures are reviewed later. For now, focus on the importance of preparing and reconciling the four financial statements recommended.

Figure 5.4 is a modification of the SRM process wheel included in Chapter 4. In Figure 5.4, the focus has changed from the general SRM process to financial risk management. The starting place in the process is measuring current business financial position using the Balance Sheet. For this discussion, we will assume the Balance Sheet is prepared using market values and that market values are held constant from the beginning to the end of the year. This eliminates some of the issues regarding asset valuation.

Understanding how and why net worth changes over time provides a great deal of management information about the ability of a business to meet family and business financial goals (Miller, Boehlje, Dobbins, 2001). All businesses have good and bad years. Historically, a single bad year may well be an aberration. Annual financial evaluations of the previous year's business operation can indicate whether a bad year was an aberration or if the business has fundamental weaknesses.

5.2.10.1 How Financial Statements Measure Financial Health

We will start this review with a short discussion on an incorrect application of terminology. Terminology is often used incorrectly when discussing financial information and confuses the issue of measuring financial performance. However, it is critical to understand how the nature of a financial transaction and the financial statements on which the transaction is reported determine the terminology used to describe the transaction.

The classic example is use, or misuse, of the words *income* and *expense*. Producers often use the word expense to describe the amount of any check written; they view any cash outflow as an expense. Think about writing a check for a loan payment. A single check typically includes both the principal and interest portion of the loan payment. The total check is a cash outflow but only the interest portion of the check is an expense (tax deductible) to the business. When recording this transaction in a ledger, paper or computerized, producers often list the entire amount as an expense to the operation. Inflows and incomes are also confused. Loan proceeds deposited into a checking account are an inflow into the business, but they are not income. Producers routinely record the deposit of loan proceeds as income. For someone focused on cash flow, this produces the desired result in the record-keeping system. The inflow from the loan proceeds shows up on the Cash Flow Statement even though it may be incorrectly labeled as income. An income statement produced from a computerized record-keeping system may also incorrectly show this as income if it was categorized incorrectly when recorded.

Using the inflow/income and outflow/expense terminology loosely contributes to incorrect reporting, which leads to a poor understanding of how financial information should be prepared and interpreted. The Cash Flow Statement and Income Statement both contain different information and, therefore, tell a different story about the business's financial position and performance. Preparing a complete and accurate set of financial statements is critical, but the agricultural industry generally does not use Income Statements, which leads to a greater degree of financial misinformation.

The format described in Figure 5.6 allows the reader to compare information contained in the recommended financial statements. Column 1 in Figure 5.6 summarizes information contained in the Balance Sheet. This column stands alone as a summary list of assets and liabilities. Rows in each of the other three columns are meant to be compared across the columns. For example, Cash Business Income is listed in all of the last three columns, indicating that each of these three statements contains cash income, which is also a cash inflow and considered business income. The same font is used across columns to indicate that information included in the Cash Business Income calculation is identical in all three statements, columns 2, 3, and 4. Depreciation, included in the two income statements, is not included on the Cash Flow Statement. The same font used for the label (Depreciation) indicates this expense is calculated in the same way for both forms of income statements. In other rows, like Total Business Income, columns 3 and 4, have the same label but a different font, indicating that while the label is the same, the information used to calculate total business income in a particular column is different, and hence interpretation of that number will be different among the statements. On the same row as Total Business Income, the Cash Flow Statement shows a Total Cash Inflows calculation. Obviously, total cash inflows are not the same as total business income.

The four basic financial statements previously recommended (Balance Sheet, Cash Flow Statement, Accrual Adjusted Income Statement, Statement of Owner Equity) did not include a Cash Basis Income Statement, column 3, Figure 5.6. It is included here for comparison purposes as this form of income statement is the most commonly used today. While included for discussion and comparison purposes, we want to emphasize that this form of an income statement should not be used for anything but help with preparing estimates of taxable income.

The SOE is not included in Figure 5.6. This statement uses information from the other three statements and reconciles beginning and ending net worth listed on the Balance Sheet. Reconciling net worth helps ensure the financial information in the other statements is an accurate picture of the financial business position and performance.

To summarize our financial health discussion, it is the ability of the business resource base to produce a positive net business income that is large enough to generate earned growth in equity. However, regardless of the size of the business, the resource base may be generating the maximum possible profit and contributing as much as it can to the business's financial health. That does not mean it is capable of supporting some type of family structure overlain on top of the business resource base or incorporated into the business ownership structure, as the case may be for a corporate form of ownership. If the business resource base is small and cannot produce a profit, it generates a negative net business income, although it may still be producing at its economically optimal level. It may simply be that the cost of capital assets, machinery, buildings, land, and so forth, is large enough to prevent the resource base from earning a profit. In this case, financial health may be defined as earning as small a negative net farm income as possible. Owners must then rely on contributions from off-farm sources to support the business operation.

5.2.10.2 Financial Ratios and Measures

We have discussed the general process of measuring business financial position and performance. Now let's turn to the ratios and measures calculated from financial statements. Is there a single measure that can explain financial position and performance? If something less than a complete set of financial statements is used, what financial measures are missed or calculated incorrectly? How do all the issues discussed for each of the financial statements, like ignoring deferred taxes on the Balance Sheet, affect calculated values of financial ratios?

Unfortunately, there is not a single measure derived from the financial statements that will adequately evaluate all possible situations. Business and family goals change through time and may be in unison or at odds with ensuring good business financial health. A family that is bringing children back to the operation may increase the debt load on the operation for expansion and to provide additional family living for a few years. Another operation with only one family that is nearing retirement may increase cash withdrawals to build retirement funds. Business owners knowingly make decisions that adversely affect the financial health of the business because those decisions help meet the combined goals for the family and business. Care must be exercised so the financial health of the business does not deteriorate to a point where it cannot recover from short-term management strategies. Evaluation of financial ratios must include knowledge of combined family and business goals.

Preparing annual financial statements provides information to calculate the "Sweet Sixteen." The Sweet Sixteen are a set of ratios and dollar amounts calculated from three of the four basic financial statements. While there have been hundreds of ratios and measures developed over the last few decades, the FFSC selected sixteen measures that adequately reflect the financial position and performance of an individual operation. The FFSC also proposed a standardized process of calculating each of these measures. Lenders, producers, accountants, and others still use many ratios and other measures not on the Sweet Sixteen list and calculate some of the Sweet Sixteen measures differently than the FFSC. Discussion in this text will focus on the Sweet Sixteen calculations using FFSC-recommended procedures.

The Sweet Sixteen measures are divided into the same five categories previously discussed for financial analysis: liquidity, solvency, profitability, repayment capacity, and financial efficiency. Figure 5.9 contains these measures by category.

FFSC definitions for each of these categories are:

1. *Liquidity* measures the ability of a farm business to meet financial obligations as they come due in the ordinary course of business, without disrupting the normal operations of the business.
2. *Solvency* measures the amount of borrowed capital, debt, leasing commitments, and other expense obligations used by a business relative to the amount of owner equity invested in the business. Solvency measures provide an indication of the ability to repay financial obligations if all assets were sold that day. Solvency also signals the ability to continue operations as a viable business after a financial adversity, which typically results in increased debt and reduced equity.
3. *Profitability* measures the extent to which a business generates a profit from the use of land, labor, management, and capital.
4. *Repayment capacity* measures the ability of a borrower to repay term farm debt from farm and nonfarm income. Principal payments on term loans must come from net income (with depreciation added back) after owner withdrawals, income taxes, and Social Security taxes.

Liquidity	Beginning	Ending
Current Ratio	6.44	6.61
Working Capital	$198,452	$204,544
Solvency		
Debt/Asset Ratio	18.24%	0.178
Equity/Asset Ratio	81.76%	0.822
Debt/Equity Ratio	0.22	0.216
Profitability		Ending
Rate of Return on Business Assets		1.32%
Rate of Return on Business Equity		0.32%
Operating Profit Margin Ratio		0.06
Net Business Income		$29,151
Repayment Capacity		
Term Debt and Capital Lease Coverage Ratio		1.28
Capital Replacement and Term Debt Repayment Margin		$6,742
Financial Efficiency		
Asset Turnover Ratio		0.23
Operating Expense Ratio		0.78
Depreciation Expense Ratio		0.10
Interest Expense Ratio		0.05
Net Farm Income From Operations Ratio		0.08
	Check Sum	100.00%

FIGURE 5.9 Sweet Sixteen ratios and measures by category for EWS Farms.

5. *Financial efficiency* measures the intensity with which a business uses its assets to generate gross revenues and the effectiveness of production, purchasing, pricing, financing, and marketing decisions.

The ratios and measures in each category draw on information from the Balance Sheet, income statement, and portions of the Cash Flow Statement. Figure 5.10 is a cumulative indicator of which financial ratios and measures can be calculated from each of the financial statements.

Preparing just a Balance Sheet allows calculation of both the liquidity and solvency measures recommended by the FFSG. Preparing a Cash Flow Statement does not allow calculation of any additional measures; however, the Cash Flow Statement does include information that can be used to help calculate the Sweet Sixteen and prepare the ending Balance Sheet. Preparation of the Cash Flow Statement for the purpose of calculating the Sweet Sixteen is not required, as individual pieces of information found on this statement can be gleaned from other sources and used to calculate some of the measures. An accurate Accrual Adjusted Income Statement must be prepared.

Preparing an Accrual Adjusted Income Statement allows calculation of all of the remaining ratios under the categories of profitability, repayment capacity, and financial efficiency. These categories combine information from the income statement and the Balance Sheet. The reason for using the Accrual Adjusted Income Statement as the

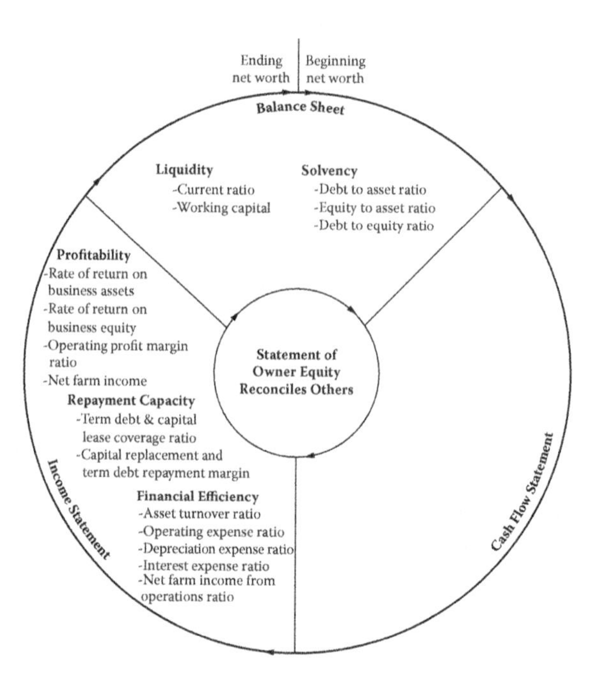

FIGURE 5.10 Financial statements required to calculate the Sweet Sixteen.

starting point is that it measures the business resources' ability to produce net business income. Neither the Cash Flow Statement, nor the Schedule F, measures this ability.

The Balance Sheet and Cash Flow Statement are typically the two statements required in agriculture lending and financial analysis. If an Accrual Adjusted Income Statement is required to calculate many of the Sweet Sixteen ratios and measures, one must ask, "How are these measures being calculated for agricultural operations today?" The values of the ratios and measures suggested by the FFSC but calculated using substitute information, Schedule F Net Income, are at best suspect and at worst misleading to the point of financial disaster.

While the Cash Flow Statement is not required to calculate any of the Sweet Sixteen ratios and measures, it provides information about the ability of an operation to meet its cash needs. Producers and lenders rely on the Cash Flow Statement for information about a firm's ability to pay all cash operating expenses and other outflows, including operating and term debt principal payments. It is important to note that repayment capacity measures included in the Sweet Sixteen, use Net Farm Income from Operations, derived from an Accrual Adjusted Income Statement, as the starting value to calculate repayment capacity measures—not Net Cash Flow. Net Farm Income from Operations is adjusted by several other values to arrive at the Term Debt and Capital Lease Coverage Ratio and the Capital Replacement and Term Debt Repayment Margin.

Financial ratios for an individual operation are typically judged by benchmarking the operation's values to industry-accepted standards for a given ratio. This information can be presented in many different ways. Figure 5.11 is an illustration of a

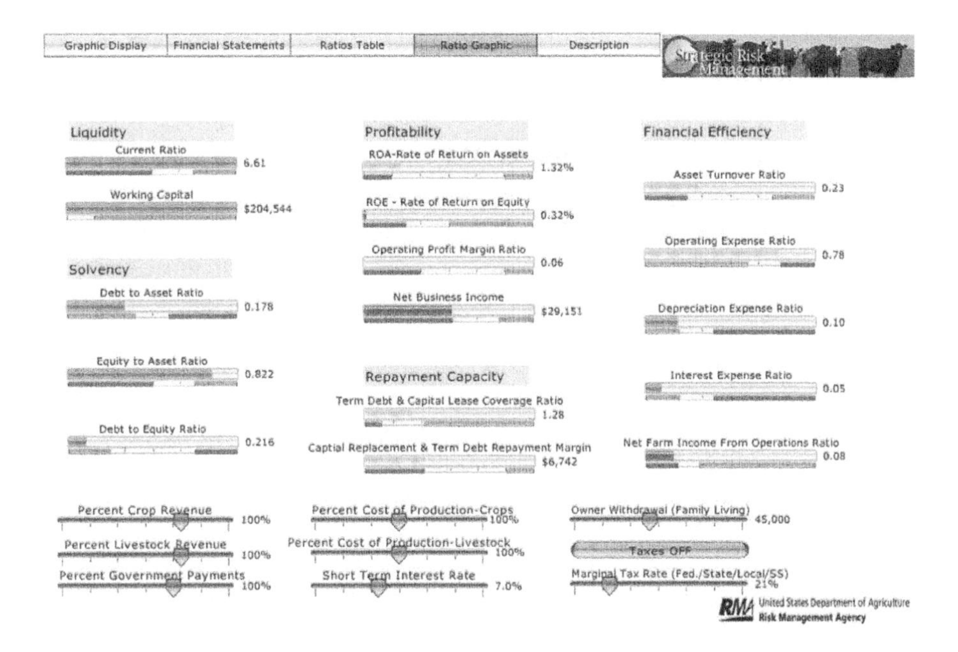

FIGURE 5.11 Ratio alert levels for EWS Farms. (From the Risk Management Agency, U.S. Department of Agriculture.)

graphic display of the ratio values shown in Figure 5.9. The graphic display is from the Flash version of the EWS Farms RDFinancial file. This version of the file allows the user to move sliders and explore a selected set of changes on the business's financial condition. Figure 5.11 also shows representative *alert levels* in the agricultural sector for each of the Sweet Sixteen ratios and measures. You will note that several of EWS Farms ratio values are in the yellow or red range. A quick analysis indicates that measures of financial *position* are solidly in the green, but measures of financial *performance* are mostly in the red and yellow range. This indicates where the management team for EWS Farms should focus their efforts for tactical planning to ensure long-term survival.

5.3 PART 3: FINANCIAL RISK MANAGEMENT TOOLS

Our discussion about the financial statements raised a lot of issues that must be handled when preparing detailed statements. Many of these issues are simplified for the user by using available tools that address the issues discussed for particular statements. Examples are statements that show both cost and market value on the Balance Sheet and automatically make accrual adjustments from balance sheet entries to the Accrual Adjusted Income Statement.

Financial analysis tools are available from many land-grant universities and commercial agricultural software vendors. Commercial vendors provide both complete day-to-day record-keeping activities for financial and production records and software that can be used to analyze the financial implications of individual business management decisions. Our focus here will be on software, mostly Excel™ spreadsheets available from land-grant universities.

5.3.1 Complete Financial Statement Preparation Tools

Listed below is a select set of universities and Web sites that provide access to the complete financial analysis tools they provide. This is not meant to be a comprehensive list. Each of these tools provides the type of analysis discussed in this chapter. A Web link and short description is provided for each software tool listed. (The Web links were active as of January 2009.) Our recommendation here is similar to our previous discussion. Do not let the details keep you from doing something that may prove crucial to your long-term survival. We will start with complete financial statements that measure business position and performance.

Montana State University: The following tools are available at http://www.montana.edu/softwaredownloads/financialmgtdownloads.html.

- **RDFinancial.xls:** RDFinancial contains a condensed set of the four basic financial statements and is intended to help explore financial management concepts. RDFinancial.xls is available at the Montana State University Web site above, or an Excel and Flash file version for EWS Farms can be downloaded at http://www.srmprocess.com/.
- **Financial Statements.xls:** This Excel spreadsheet combines beginning and ending two-column Balance Sheets listing both cost and market values, a Cash Flow Statement, an Accrual Adjusted Income Statement, a statement of cash flows, an owner equity statement, a valuation equity statement, a set of schedules for the Balance Sheets and Cash Flow Statements, deferred income tax calculations for the Balance Sheets, and the Sweet Sixteen ratio calculations based on both cost and market values in the Balance Sheets. The Cash Flow Statement included in this package of financial statements can be used as a stand-alone financial statement. The income statement is presented in a two-column format, comparing a cash-basis income statement to an Accrual Adjusted Income Statement. The statement of cash flow is another useful format for analyzing cash flow information that was not discussed in this chapter. The owner equity and valuation equity statements contain the information we have discussed in our condensed SOE.

FINPACK from the Center for Farm Financial Management: The following tools are available at http://www.cffm.umn.edu/.

- **FINPACK:** FINPACK software helps organize and analyze the current financial situation of your farm, answering the question, "Where am I?" FINPACK will help you explore alternatives within your agricultural business, assisting in answering the question, "Where do I want to be?" After projections are analyzed, FINPACK provides you with the information to make better management decisions about your business.

- This software comes in several versions. The *FINPACK Producer* version helps evaluate your financial situation, explore alternatives, and make informed decisions about the future direction of your farm. It is not a record-keeping system. Instead, FINPACK gives you the tools you need to effectively use your records in managing your farm or ranch. *FINPACK for Ag Professionals* helps evaluate your client's financial situation, explore alternatives, and recommend management strategies. *FINPACK for Ag Lenders* is the choice of lenders who want to help their customers grow and gain financial success. Make timely and thorough credit evaluations using all the planning and analysis components agricultural professionals have come to rely on, plus features for risk rating and credit scoring, collateral analysis, FSA guarantees, and loan presentations.

University of Illinois: http://www.farmdoc.uiuc.edu/fasttools/index.asp.

- **FAST Tools:** The financial statements tool is only one of many available at this Web site. The financial statements tool generates farm financial statements for a two-year period and evaluates financial performance with ratio analysis. Reports produced include beginning and ending Balance Sheets, an income statement, statement of cash flows, Statement of Owner Equity, and financial ratios.

Oklahoma State University: http://www.agecon.okstate.edu/iffs/.

- **IFFS:** IFFS generates projected or actual financial statements useful in decision making and evaluating performance. Projected statements provide insight into anticipated credit needs, cash flows, income levels, and changes in financial ratios, all of which is important information for both potential new producers and continuing operations. Historical financial statements provide important benchmarks, document financial progress, and help identify business strengths and weaknesses. IFFS is a set of interdependent Excel-based workbooks and budget files: Crop and Livestock Budgets (CLBUD), Additional Information (AI), and Multiple-Year Integrated Statements (MULTSTAT). The software facilitates data entry through point-and-click options, drop-down lists, and on-screen "help," while also maintaining the keyboard commands of original versions of the software. Customized enterprise budgets, a monthly Cash Flow Statement, debt worksheet, Balance Sheet, income statement, and financial measures can be generated.

5.3.2 Risk Navigator SRM Excel Tools

The Risk Navigator SRM site contains several tools related to individual financial statements and financial analysis. These tools include *BalanceSheet.xls*, an Excel spreadsheet that allows a user to prepare a single Balance Sheet. This can be for the beginning or end of the year. The *Simple Cash Flow.xls* spreadsheet allows preparation of a cash flow projection for the coming year or a historical cash flow analysis. The *Cash Flow and Income Statements.xls* allows preparation of a Cash Flow Statement, an Accrual Adjusted Income Statement, and a statement of cash flows. These statements are driven by a set of schedules used to enter detailed information for the cash flow and income statements. Both a cash basis and Accrual Adjusted Income Statement are prepared.

The Risk Navigator SRM site also contains the *Ratios EWS Farms.xls* spreadsheet. This spreadsheet is loaded with financial information from the RDFinancial spreadsheet used in this chapter, but is a stand-alone version of the ratio calculations. The user is allowed to enter cost and market values for assets and other financial performance measures from any operation, and the spreadsheet calculates the Sweet Sixteen ratios based on the data entered. This tool, like RDFinancial, displays the full formula used to calculate each of the 16 ratios and measures for both cost and market value-based calculations.

Other financial analysis tools are available on the Internet from land-grant universities. We encourage you to explore the sites listed above as well as others you may be familiar with.

5.3.3 Risk Navigator SRM Flash Files

In addition to the tools listed above, several Excel spreadsheets have been converted to Macromedia Flash files. These are designed to be more user friendly than Excel and don't require anything but a Web browser. Figure 5.11 above is an example of one page in the EWS Farms Flash file. This file is also available for download at the Risk Navigator SRM Web site.

5.3.3.1 Sweet Sixteen Ratio Analyzer

Our previous discussion indicated that there are several issues with the way financial statements are prepared and how these can affect interpretation of results. The Sweet Sixteen Ratio Analyzer is a tool that allows the user to explore changing the values from the EWS Farms Financial Statement to see how a change affects one or more of the ratios. This tool also allows the user to enter all of the values from their own financial information to calculate their own ratios. Figure 5.12 shows the profitability ratios and how they are calculated for EWS Farms. The data used to calculate these ratios are from the Accrual Adjusted Income Statement and average values calculated from the beginning and ending Balance Sheets. This tool is available on the Risk Navigator SRM Web site.

5.3.3.2 Deferred Tax/Missing Data Error Analyzer

The RDFinancial statements in this chapter did not include deferred taxes for accrual income, the sale of all current assets, or the sale of all non-current assets. This is typical of how Balance Sheets are prepared, unless they are prepared specifically to estimate the results of a liquidation. As discussed earlier, any type of information left off the Balance Sheet or any of the other financial statements can have a significant impact on the financial health picture painted by the statements themselves and on the financial ratios calculated from values on the statements. Since deferred taxes are typically

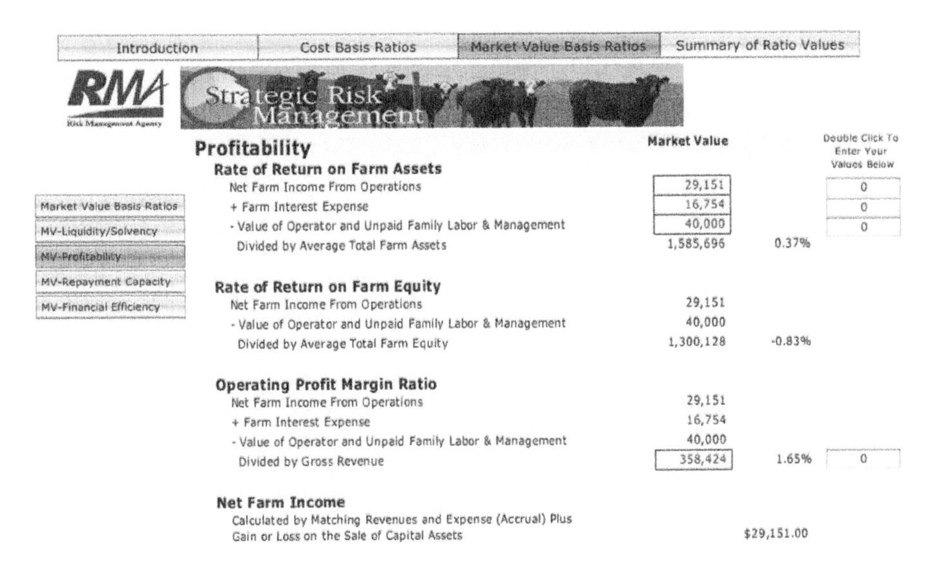

FIGURE 5.12 Profitability ratios with assets at market value for EWS Farms, ending Balance Sheet. (From the Risk Management Agency, U.S. Department of Agriculture.)

missing from the financial statements, the Deferred Taxes/Missing Data Analysis Tool helps show the relatively large impact that deferred taxes can have on the Sweet Sixteen.

This tool also allows the user to enter their own asset values, using market value and information from the Accrual Adjusted Income Statement to calculate their Sweet Sixteen financial ratios. This tool is equipped with sliders that allow the user to set the marginal tax rate for ordinary income and the capital gains tax rate for assets that are typically listed in the non-current section of the Balance Sheets. The sliders' initial values are set to zero, but Figure 5.13 shows the sliders set to 28% and 15% tax rates for current assets and non-current assets, respectively. When the sliders are in the 0% tax bracket position, all of the ratios in the right-hand column are equal to the ratios in the middle column. Moving the slider changes the income tax calculated for the current and non-current assets, assuming all assets are liquidated.

This tool is not meant to provide accurate estimates of the tax liabilities that may occur on the EWS Farms operation or your operation if you enter all of your values. It is only meant to help illustrate how missing data, in this instance deferred taxes, can have a significant impact on the results of a financial health checkup. The Financial Statements.xls spreadsheet discussed above provides a much more rigorous estimate of deferred taxes.

5.3.4 Summarizing All Financial Information: Credit Scoring

As we have explored the financial health of EWS Farms, we reviewed the financial statements necessary and the ratios and measures calculated from these statements to get a picture of overall financial health. The Sweet Sixteen is a relatively small set of the available financial ratios used by the agricultural industry. With sixteen

Introduction | Market Value Basis Ratios | Summary and Deferred Tax

RMA
Risk Management Agency

Strategic Risk Management

Ending Balance Sheet Market Values for Current	
Assets	Liabilities
234,900	93,925

Ending Balance Sheet Market Values for Current	
Assets	Liabilities
1,330,500	405,989

Ending Deferred Tax Liability Values
$220,626

Ending Net Worth WO/Tax	Ending Net Worth W/Tax
1,292,203	1,071,577

Tax Rate on Ordinary Income 28%

Capital Gains Tax Rate 15%

FFSC Sweet Sixteen Ratios From Ending Balance Sheet

	Cost Basis	Market Value	Market Value With Def. Taxes
Liquidity			
Current Ratio	3.87	6.61	2.57
Working Capital	$104,543	$204,543	$147,066
Solvency			
Debt/Asset Ratio	76.41%	17.77%	31.81%
Equity/Asset Ratio	23.59%	82.23%	68.19%
Debt/Equity Ratio	3.24	0.22	0.47
Profitability			
Rate of Return on Farm Assets	1.54%	0.37%	0.37%
Rate of Return on Farm Equity	-11.16%	-0.83%	-1.01%
Operating Profit Margin Ratio	1.65%	1.65%	1.65%
Net Farm Income	$29,151	$29,151	$29,151
Repayment Capacity			
Term Debt and Capital Lease Coverage Ratio	1.28	1.28	-7.75
Capital Replacement and Term Debt Repayment Margi	$6,743	$6,743	($213,883)
Financial Efficiency			
Asset Turnover Ratio	93.64%	22.60%	22.60%
Operating Expense Ratio	77.57%	77.57%	77.57%
Depreciation Expense Ratio	9.63%	9.63%	9.63%
Interest Expense Ratio	4.67%	4.67%	4.67%
Net Farm Income From Operations Ratio	8.13%	8.13%	8.13%
Check Sum, Excludes Asset Turnover Ratio	100.00%	100.00%	100.00%

FIGURE 5.13 Deferred tax and missing data analysis tool. (From the Risk Management Agency, U.S. Department of Agriculture.)

different evaluations of financial health, is there something that we can really hang our hat on that says this is the one that reflects overall financial health. No, there is not just one, but an additional evaluation of financial health is provided by credit scoring models. Credit scoring models attempt to provide one single measure that evaluates the overall financial health of an operation. As you might expect, there are many of these models that use different approaches to providing a single score. Credit scoring models can be based on financial information, character assessments of the owner/operator, or some combination of both. These models can also vary by the type of loan an individual operation is trying to secure, such as long-term loans for capital purchases versus short-term operating credit. Credit scoring models developed using statistical analysis of financial data from actual operations can provide an objective rating of an operations financial health. Once developed, a credit scoring model could also be used to rate an individual operation against the average score or range of scores for other operations of similar size and characteristics.

One such model was developed with the help of and for the Sixth Farm Credit District (Splett, Barry, Dixon, and Ellineger, 1994) using a select set of the Sweet Sixteen ratios and measures recommended by the Farm Financial Standards Guidelines. Figure 5.14 shows the Term Loan model applied to EWS Farms.

The detailed calculations show how each selected ratio and measure included in this model was weighted and the potential impact that each has on the credit score for a long-term loan, 1.80 for EWS Farms. The Operating Loan Credit Score is shown as 2.20 (Figure 5.15). The credit scoring scale, developed using the Sixth Farm Credit District data set, shows the break points in the overall scale and how each break point relates to the class of loan and its associated risk. Not only are the

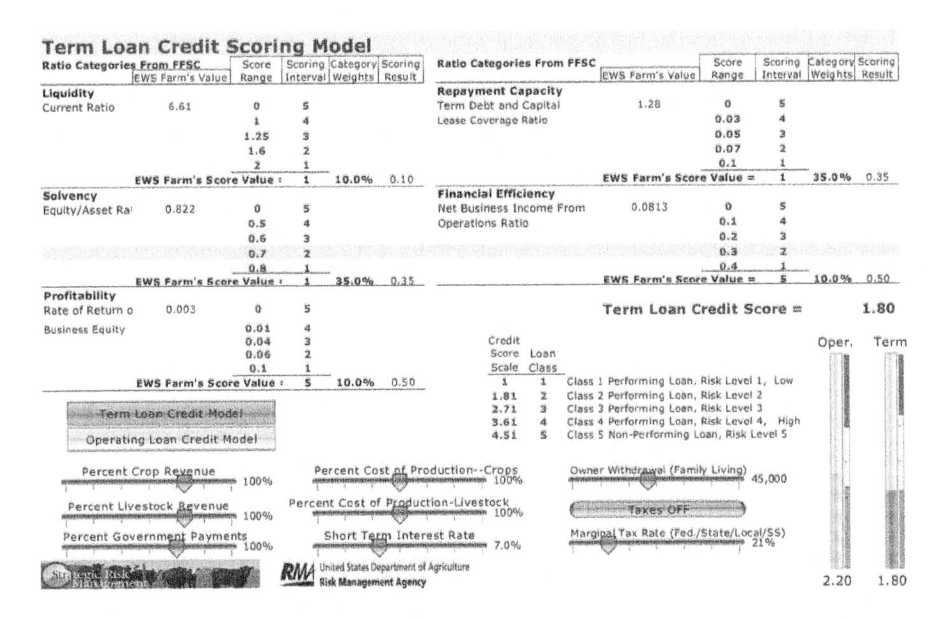

FIGURE 5.14 Term Loan credit scoring model. (From the Risk Management Agency, U.S. Department of Agriculture.)

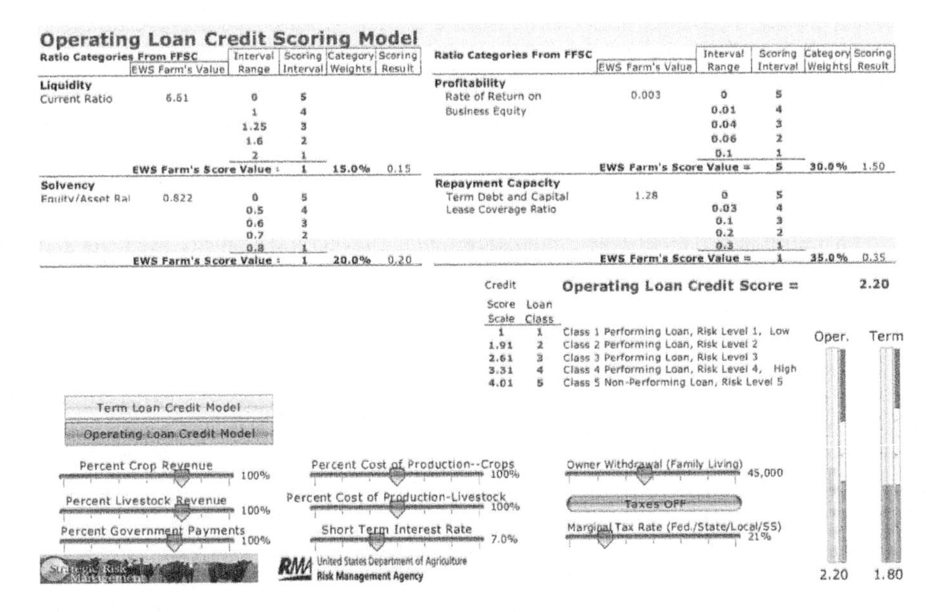

FIGURE 5.15 Operating Loan credit scoring model. (From the Risk Management Agency, U.S. Department of Agriculture.)

models different, but the break points for the credit score are also different for each model as are the weights applied to the individual ratio and measures selected for use in each model.

EWS Farms' credit score for the Term Loan (1.80) is just below the break point (1.81) between a class one and class two loan. The credit score for the operating loan model (2.20) is above the break point (1.91) between the class two and class three loans. In general, however, EWS Farms is in good financial condition and would not be considered a high risk for either type of loan, which is the same conclusion made in our previous discussions.

By now, it may seem that it would sure be nice if all this information could be crunched down into just one number using a uniform set of calculations for all possible evaluation criteria, and you are right, it would be nice. Unfortunately, given the variety in types of agricultural operations—beef, small grains, dairy, hogs, orchards—and the types of financing arrangements they need, this is not possible. Note that only a few of the Sweet Sixteen measures in our previous discussions were used in either of the two models presented here and there were no "characteristics of the individual owner/operator" included in either model. These characteristics might include a yes/no or numerical rating on an individual's strengths and weakness with respect to integrity, education, and training in each of the five risk areas (production, financial, marketing/price, institutional, human resources) and other character traits. The rankings of selected characteristics would then be incorporated into the overall credit score for a particular model.

5.3.5 Summary

Many producers consider financial analysis a tedious process and prepare financial statements solely for tax purposes, but records used for taxes are inadequate measures of financial performance. While a tedious process, producers should commit to learning the record keeping and data preparation procedures required to complete an adequate set of financial statements. Continuing to operate with minimal records, making management decisions based on incomplete financial statements is truly a risky business.

The ability to use tactical financial planning comes with the ability to create and use the recommended financial statements. If this cannot be accomplished, the strategic and operational portions of the financial planning process, Figure 5.4, are going to be very difficult, if not impossible. Planning based on poor financial information may be in error and result in a business and family financial disaster.

REFERENCES

Coble, K. H., T. O. Knight, G.F. Patrick, and A. E. Baquet. 1999. *Crop producer risk management survey: Preliminary summary of selected data: A report from the Understanding Farmer Risk Management Decision Making and Educational Needs Research Project.* Mississippi State University, Information Report 99-001.

Ellinger, P. 1999. Cash or accrual income—does it really make a difference? *Ag Lender*, March.

Ellinger, P. 1998. Ratio Analysis: Comparative analysis: Guidelines for liquidity and solvency measures, *Ag Lender*, March.

Farm Financial Standards Council (FFSC). 1997. *Financial guidelines for agricultural producers: Recommendations of the Farm Financial Standards Council*, http://www.ffsc.org.

Frey, T. L., and D. A. Klinefelter. 1980. *Coordinated financial statements for agriculture*, Second Edition. Skokie, IL: Agri Finance.

Griffith, D. Financial Statements xls, http://www.montana.edu/extensionecon/financialmgt downloads.html.

Griffith, D. 1990. *Preparing and interpreting a balance sheet*. Montana State University Extension Bulletin EB 92. Bozeman, MT: Montana State University.

Internal Revenue Service (IRS). 2007. *Publication 225: Farmers tax guide*. Washington, DC: Government Printing Office.

Langemeier, M. R. 2004a. *Balance sheet—A financial management tool, MF-291*. Kansas State University Agricultural Experiment Station and Cooperative Extension Service, October.

Langemeier, M. R. 2004b *Financial ratios used in financial management*, MF-270. Kansas State University Agricultural Experiment Station and Cooperative Extension Service, October.

Langemeier, M. R. 2004c. *Income statement—A financial management tool*, MF-294. Kansas State University Agricultural Experiment Station and Cooperative Extension Service, October.

Miller, A., M. Boehlje, and C. Dobbins. 2001. *Key financial performance measures for farm general managers*. Purdue University, Publication ID-243.

Mishra, A. K., H. S. El-Osta, M. J. Morehart, J. D. Johnson, and J. W. Hopkins. 2002. *Income, wealth, and the economic well-being of farm households*. Economic Research Service, U.S. Department of Agriculture. Agricultural Economics Report No. 812, July.

Niswonger, R. C., and P. E. Fess. 1969. *Accounting principles*. Cincinnati, OH: South-Western Publishing Co.

Splett, N. S., P. J. Barry, B. L. Dixon, and P. N. Ellinger. 1994. A joint experience and statistical approach to credit scoring, *Agricultural Finance Review* 54.

FURTHER READING

Ellinger, P. 1997a. Ratio analysis: Look behind the numbers, *AgriFinance*, July.

Ellinger, P. 1997b. Ratio analysis: Look among the numbers, *AgriFinance*, November.

Financial Accounting Standards Board (FASB), http://www.fasb.org.

Griffith, D, *A simple cash flow statement*, http://www.montana.edu/extensionecon/financialmgtdownloads.html.

US Department of Agriculture, Economic Research Service. Accessed June 10, 2009 from http://www.ers.usda.gov

6 Step 2
Determine Risk Preferences

Dana Hoag and Catherine Keske

CONTENTS

Although agricultural production is a risky business, producers can reduce potentially devastating losses if they are willing to sacrifice some of their financial rewards. Would you invest in a new crop variety that could make you $100 per acre if there was also a chance of losing $50 per acre? How would your answer change if you could make $500 instead of $100? Each farm manager might answer this question differently. After all, everyone has a different view about what constitutes a "risky" decision, and everyone has a different preference for the amount of risk he or she is willing to face.

How we feel about risk influences the business decisions we make, from straightforward choices like selection of crop variety to more complex events like buying a new farm or adding another operator to the business. These decisions may determine how wealthy we become or how well we sleep at night. It's important to know your personal risk preference and how it influences your decision making. Step 2 in the SRM process will help you become better acquainted with your risk preferences. When combined with your financial health analysis from Step 1, these preferences will guide your choice of goals in Step 3.

Generally speaking, greater risks lead to greater wealth. Many people accept the risks associated with starting a new business or investing in a new technology because they realize that risk taking often results in a better long-term payoff. If this weren't true, no one would take risks. Some people even seek out risks for the thrill of a big payoff, though most of us don't enjoy the emotional highs and lows associated with risk taking or the inconvenience risks can cause. Instead, we seek to avoid risk, or at least reduce risk to an acceptable level. The threshold of what is deemed to be an acceptable level of risk varies considerably among individuals. In this chapter, you will learn how to measure your risk attitude and determine what you find to be an acceptable level of risk and return.

Like the previous chapters, this chapter is divided into three distinct parts. The first part examines and measures the risk tolerance levels for Aaron Sprague and EWS Farms. Part 2 discusses the mathematical principles and formulas used to define risk preferences. The three kinds of risk preference or tolerance (risk averse, risk neutral, and risk loving), different approaches for measuring risk tolerance, and the associated strengths and weaknesses of each approach are described as well. Part 3 defines the Risk Navigator tool designed for this chapter, the Risk Preference Calculator, and provides you the opportunity to determine your own risk tolerance.

6.1 PART 1: THE EWS FARMS CASE STUDY

6.1.1 Objectives

This chapter has a very straightforward objective: *Determine your personal preference, or tolerance, for risk.*

While the objective is simple, the process is very difficult. There are several methods that have been developed to measure individual risk preferences. Our objective is to use more than one method to increase the chance of an accurate reading.

6.1.2 Strategy

There are two main schools of thought about determining risk attitudes; both approaches are used in this chapter. The first approach is based on psychology. Respondents are asked how they might react to situations that involve uncomfortable choices. (Later in the chapter, you'll find a psychology-based risk preference quiz.) The second technique reflects an economic or financial approach for measuring risk preference. Respondents complete an inventory of questions where they make choices that reflect risk trade-offs and financial gains and losses. The Risk Navigator SRM tool uses both methods to generate a finite numerical score and ranks the respondent on a continuum from risk loving to highly risk averse. Aaron Sprague's scores are shown in the following implementation section.

As with previous chapters, we provide a comprehensive discussion in Part 2 to describe the complex calculations used to determine Aaron Sprague's risk preference scores. We explain the mathematical formulas used to determine risk preferences to help readers develop an intuitive understanding of how risk attitude affects the premium Aaron is willing to pay to avoid risk and how this knowledge can help him to make better risk-related decisions in the future. For example, Aaron will have more information to decide whether he should sell some of his crop at harvest or store the crop for a possible price increase. Be aware that the information presented in Part 2 is technical and the material may overwhelm readers unfamiliar with algebra and some calculus. Regardless of your background, don't let the complexity of material deter you from calculating your own risk preference measurement with the Risk Navigator SRM tools in Part 3. It is simple.

6.1.3 Implementation

Implementation of Step 2 is straightforward. First, Aaron completes the psychology-based risk tolerance survey presented in Part 3. For example, question 1 asks: "In general, how would your best friend describe you as a risk taker?" He checked box "b: Willing to take risks after completing adequate research."

Aaron's quiz score of 31 reveals that he has an "above average tolerance for risk." His score isn't surprising simply because he is fairly young; risk aversion increases with age and younger people are able to tolerate more risk. Aaron also has a great deal of confidence when making decisions because of his own personal investment in education.

Next, Aaron takes the risk preference inventory with the Risk Navigator SRM tool, Risk Preference Calculator. He answers questions about how much he would pay for bets that have a 50-50 chance of yielding a good or bad outcome. Based on this exercise, Aaron finds that he has a risk preference score of 0.52, which is somewhat risk averse. This result means that Aaron is very willing to take on risks, but he is not a risk lover. In other words, Aaron would spend some effort or money to avoid risks when appropriate, but his personal scores indicate he is generally willing to take on risks. To take the quiz yourself, jump ahead to Part 3.

6.2 PART 2: THE FUNDAMENTALS OF DEFINING AND MEASURING RISK ATTITUDE

The term *risk attitude* simply means there is a fear/greed trade-off (FinaMetrica, 2008) between making money and avoiding potential unfavorable consequences as a result of taking risks. There are three main attitudes a person can have toward risk (Hardaker et al., 2004). An individual who is afraid of taking risks is known as risk averse. This individual prefers an investment with a lower, but certain, expected payoff to an investment with a higher, but uncertain, payoff. An individual is said to be risk neutral if he or she cares only about the expected payoff of an investment and not the risk that must be taken to achieve the investment goal. Such an individual will neither actively take risks nor pay to avoid them. This individual would self-insure against loss, for example. An individual who actively engages in risky investments is referred to as risk seeking or risk loving.

We can use a decision tree of Aaron's preferences to further explain this concept (see Chapter 9 for a formal discussion about decision trees). Aaron has an opportunity to store some of his corn crop when he harvests it, with the hope that the future price

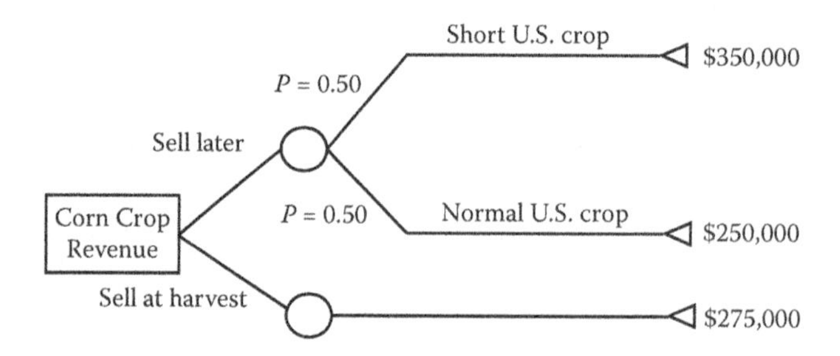

FIGURE 6.1 A decision tree for holding or selling corn crops.

goes up. The alternative choice is to sell the crop on the cash market at harvest for an expected $275,000. There is a chance that prices will go up if there is a short U.S. crop, and he could increase his revenue to $350,000 if he stores his corn. However, there is an equal chance that there will be a normal U.S. crop and that he would have to sell for $250,000—less than he could have received at harvest. As shown in Figure 6.1, there is one decision (sell later vs. sell at harvest), and three possible financial outcomes, a certain $275,000 or a risk between $250,000 and $350,000.

Based upon Aaron's risk preferences and the concept of expected value (EV), we can project at what point he should sell his crop. The EV is the sum of the probability of each possible outcome multiplied by its payoff (see Chapter 10 for more details about probability). In this case, the EV is ($350,000 × 0.5) + ($250,000 × 0.5), which equals $300,000. These probabilities are also reflected in Figure 6.1 by the notation $P = 0.50$. Aaron's choice is to either hold the crop for an expected value of $300,000 or to sell it for a sure thing of $275,000—by accepting $275,000 instead of an average of $300,000 over time. If Aaron accepts a risk-free opportunity to sell today, he is considered risk averse because he would lose an average of $25,000 per year over time in order to avoid risks. If he chooses to hold his crop, he is less risk averse (either risk neutral or risk loving) because he is willing to take on the risk that he could lose income in the hopes that he may make even more money than he is certain to receive if he sold his crop at harvest. Aaron is considered risk neutral if he is indifferent to risk and simply chooses the action that will produce the highest expected value (EV). Aaron is a risk lover if he is willing to pay more than $300,000 to hold the crop just so he could experience the risk.

As we showed with the risk inventory, Aaron is almost risk neutral. He selects the option that provides the highest expected value, and he holds his crop in the hopes that there is a short U.S. crop this year.

6.2.1 Defining Risk Preference Using a Utility Function

Risk preference is a measure of the amount of risk we are willing to take to achieve an investment goal. This is also known as *risk tolerance*. Risk preference scores help define what constitutes acceptable versus unacceptable levels of risk. While risk preference is beneficial for investors to understand, the formulas are rather complex. In this section, we review the theory and formulas used to create the algorithms used in Part 3, Risk Navigator SRM tools.

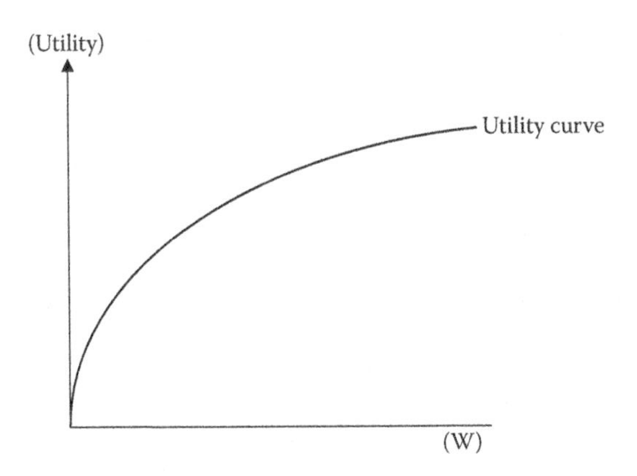

FIGURE 6.2 Risk and the utility function.

An individual's risk preference depends upon his wealth, how much satisfaction he derives from his wealth, and the rate of satisfaction he derives from increases in wealth. This can be measured by utility (U), a scale of satisfaction. The amount of utility a person receives from wealth (W) is written as U(W). A utility function, shown in Figure 6.2, maps utility (the dependent variable) as a function of wealth. In this case, utility is scaled from 0 to 100, where 0 reflects no satisfaction and 100 is maximum satisfaction. The curve is bowed outward (concave) because each additional unit of (W) adds less satisfaction than the previous unit. When wealth is small, an additional dollar can provide a lot of satisfaction. But as you grow wealthier, the last dollar received doesn't provide nearly as much satisfaction as that first dollar. Think about sitting down to eat a bag of popcorn. The first bite gives you more satisfaction than the last.

Risk attitudes are determined by the shape of the utility function shown in Figure 6.2, which reflects how quickly satisfaction diminishes with additional wealth. The slope for risk-averse, risk-neutral, and risk-seeking individuals is positive. This is intuitive because more wealth or profit is preferred to less. Mathematically, the slope of the utility function can be found by taking its first derivative:

$$U'(W) > 0$$

where U is equal to utility, which is a function of wealth or income (W). The positive sign (> 0) means that more wealth is preferred to less.

The second derivative of the utility function indicates the rate at which each additional dollar provides changing satisfaction. The second derivative indicates risk attitude:

$$U''(W) < 0 \text{ implies risk aversion.}$$

This concave, or humped, function is shown in Figure 6.2. This means an additional dollar of wealth provides incrementally less satisfaction than the previous dollar.

$$U''(W) = 0 \text{ denotes risk indifference (a linear function)}$$

This means that the next dollar of wealth provides exactly the same satisfaction as the previous dollar.

$$U''(W) > 0 \text{ reflects risk-seeking behavior}$$

The convex (U-shaped) function indicates that an additional dollar of wealth provides incrementally more satisfaction than the previous dollar.

Risk-aversion levels are not constant for most people, and studies show that most people are risk averse to some degree (Meyer and Meyer 2006; Harwood et al., 1999; Kimball, Sahm, and Shapiro, 2007). People change their preferences for all sorts of reasons, including the context of the investment. For example, you might be risk loving and buy a lottery ticket if the cost to you is only $1, but you'd become risk averse if you had to spend $100 for a ticket with the same proportional chances of winning. You might be risk averse for a bank investment if your retirement savings are at stake, but risk loving for business investments that involve increasing discretionary income.

Next, we explain risk preference through the concepts of risk premium (RP), certainty equivalent (CE), and expected value (EV). Risk premium (RP) is the amount of money someone is willing to pay (or sacrifice) to avoid (assume) a risky alternative. Insurance is a good example of an RP because the insured pays to avoid risk. How much a person would pay to insure against risk is important because it determines the profit that can be made selling someone insurance. Insurance companies invest in research to determine risk premiums and compare it to the cost of providing the insurance. Think about what you pay per period for automobile, life, health, fire, or other insurance. If you pay $1,000 per period, you know the insurance company expects to pay less than $1,000 in claims per period, averaged over a number of periods, if it wants to make a profit. The RP is the difference between what the company expects to pay and what you pay. For example, if the company expects to pay $800 on average per period for a claim, your RP is $200. In other words, you give up $200 per period to transfer your risk to the insurance company by paying $1,000 per period instead of the true cost of $800.

A related concept, certainty equivalent (CE), is defined as the "certain" amount that a person would have to be paid to give up the potential payoff from a risky activity. The risk premium (RP) between two choices is equal to the expected value (EV) of the risky event minus its certainty equivalent (CE). In the insurance example above, the CE is $1,000. That is, you would be equally happy either paying $1,000 (by paying the insurance premium up front regardless of whether or not you have an accident) or accepting the risk on an accident, which the insurance company has determined at an average expected cost of $800 per period (RP($200) = EV(–$800) – CE(–$1,000)).

Consider Aaron's marketing opportunity; if Aaron chose to sell his crop at harvest, his CE would be $275,000 and his RP would be at least $25,000:

$$\text{RP } (\$25{,}000) = \text{EV } (\$300{,}000) - \text{CE } (\$275{,}000)$$

Recall that we calculated Aaron's EV a few pages earlier. If Aaron chose to sell his crop at harvest (instead of holding his crop, as previously discussed), his RP would be at least $25,000. We wouldn't really know how much more he would be

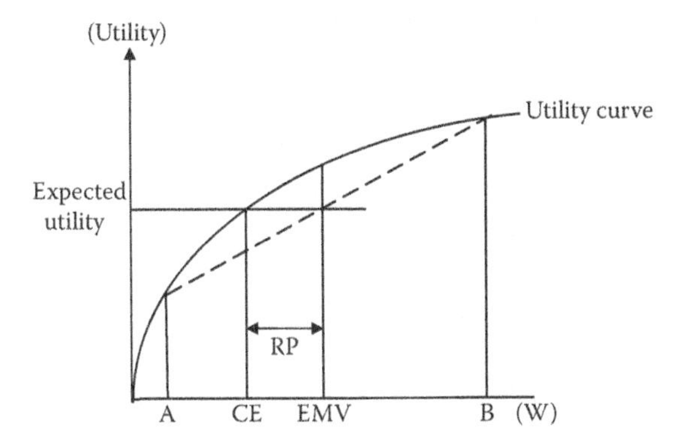

FIGURE 6.3 Expected monetary value, risk premium, and certainty equivalent.

willing to give up to avoid risk until we looked at the slope of his utility curve. We examine this in Figure 6.3.

For the sake of simplicity, Aaron has a 50-50 chance of earning two amounts, shown as A and B in Figure 6.3. The EV is halfway between at EMV (Expected Monetary Value). EMV is the weighted average; that is, the sum of each outcome multiplied by its probability. In this case, EMV = 0.5A + 0.5B. For a risk-neutral person, the EMV would have a utility on the dotted line. Given that the utility is bowed and not straight, the expected utility (EU) must be measured horizontally from where EMV hits the dotted line. By definition, the utility, or satisfaction, of the CE is equal to the utility of the gamble to hold the crop. That is, the EU of the gamble is equal to the utility of the CE, which is not on the dotted line connecting A and B when utility is bowed. The horizontal line between the EMV and CE represents the RP. A more risk-averse person would have greater bowing in their utility function and would be willing to pay a larger premium to avoid risk than someone with less bowing in their function. Estimating the bow in a person's utility function is how economists measure risk preference.

Practically speaking, risk tolerance measures the curvature of the utility function. A measure of this curvature was defined by Pratt (1964) and Arrow (1965), who coined the term *coefficient of absolute risk aversion*, ra(W):

$$ra(W) = -U''(W)/U'(W)$$

The coefficient of absolute risk aversion is the negative of the second derivative divided by the first. This coefficient is positive for risk-averse individuals, zero for risk-neutral individuals, and negative for risk-loving individuals. A risk-averse decision maker's utility function is concave (bowed upward), as described above in Figure 6.3. As shown in Figure 6.4, a risk-neutral decision maker's utility function is flat or straight; there is no difference between the CE and EMV. A risk-lover would have a convex (bowed down) utility function and would be willing to pay more than EMV for a chance at risk.

Since "r" changes with the size of the gamble, the concept of a coefficient of relative risk aversion (RAC) was created simply by multiplying the coefficient of absolute risk aversion by wealth:

$$RAC = rr(W) = W\ ra(W) = RAC\ (W)$$

The RAC equals 0 for someone who is risk neutral. It varies from about 0.5 to 4.0 for risk-averse people as suggested below by Anderson and Dillon (1992):

RAC (W) = 0.5: Hardly risk averse at all
RAC (W) = 1.0: Somewhat risk averse (normal)
RAC (W) = 2.0 Rather risk averse
RAC (W) = 3.0: Very risk averse
RAC (W) = 4.0: Almost paranoid about risk

Many researchers have looked at measures of risk aversion. Perhaps the most popular measure, relative risk tolerance (RRT), is directly related to the RAC. Many people use the terms risk tolerance and risk preference interchangeably. There are some subtle differences, but RRT and RAC are, for practical purposes, the inverse of each other. Both provide straightforward measures of risk aversion, and either scale is appropriate. The scale of RRT spans from 0.25 (1/4 compared to RAC) for a person who is almost paranoid about risk to 2 (1/0.5) for someone who is rarely risk averse to 10 (1/.1) for someone who is almost risk neutral.

One confusing point about risk tolerance is that the term is also used interchangeably to refer to the amount of risk a business or person can tolerate. Risk tolerance is actually different from the concept of risk preference—while RRT is not so different. It is based on the amount ($X) where a person would be indifferent between

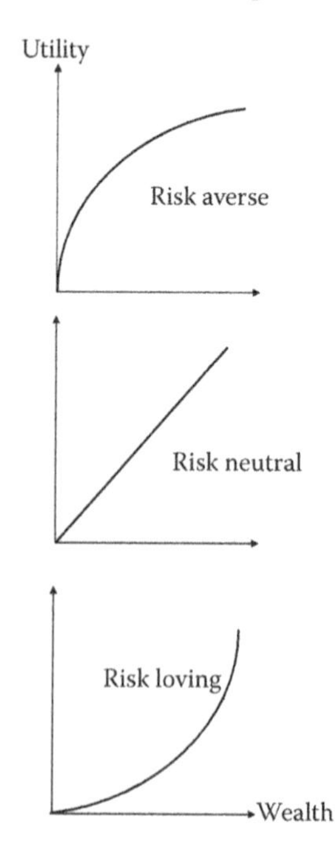

FIGURE 6.4 Utility functions for risk-loving, risk-neutral, and risk-averse decision makers.

an equal chance of receiving \$X and losing \$X/2. This calculation is explained in the next section, but let's look at an example here. Myerson (2005) shows that risk tolerance for a 50-50 bet of receiving either 0 or 20,000 is \$15,641 for a person who would pay \$7,000 to take the bet. Risk tolerance increases to \$99,833 for someone willing to pay \$9,900. The EV of this bet is \$10,000. A person willing to pay \$9,900 is more risk tolerant than someone only willing to pay \$7,000; that much is clear. But interpreting what \$15,641 or \$99,833 of risk tolerance means is not very intuitive. It is perhaps more intuitive to know that the risk premium is \$3,000, where the risk tolerance was \$15,641, and is \$100 where the risk tolerance was \$99,833.

While it may be less intuitive than risk preference, it turns out that risk tolerance can be very helpful for computing the RRT (and, by inversion, the RAC). A Harvard University study by Howard suggests guidelines to estimate risk tolerance, R, in terms of net income, total sales, and equity: R = 1.24 multiplied by net income, 6.4% sales, or 15.7% of equity (Harwood et al., 1999). Knowing how to relate R to observable business variables provides a practical link for computing RRT because the RAC can be computed from R as shown later.

6.2.2 Functional Forms of Utility Functions

The functional form of the utility function can have a big impact on how it represents utility (Hardaker et al., 2004). While there are many functions and nuances that could be presented, in this section we focus on what is used in Risk Navigator SRM. The most common form used is the negative exponential utility function:

$$u(w) = 1 - \exp^{(-\alpha w)}$$

The coefficient of absolute risk aversion for this functional form equals:

$$ra(W) = -U''(W)/U'(W) = \alpha$$

This function is popular because it is mathematically well behaved and relatively easy to use. The coefficient of absolute risk aversion is constant and positive over all levels of wealth and income. This implies that the negative exponential utility function exhibits constant absolute risk aversion. This functional form appears in many contexts. The RAC is simply the absolute value of the constant multiplied by w. For this function, an individual's RP does not depend upon his initial amount of wealth. For example, you might be indifferent between doing nothing and buying a business that has an equal chance of earning \$5,000 or losing \$4,000 (Myerson, 2005). You are considered risk averse since this bet carries an EV of \$500. Many researchers suggest that you might take that bet if you had, for example, just inherited \$100,000 (Fleisher, 1990; Myerson, 2005), meaning that wealth does not matter. You would be willing to take on more risk because you had more wealth. This functional form is widely used despite this questionable property. An expanded discussion about constant and variable risk aversion can be found in Chapter 13 of Clemens and Reilly (2001) or in Myerson (2005).

There are other utility functions to consider. For the sake of showing alternatives, the power utility function is expressed as:

$$u(w) = \alpha + \beta w^{\gamma}$$

where α, β, and γ are parameters and the coefficient γ is restricted between 0 and 1 ($0 < \gamma < 1$).

The absolute risk aversion coefficient for this function is:

$$ra(W) = -U''(W)/U'(W) = -(\gamma - 1)w^{-1}$$

The power utility function exhibits decreasing absolute risk aversion (DARA) since it is positive ($\gamma < 1$) and decreases while wealth increases. This property makes the power utility function very attractive since risk aversion generally decreases with an increase in wealth.

Another function is the logarithmic utility function:

$$U(W) = \ln(W)$$

The absolute risk aversion coefficient for this function is:

$$ra(W) = -U''(W)/U'(W) = w^{-1}$$

The logarithmic utility function also exhibits decreasing absolute risk aversion (DARA) because it is positive and decreases while wealth increases. This property also makes the function well behaved.

6.2.3 Eliciting a Utility Function

Researchers use the utility functions discussed above to explain observed behaviors and to draw conclusions about how individuals might respond to risk-related decisions. Practically speaking, an individual does not know their own utility function, nor can it be easily revealed. Researchers are learning more about how to elicit a person's utility function, however. For a more thorough discussion, refer to Clemens and Reilly (2001) or Hardaker et al. (2004).

The method we favor is eliciting utility by asking people about their CE. Underlying this notion is an exponential utility function that can be expressed as:

$$U(X) = 1 - e^{-xr}$$

Where r is the coefficient of absolute risk aversion and X is profit or income; this function makes it very easy to compute RACs (r multiplied by X).

Hardaker et al. (2004) describe a method called Equally Likely Certainty Equivalent (ELCE), which relies on plotting a series of certainty equivalents for risky prospects. For simplicity, we plot utility in Figure 6.5 between 0 and 1, where 1 is the highest utility possible. The method first plots the utility of 0 (0,0) and then 1 (400,1), as shown in Figure 6.5. In this case, 400 is assumed to be the highest level that X can assume. The method proceeds by determining points between 400 and 0. For example, Aaron is asked how much he would pay to play the following lottery:

Win $400 with a probability of 0.5 in the best case

and

Win $0 with a probability of 0.5 in the worst case

A risk-neutral person would pay the EV of $200. Graphically, this falls halfway between the line segment from the 0 and 400 endpoints. A risk-averse individual would pay less than $200. Suppose Aaron said he would pay $120. He is risk averse; his CE is $120 and his RP is $80. We can plot another point at 120 on the x-axis in Figure 6.5. Utility for this point is 0.5, as shown below:

$$U = 0.5\ U(400) + 0.5\ U(0) = 0.5(1) + 0.5(0) = 0.5$$

The ELCE process continues by asking about the CE for a 50-50 bet between 0 and 120, where utility is 0.25, and between 120 and 400, where utility is 0.75. The process can continue by filling in as many points as you wish. Eventually, this traces out a person's utility function as shown in Figure 6.5. The Risk Navigator Risk Preference Calculator shows the utility function as you indicate your preferences for each lottery and computes the utility function with regression analysis.

It is easy to estimate relative risk tolerance directly from the exponential utility function equation since utility can also be expressed as:

$$U(X) = 1 - e^{-x/R}$$

R reflects *risk tolerance*. As explained earlier, R can be computed from business information such as gross sales. A small "R" indicates the function is more bowed and the person is more risk averse. Someone who is not very risk averse can tolerate more risk; he or she has a larger R. R is approximately equal to the maximum amount, X, where a person is indifferent between a lottery with an equal chance of receiving X and losing X/2.

The two expressions above, with r and R, allow us to show both risk preference and how risk tolerant a person is using the RAC or RRT. In theory, you could measure

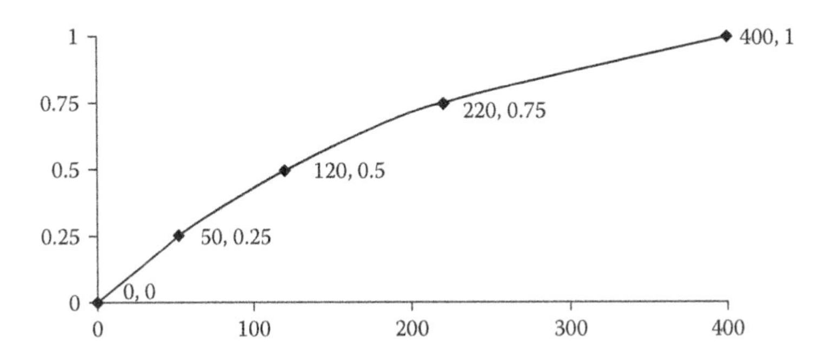

FIGURE 6.5 An example of eliciting a utility function.

a person's tolerance, R, directly by asking him about the decisions he would make with regard to bets (Clemens and Reilly, 2001). In concept, a person would be asked to determine the maximum amount, X, that he would pay for the bet of winning X or losing X/2. For example, suppose that Aaron had answered X = $275,000 (which we enter as $275, shown in the equation below) regarding his decision whether to hold or store his crop. Because $275,000 is close to the EV of $300,000, Aaron is not very risk averse. That is, he can tolerate taking a chance of losing up to $275,000. Aaron's utility function is:

$$U(x) = 1 - e^{-x/275}$$

6.2.4 Finding Risk Premiums and Certainty Equivalents

As we discussed in the previous section, risk tolerance can be measured once the utility function is identified. We used the equation estimated above to find Aaron's expected utility:

$$U(x) = 1 - e^{-x/275}$$

Let's return to Aaron's decision to store his corn. Suppose that if he stores his crop, Aaron faces the following probabilities for $250, $260, $310, and $350 (measured in thousand dollars):

$250 with a probability of 0.2
$260 with a probability of 0.2
$310 with a probability of 0.3
$350 with a probability of 0.3

To determine Aaron's expected utility of crop storage, take the sum of the utility of the payoffs multiplied by the probability of each of these payoffs:

$$EU(\text{crop storage}) = 0.2\ U(250) + 0.2\ U(260) + 0.3\ U(310) + 0.3\ U(350)$$

The utility payoffs are plugged into the exponential utility function (shown above) to obtain respective utility values. That is, plugging 250 into the utility function is equal to 0.597. This implies:

$$EU\ (\text{crop storage}) = 0.2(.597) + 0.2\ (0.611) + 0.3\ (0.676) + 0.3\ (0.720) = 0.661$$

The CE now can be found by solving the following identity:

$$0.661 = 1 - e^{-x/275}$$

which implies:

$$0.339 = e^{-x/275}$$

and taking natural logs on both sides of the equation leaves:

$$\ln(0.339) = -x/275$$

which implies:

$$x = \$297.10.$$

Since the utility of the CE is equal to the EU of the risky venture, the CE is therefore equal to $297,100. Remember, the RP is equal to EV – CE. The RP is then $2,900 as follows:

$$RP = \$300,000 - \$297,100 = \$2,900$$

The $2,900 RP is a very small number relative to the size of the gamble, once again confirming that Aaron is not very risk averse. Aaron is willing to lose no more than about $2,900 to gain certainty.

6.2.5 Concluding Comments about Measuring Preferences

It is clear that measuring risk preferences requires an understanding of several complex mathematical concepts; however, it is important to point out that mathematical rigor isn't always required to calculate risk preferences. In fact, sometimes little value is added when quantifying risk preferences. In these cases, determining the decision maker's preferences probably isn't even warranted. This may be the case when the risk premium is relatively small. That is, if you are willing to pay only a small premium to avoid risk, say 1% of EV, then it might not be worth conducting a complicated risk analysis. Aaron, for example, only had a $2,900 RP on a $300,000 investment. The risk ranker tool described in Chapter 11 helps producers determine when risk premiums are too small to consider. It is also possible to use techniques that avoid eliciting risk altogether. Information about statistical methods that rely on observing behavior can be found in other literature. These methods, called *stochastic efficiency*, are also discussed in Chapter 11.

Results also may be influenced by framing, perception, and measurement. In 1981, two psychologists, Kahneman and Tversky, showed that risk attitude changes according to the way a problem is framed (Clemens and Reilly, 2001). A person might respond differently to a question posed in the negative than to one posed as positive: "You would lose this much money if …" versus "This is how much you would gain if …" Framing problems are usually overcome by carefully considering the wording of a question or asking it in multiple ways.

Perception can affect the utility measurement if respondents don't interpret a question as it was intended. For example, people who are morally opposed to gambling have had problems complying with the CE method because they don't like answering questions about a gamble. That's why we use an example of growing onions instead of an outright gamble when we use ELCE method in our Risk Navigator tool. Another perception problem is anchoring. For most people, the ability to answer a question accurately is limited because they are anchored on prior knowledge. A farmer, for example, might have difficulty estimating yield for a new seed variety if the yield was substantially different from what he or she experienced in the past. Some perception problems can be overcome by carefully framing questions. Other perception problems can be overcome by using different methods. For example, if a person has an aversion to gambling, the researcher can ask about probability equivalents instead of certainty equivalents (Clemens and Reilly, 2001).

Measurement is a much more complicated problem. We have already discussed three different functional forms for utility. Each of these functions has limitations and no one function overcomes all problems. The negative exponential function, for example, imposes constant risk aversion. (Note: Most of these problems are too complex to discuss here, so refer to the reference list literature for more information.)

Finally, it's worth noting that everything can't be structured. These methods are capable of framing a problem in realistic terms, but there are always exceptions. Even a risk-averse person might be motivated to make a risk-loving choice. Suppose that your banker, for example, said she will call in your loan if you don't make a certain profit mark this year. You might choose to take a very risky action because playing it safe is actually riskier than failing to generate revenue.

6.3 PART 3: RISK NAVIGATOR MANAGEMENT TOOLS

The previous section provides an in-depth explanation about how to compute risk preference and tolerance. In this section, we discuss the benefit of using our online tools to make these calculations simply and automatically with only a little information from you.

The Risk Navigator Risk Preference Calculator estimates risk preferences automatically, so you don't have to endure all of the hand calculations. This tool provides three distinct methods to estimate your risk preferences:

1. The first method, "Take a Quiz," is based upon the psychology discipline's approach to calculating risk preferences.
2. The second method estimates your RAC and RRT using the ELCE method. Click on "Estimate My Risk Preferences."
3. The third approach incorporates the Harvard study estimates to determine Standard Risk Tolerances, which reflect how much risk tolerance you might have, given what researchers have found for typical businesses around the country.

You can choose one or more approaches, as you like. Each of these approaches is discussed in the next section.

6.3.1 Risk Preference Calculator: Take a Quiz

When you open the Risk Preference Calculator, you will see the main menu tabs off to the left and a description of the three methods on the right. Choose any or all

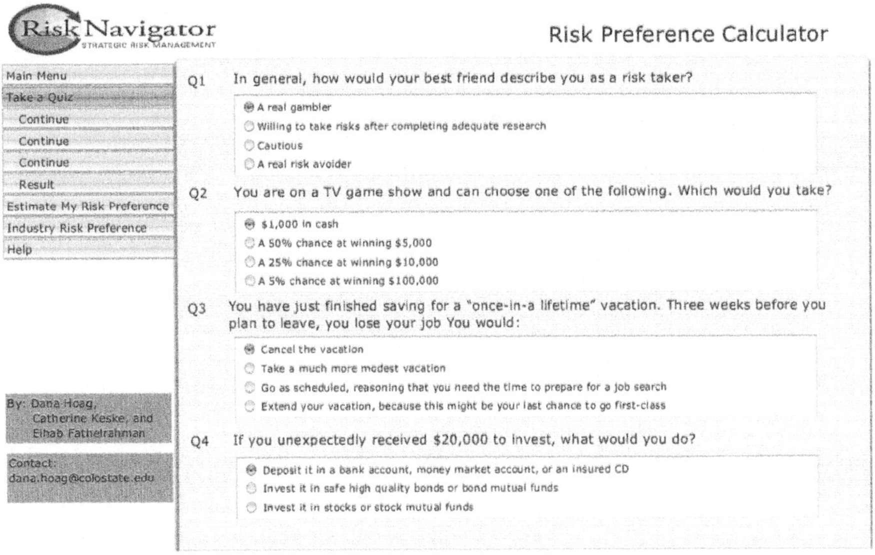

Risk Navigator
STRATEGIC RISK MANAGEMENT

Risk Preference Calculator

Main Menu
Take a Quiz
Continue
Continue
Continue
Result
Estimate My Risk Preference
Industry Risk Preference
Help

By: Dana Hoag, Catherine Keske, and Elhab Fathelrahman

Contact: dana.hoag@colostate.edu

Q1 In general, how would your best friend describe you as a risk taker?
- A real gambler
- Willing to take risks after completing adequate research
- Cautious
- A real risk avoider

Q2 You are on a TV game show and can choose one of the following. Which would you take?
- $1,000 in cash
- A 50% chance at winning $5,000
- A 25% chance at winning $10,000
- A 5% chance at winning $100,000

Q3 You have just finished saving for a "once-in-a lifetime" vacation. Three weeks before you plan to leave, you lose your job You would:
- Cancel the vacation
- Take a much more modest vacation
- Go as scheduled, reasoning that you need the time to prepare for a job search
- Extend your vacation, because this might be your last chance to go first-class

Q4 If you unexpectedly received $20,000 to invest, what would you do?
- Deposit it in a bank account, money market account, or an insured CD
- Invest it in safe high quality bonds or bond mutual funds
- Invest it in stocks or stock mutual funds

FIGURE 6.6 Risk Preference Calculator: Take a Quiz.

of the methods. We'll start here with the first option, "Take a Quiz," as shown in Figure 6.6. This tool uses a quiz developed by J. E. Grable and R. H. Lytton (1999). You can also find the quiz at the Web site: http://njaes.rutgers.edu/money/riskquiz/default.asp.

When taking the quiz, you have to click on "Continue the Quiz" and then on "Results" to complete the survey. When Aaron is done, his score is 31, "above average tolerance for risk," as shown below:

0–18 Low tolerance level
19–22 Below-average tolerance for risk
23–28 Average/moderate tolerance for risk
29–32 Above-average tolerance for risk
33–47 High tolerance for risk

There is also a much more involved quiz provided at the Web site MyRiskTolerance.com (http://www.myrisktolerance.com) by FinaMetrica®. There is a modest fee to take this exam, but it is highly tested and normed across many people.

6.3.2 Risk Preference Calculator: Estimate My Risk Preference

Clicking on "Estimate My Risk Preference" will bring up the screen shown in Figure 6.7. Using the ELCE method, you are asked to indicate how much you would pay (your CE) for four different choices. We can make a specific, personalized estimate of your risk preference and tolerance based on these questions. The first choice is for a 50-50 chance of getting 0 or $100,000. The EV is $50,000. To make the

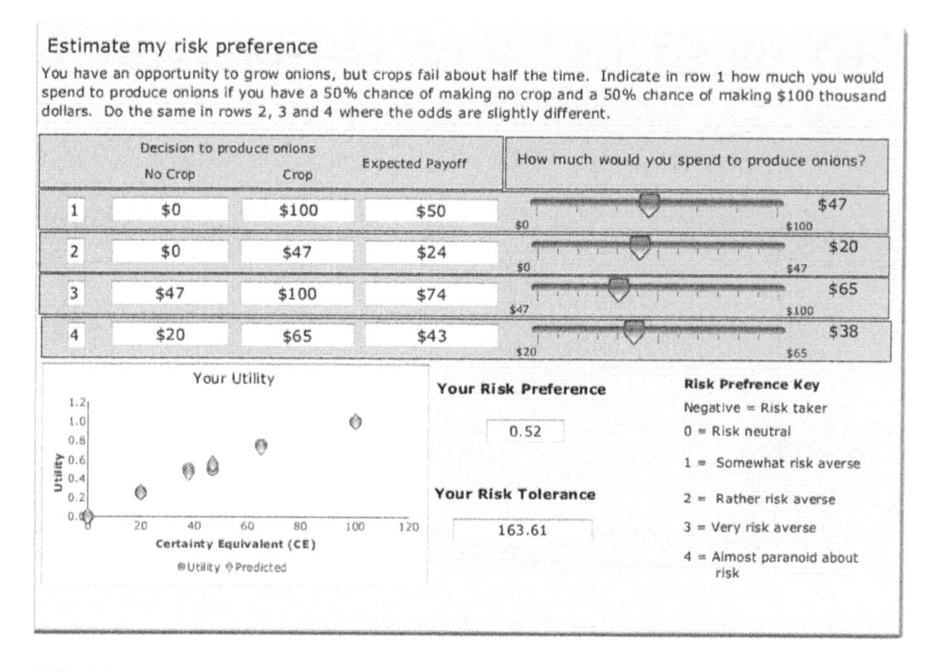

FIGURE 6.7 Risk Preference Calculator: Estimate My Risk Preference.

program more interesting, we have asked you to think about the choice being about growing onions. Onions are difficult to grow where you live. You make a marketable crop only half the time, but it can still be profitable because onions fetch a high price. Your profit will depend on whether you make a crop and the price of onions. At the highest price of onions, there is a 50% chance of earning $100,000 and a 50% chance of having no crop to sell. You have a $50,000 expected value. Risk Preference Calculator asks you how much you would be willing to pay to plant onions under these conditions. If you were willing to pay $45,000, you would be risk averse, since the EV is $50,000. If you would pay more than $50,000, your preferences are risk loving. If you would pay $50,000, your preferences are considered risk neutral. Most people would pay something less than $50,000 to take this opportunity.

Note that Aaron slides the bar for this bet on the top row over to the number $47,000. His CE is $47,000 and his RP is $3,000. That is, he would pay up to $47,000 for the chance at this investment, when it has an EV of $50,000. He has to be compensated by $3,000 to take on the risk. You can think about this problem in any terms you like; it does not have to be onions. The example is just to make you more likely to be realistic in your bids, so pick something that will make you be realistic.

Navigator uses information from previous turns to guide you to the subsequent alternatives. For example, the CE from the first question is used to structure the next investment on the second row. In Aaron's case, the second investment is a 50-50 bet between 0 and $47,000. (Remember, this $47,000 is his CE from his first answer.) The EV is $24,000. He slides the bar over to $20,000. Aaron proceeds to slide the bar to his CE on all four rows. Each row represents a different price for onions, with a 50% chance of harvesting a crop.

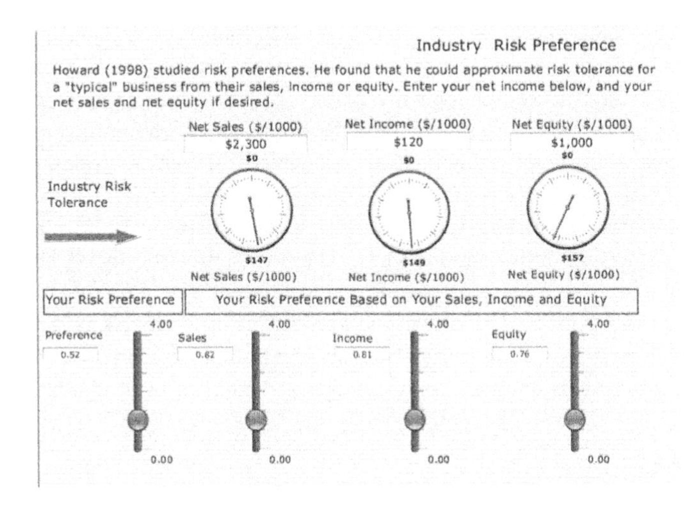

FIGURE 6.8 Risk Preference Calculator: Industry Risk Preference.

Once Aaron has entered his preferences, Navigator computes his relative risk preference, shown on the thermometer, and his risk tolerance too. Aaron's RAC is 0.52 and his RRT is the inverse, 163.61. According to the scale on the right, he is about halfway between risk neutral and somewhat risk averse. Let's just call him somewhat risk averse.

6.3.3 Risk Preference Calculator: Industry Risk Preference

The last tool we will use is the Industry Risk Preference Calculator. This tool allows you to compare your personal risk preference score with other business operators that have similar net sales, net income, or net equity. Navigator computes your risk tolerance, the amount you would pay, $X, for an even bet between getting X and losing X/2. Aaron's personal risk tolerance of 163,000 was computed in the previous tool. Based on the Harvard study, risk tolerance is 1.24 multiplied by net sales, 6.4% of net income, or 15.7% of net equity for the businesses studied; enter net income and at your option, net sales and net equity, and Risk Navigator does the rest.

As shown in Figure 6.8, a business operator with an income of $120,000 would typically have a risk tolerance of $149,000 (shown just below the circle). As discussed in the previous section, risk tolerance is difficult to interpret. So Risk Navigator prints the corresponding risk preference below in a thermometer. Given the values Aaron put in, others with similar sales, equity, and income would have a risk preference of about 0.81. This compares similarly to Aaron. You can use this tool to determine how well you match your cohorts. Are you typical, or perhaps more or less risk averse than other farmers with similar financial attributes?

6.4 CONCLUSION

Everyone looks at risk differently. These preferences dictate our choices in business—from taking advantage of a unique opportunity to finding ways to reduce everyday risks. Knowing your risk preference will help you evaluate management decisions and decide how much time you want to invest in managing these risks. The tools described in this chapter are useful in determining risk-preference and risk-tolerance levels. When combined with the financial information from Chapter 5, knowledge about your risk preference is invaluable in helping set operational goals in the next chapter.

Finally, we offer a word of warning. These estimates are just that—estimates! It can be interesting and informative to know where your preferences lie, but estimates should be thought of as a ballpark of how you feel, rather than a precise measure. We offer methods in Chapter 11 to allow you to avoid using your risk preference score when possible. However, sometimes knowing precisely how you feel about avoiding risk is required if one is to choose one management practice over another.

REFERENCES

Anderson, J., and J. L. Dillon. 1992. *Risk analysis in dryland farming systems*. Farming Systems Management Series, No. 2. FAO Rome.

Arrow, K. J. 1965. *Aspects of the theory of risk-bearing*. Helsinki: Academic Bookstore.

Clemen, R. T. and T. Reilly. 2001. *Making hard decisions with Decision Tools®*. Pacific Grove, CA: Duxbury.

FinaMetrica. 2008. MyRiskTolerance.com. May.

Fleisher, B. 1990. *Agricultural risk management*. Boulder, CO: Lynne Rienner Publishers.

Grable, J. E., and R. H. Lytton. 1999. Financial risk tolerance revisited: The development of a risk assessment instrument. *Financial Services Review* 8: 163–181.

Hardaker, J. B., R. M. Huirne, J. R. Anderson, and G. Lien. 2004. *Coping with risk in agriculture*, 3rd ed. Wallingford, UK: CABI Publishing.

Harwood, J., R. Heifner, K. Coble, J. Perry, and A. Somwaru. 1999. Managing risk in farming: Concepts, research and analysis. Market and Trade Division, Economic Research Service. United States Department of Agriculture. Report No. 774.

Kimball, M., C. Sahm, and M. Shapiro. 2007. Imputing risk tolerance from survey responses. Working Paper No. W13337, National Bureau of Economic Research (NBER).

Meyer, D., and J. Meyer. 2006. *Measuring risk aversion*. Hanover, MA: Now Publishers, Inc.

Myerson, R. 2005. *Probability models for economic decisions*, Duxbury Applied Series. Belmont, CA: Thomson Brooks/Cole.

Pratt, J.W. 1964. Risk Aversion in the Small and the Large. *Econometrica* 32: 122–136.

7 Step 3
Establish Risk Goals

John P. Hewlett

CONTENTS

Alice: "… which way I ought to go from here?"
Cheshire cat: "That depends a good deal on where you want to get to."
Alice: "I don't much care where—"
Cheshire cat: "Then it doesn't matter which way you go!!"

—Lewis Carroll, *Alice's Adventures in Wonderland*, 1865

7.1 PART 1: THE EWS FARMS CASE STUDY

An old adage states, "If you don't know where you're going, any road will get you there." Having clear, well-defined goals can help focus energy and effort. While this may not guarantee success, it does make it more likely. Developing plans for managing risk increases your chances of success.

7.1.1 Objective

The Strategic Risk Management (SRM) process is designed to help farms and ranches become more successful. Managing risk is similar to managing other dimensions of an agricultural operation. Setting goals and objectives is one means for describing what that future looks like, and the process of drafting the goals can be very helpful in determining what exactly is required to get from here to there. In other words, thinking through the steps can help the people involved see what resources will be required, what changes must be made in current operations, and what some of the milestones might be along the way.

7.1.2 Strategy

The strategy begins with a vision for the future, which is often outlined as a mission statement or a description of the operation's purpose. The mission statement should describe what the management team sees the operation becoming for the individuals, the family, and for the team. It should specify what the operation will focus on in the long run. Written mission statements help build strategic goals that work for the operation.

The next step is to set both business and personal goals for the operation. This step may include the goals of management, as well as key personnel. Goals can describe what the operation should be in 10 to 20 years, where management wants to be personally in 5 years, or the kind of education the owners would like to provide for their children. These sorts of goal statements are required if the people involved or the business as a whole is to ever reach the desired destination.

The tactical objectives provide the framework for achieving strategic goals. They identify a sequence of events or accomplishments required in order to reach long-term strategic goals. Tactical objectives generally describe *how* the strategic goals will be achieved and the order in which those events need to occur. They help to outline what needs to be accomplished in the near term as well into the future.

The operational plans are concerned with describing the specific steps and timetable required for accomplishing the tactical objectives. Operational planning deals with the *how* and *when* of the process. This step refers to planning the activities that must be accomplished in order to achieve the tactical objectives and the timing of these activities during the coming year. Operational plans often include a listing of the action steps, a timeline for completion, who is responsible for completion of each step, and some indicator to show the step has been completed.

7.1.3 Implementation

EWS Farms has involved all the family members of their Holyoke-based operation to draft a mission statement, goals, objectives, and plans for the future. As mentioned previously, family is an integral part of the business because the operation is family owned and operated. The two most current generations of the family are now charged with management of the operation. Russell and his wife Kimberlee are currently in charge of the decision making for the operation. They also have five children ranging in age from 19 to 28: daughters Desiree and Brianne, who are the oldest and youngest, as well as sons Aaron, Russell, and Dustin.

The mission statement developed by the family for EWS Farms describes the values, aspirations, and dreams of each family member. Although putting a mission statement on paper takes the involvement of everyone and many hours in conversation, the resulting document can serve as a source of strength to all who contribute.

EWS Farms Mission Statement

We want this venture to be first and foremost a family-building and supportive venture that highlights each family member's strengths by providing business roles that fit well with each person's natural abilities and interests. We want this business to maintain flexibility that will allow each family member to contribute in the way and level that fits their current life circumstances at any given point. We hope for this business to provide long-term financial stability for our entire family. We are excited about the lifestyle and quality of life an agriculturally based business will bring for our families. Through this venture, we hope to be better able to instill essential values in our children—values like the importance and impact of a large family support structure, hard work, flexibility, problem solving, effective and respectful conflict resolution, and most importantly, the strong Christian values like faith, hope, love, and God's enduring grace. This opportunity is also designed to allow family members who want to live and raise children in a small rural community—our childhood home—to be able to do so financially and professionally. When a dynamic family comes together for a common purpose, with a common vision, the possibilities are endless, and we want to see where those possibilities will lead each of us.

7.1.3.2 EWS Farms Strategic Goals

The Spragues developed a set of strategic goals, tactical objectives, and operational plans designed to make their mission statement a reality. A complete listing of the goals, objectives, and plans is provided in the appendix at the end of this chapter. Here, in the interest of space, we present just the strategic goals.

Strategic Goal 1 (Financial): Ensure short- and long-term financial success by maintaining business profitability, while expanding the overall business financial resource base.

Strategic Goal 2 (Family): Continue to live, work, and grow our families in a rural, agricultural environment. Encourage individual development and exploration in a manner that is consistent and flexible in order to allow all individuals to reach their full potential.

Strategic Goal 3 (Organizational): Continue to pursue organizational structures that fit the family dynamics of the operation, as well as allow for strategic goal attainment. Also, increase the business activities efficiency of the operation.

Strategic Goal 4 (Integrated Farm Management): Manage our farm as a co-integrated unit, while providing a step-by-step process for developing a strategic risk management plan.

7.2 PART 2: BASIC FUNDAMENTALS

The third step in the SRM process is to establish risk goals. Most individuals, businesses, and organizations have some idea of where they are headed; fewer have formal, written plans. An even smaller number have well-articulated, written plans that are shared with everyone concerned, that provide a basis for monitoring business activity, that give input to management decision making, and that offer insights into appropriate contingency responses when external forces require midcourse corrections.

7.2.1 Strategic Management

Strategic management is a technique for applying the resources available to achieve a long-term goal. In a military sense, a strategy is a maneuver designed to deceive or surprise an enemy. It is an approach directed at winning. As such, this management style generally requires all levels of management and labor to be aligned with the general strategy to ensure success. Another term often used to describe this approach is *integrated management*, which implies that all levels of the organization—its goals and objectives and its workers—are integrated and focused on reaching a particular outcome. Gable (1998) lists the following words or ideas to describe strategy: Agility, Precision, Surprise, Insightfulness, Efficiency, Cleverness, Imagination, Innovation, Using the resources at hand, Being prepared, and A plan of attack.

Coupled with the idea of strategy is the notion that a desired outcome is the focus of the individual or business activities. A step-by-step approach is used to reach that desired outcome. Moving toward the desired outcome in a stepwise manner implies an order; some activities necessarily come first and others follow. Under the SRM process, these steps are defined by well-written goals, objectives, and plans; details are provided later in this chapter.

Systems thinking is another term used to describe a management approach that acknowledges that all component parts of business planning and activities are interlinked and interrelated. The success of the entire unit depends on the successful operation of the component parts. Day-to-day plans must be well carried out to reach the corresponding objectives and goals. Further, as a system, the process for planning and implementation is carried out in an orderly fashion and is ongoing into the future.

7.2.2 Why Plan?

The benefit to planning is not obvious to everyone, even when it involves major events of the business and/or family life cycle. Estate transfer, the eventualities of an illness, the addition of new family members, and the evolution of business goals are all shifts in the internal business environment that may require advance planning. External changes to the business, however, may have implications that are just as significant as internal changes. External factors, such as agricultural commodity prices, prevailing interest rates, fuel costs, and national agricultural policy will all influence the overall business climate. In addition, these external factors combine to form the risk environment through which the business must navigate to achieve success both as a business unit and for the individuals involved.

Planning provides a means for communicating the common vision, goals, and objectives for individuals or a business. With this common understanding, a plan allows the people involved to react similarly to changes in the environment. Change may be either positive or negative and may be either anticipated or unanticipated. Regardless of the source or nature of change, a well-orchestrated planning process and resulting planning documents can lead to a more resilient business.

Checkland (1981) defines a *formal system* as a general model of any human activity. A system is formal if, and only if, it satisfies the following criteria:

It has an ongoing purpose or mission.
The system has a measure of performance that signals progress in trying to achieve the purpose.
It contains a decision-making process.

The system has components that are also systems.
The system has components that interact, that show a degree of connectivity such that effects and actions can be transmitted through the system.
The system exists in wider systems and/or environments with which it interacts.
The system has a boundary separating it from the wider systems and environment.
The system has resources.
The system has some guarantee of continuity and will recover stability after some degree of disturbance.

Strategic planning may be differentiated from other types of planning in that it "provides a framework for managers and others in the organization to assess strategic situations similarly, discuss the alternatives in a common language, and decide on actions (based on a shared set of values and understandings) that need to be taken in a reasonable period of time." (Goodstein, Nolan, and Pfeiffer, 1993, 6). In addition, strategic planning involves the process of developing a strategy for the future, considering what other players may do, and considering alternative actions that may be required to maintain or enhance the business position.

The strategy employed should move the business toward its stated goals through the actions the business takes. The strategy and goals, therefore, come before the action. In a nutshell, this may be the most compelling reason for engaging in planning as the first step. This planning is neither forecasting nor a method of applying quantitative analysis to regular business planning. Perhaps most importantly, strategic planning does not eliminate risk; rather, strategic planning outlines an approach, a method of operating into the future that provides the individuals involved with a common yardstick to measure opportunities, to react to changing conditions, and to assess the risks presented in order to move from where they are today to where they want to be in the future.

7.2.3 Roadblocks and Stepping-Stones

Burleson and Burleson (1994) present a number of roadblocks that prevent people from reaching their goals. When setting goals, they suggest listing the roadblocks and then strategizing ways to turn the roadblocks into stepping-stones. A partial list of common roadblocks and subsequent stepping-stones is presented in Table 7.1. After reading through this chapter's material on goal setting and drafting a goal or two, come back to this page, decide what roadblocks might prevent you from reaching those goals, and thank about how they might be turned into stepping-stones.

7.2.4 Learning Cycle

Planning is part of the process in which people learn and change over a lifetime. Many individuals have researched how students and others learn; however, Kolb

TABLE 7.1
Roadblocks and Stepping-Stones

Roadblocks	Stepping-Stones
Lack of commitment	Learn to ask for help
Negative feeling	Know how to adjust the pace
Lack of funding	Think big – start small
Unmotivated	Don't quit
Lack of education/skills	Learn from others
Past failures	Learn when to delegate
Fear of injury/failure	Know your strengths
Hard to stay focused	Anticipate change
Lack of self-discipline	Develop detailed plans
Too busy/involved	Know when to stop
Poor self image	Be prepared
Fear of competition	Use a budget
Addictions to ___________	Take a chance
Depression	Do your homework first
No encouragement	Study the competition
Peer pressure	Understand your weaknesses
Impatience	Look to the future
Afraid of change	Predict success
Bitterness	Stand your ground
Stuck in a rut	Learn from past mistakes

Source: Adapted from Burleson and Burleson, 1994.

(1984) developed the Experiential Learning Cycle (Figure 7.1), which many have found helpful in understanding this process. It is important to recognize that learning may occur without following each step or by stepping quickly through the process. But effective learners will take the time necessary to complete each phase of the process. When done correctly, it helps the individual move from the current plane of understanding or ability to a higher level of functionality.

7.2.5 Attitude

Attitude is another important component to consider when preparing to shape the future. A negative attitude not only saps energy needed for other things, but it also pulls energy away from others. On the other hand, a positive attitude can help open doors, convince others ideas have merit, and assist in getting through life's rough spots. Individuals can *choose* to be a victim of circumstance or to rise above the difficulties and come out on top.

The approach used in daily life with a spouse, friends, coworkers, or the boss makes a big difference. Choosing to have a positive attitude, even in the face of adversity, not only helps keep things moving forward, but may help others try a

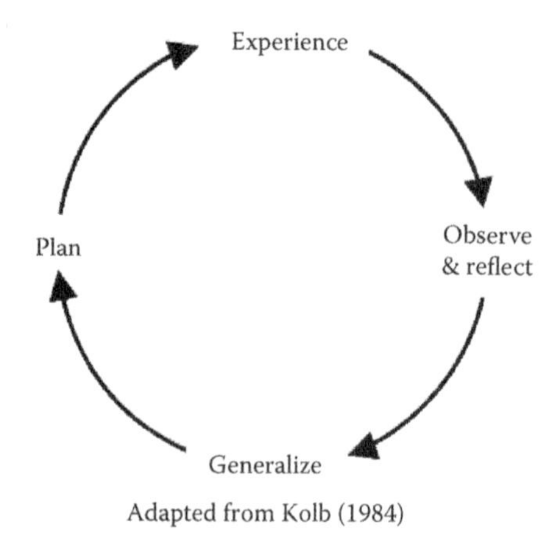

FIGURE 7.1 Experiential Learning Cycle.

little harder or put up with difficulties just a little longer. Attitude can make all the difference.

7.2.6 Time Management

How individuals manage time and other resources has a large impact on how easily they achieve stated goals. Many factors can take time away and reduce focus on the tasks at hand. Examples include: an unorganized approach, too little sleep, lack of self-discipline, the wrong tools, and too many demands. On the other hand, practices exist that may help provide better time management. These practices include: developing a project plan, setting priorities, thinking before acting, and planning for downtime. Time is just one of many resources a manager must allocate wisely in order to reach the goals of the organization. In addition, the prudent manager will already realize it is a resource that is easily mismanaged and in short supply.

In his book *The Seven Habits of Highly Effective People* (1989), Stephen Covey outlines a matrix for time management. A Risk Navigator Tool based on this matrix has been included for assessment of where individuals spend a majority of their time (Figure 7.2). Covey suggests that time management literature is moving away from scheduled, highly disciplined approaches toward self-management. He encourages individuals to shift their emphasis away from activities that are not-important-but-urgent or not-urgent-and-not-important to activities that are important-but-not-urgent. These activities are located in Quadrant II of the Time Management worksheet. Strategic goals and tactical objectives in the vernacular of this chapter would fit into Quadrant II. These directly contribute to making the mission statement of the organization become a reality.

7.2.7 Who Is on the Team?

The first step in goal setting is to determine who is on the management team—a management team roster. Is it mom, dad, daughter, and her husband? Perhaps it's

	Urgent	Not Urgent
Important	Quadrant I Crises Pressing problems Deadline-drive projects	Quadrant II Prevention, relationship building Recognizing new opportunities, planning, recreation
Not Important	Quadrant III Interruptions, some calls, some mail, some reports, some meetings, pressing matters, popular activities	Quadrant IV Trivia, busy work, some mail, some phone calls, time wasters, pleasant activities, no-brainers

FIGURE 7.2 Time Management Worksheet. (Adapted from Covey, 1989.)

a mixture of parents, uncles, brothers, sisters, and spouses all joined together in a corporation. It may be just an operator and her spouse. A Risk Navigator tool for developing your personal team roster is described later in Part 3.

Once the team is defined, every player should be involved in the goal-setting process. While this can be done without everyone, it won't be nearly as effective without all of their input. If yours is a family operation, include *all* family members. If the operation employs hired help, include the hired help and their spouses. It is important to remember that if the members of the team do not have ownership of the goals, they probably won't be working hard to help reach them.

When listing who is on the management team roster, consider the following questions (Hewlett et al., 1992, p. 8):

Who's the coach? Is it Dad, Mom, Granddad, or someone else?

Who's the quarterback? Is the quarterback Dad, Mom, the ranch manager, an uncle, or someone else?

Who are the other team players? Do the players include children, other family players, hired managers or workers, parents, stockholders, or others? Does everyone who is actively involved in the operation really want to be involved?

Are there players on the bench who would like to be in the game? Are other children, family members, hired managers/workers, parents, stockholders, or others who would like to be actively involved in the operation? Have you ever asked them if they want to be involved?

7.3 MISSION STATEMENTS

The next step in setting goals for the operation is to visualize where the operation should be in the future. The future may be 20 to 30 years down the road, or it may be just 10 years from today. Either is okay. The main objective is to form a picture of what the future should look like. Some experts claim that we can program our minds to help bring about the things we really want to achieve.

A mission statement should describe what a business or individual intends to:

- **Become**. Where is the business headed? What is it striving for? What makes the business "tick"? What inspires the people who are involved?
- **Provide**. What services or benefits does the business provide? What does it do for the people who provide the labor or management? What does it provide to the community or industry?
- **Produce**. What product or service does the organization produce? What do customers think of when they think of the product? What attributes, aesthetics, or feelings might be associated with the product(s)?

An operation can be or become:

- The lowest-cost producer in the county
- Renowned regionwide for producing high-quality breeding stock
- An operation known for high-quality hay or a source of reasonably priced livestock forage
- Known for its quality hunting opportunities or recreational experiences

An operation could provide:

- Money for retirement
- Enough income for the family to live modestly until retirement
- The lifestyle the family desires
- The kind of recreational experience potential customers are looking for

The operation may want to capture its vision in a mission statement. This should be a list of all the things the operation wants to be and all the things the operation wants to do. A mission statement should be a description of the purpose of the operation. It should describe what the management team sees the operation becoming for the individuals, the family, and for the team. A mission statement should specify what the operation will focus on in the long run.

Written mission statements help build strategic goals that work for the operation. A mission statement may be viewed as the trunk of a tree. From the mission statement springs the support and direction for the entire operation. All the operation's goals come from the mission statement or vision for the operation.

The mission of the Warbonnet Ranch is to be known as a superior range and livestock operation [what it wants to become] that produces quality organic beef [what it wants to produce] while returning a profit for the Warbonnet Ranch and its shareholders and providing ranch educational opportunities for visitors [what it wants to provide].

A mission statement also might include the driving forces present in the operating environment. These forces may reflect the business's competitive advantage or fundamental strength when compared to its competition. Tregoe et al. (1989) lists the following basic categories of driving forces:

1. Products offered
2. Market served
3. Technology
4. Low-cost production capability
5. Operations capability
6. Method of distribution/sale
7. Natural resources
8. Profit/return

A mission statement should be concise and meaningful to the people in the operation. It should clearly present the elements suggested above, while providing enough detail that it would make sense if read by someone unfamiliar with the inner workings of the business. In addition, once written it should be posted in a prominent place where it may be easily read by both outsiders and insiders. Printing the mission statement on the back of business cards, posting it on the boardroom wall, or taping it to the refrigerator are all ways of keeping the mission statement foremost in the minds of the people.

7.3.1 Values

Values play a large role in how we view what goes on around us. The old euphemism about rose-colored glasses comes from this very idea. Values are our most fundamental beliefs and they influence how we act in a given situation. Values may be held about both our personal and professional lives. When these separate sets of values are not in alignment with one another, an individual may have a hard time continuing in the situation regardless of how much he is paid.

Develop Your Own Success Map

Review the list of items below. Select the five that are most important to you. Prioritize the list and number them 1 through 5 (Adapted from Cairo, 1998).

Security
Friendship
Wealth
Retirement
Good health
Being in business for yourself
Relationship with spouse/mate
Long life
Relationship with children
Travel
Relationship with family (other)
Respect of peers
Fame
Spiritual fulfillment
Job/career
Charity/contributions to others
Power
_________________ Other
Happiness

Eighty percent of your time and energy should be devoted to the top five values you have listed. Post them where you can see them every day to keep you focused.

Mission statements, if they are to be effective, must include as many values of the people involved as possible. Take the time to learn about and include the personal values of all the individuals in the business. These values help shape the culture and custom of the business organization. Additionally, discuss and think through the values that govern the business and how they interact with its customers and input suppliers. All these values should be included in the business mission statement if possible.

7.3.2 Motivation

Cairo states that "[m]otivation is by far the step that has generated the most debate, study and theories" (1998, 20). In short, if you are rewarded or recognized for doing something, you will be motivated; if you reward others, they will be motivated. A business manager is concerned about motivating the labor force to provide the work to accomplish the tasks that must be completed. While this is important, it is also critical that the manager be motivated to provide management for the organization. Providing incentive structures (or consequences) can help to keep motivation high, even during off-peak seasons. Further, it may even help the organization reach its goals more quickly.

Outside motivation has a limited capacity for encouraging individual action. Internal motivation is more likely to generate individual action. Internal motivation arises when the values and principles of the individual are aligned with the business. Conversely, if an individual is not in alignment with the organization, this can be a source of strong *de-motivation*. For this reason, motivation and values alignment should be carefully considered when drafting mission statements and strategic goals.

7.3.3 Decision vs. Behavior Change

Change is pervasive in today's business climate. The constant evolution of technology and the enhancements it brings, as well as increased abilities to communicate, transport, and retool, have dramatically shaped the past 15 to 20 years. However, change generally moves at a slower pace when it comes to human systems. Smith states, "Study after study shows that up to four out of five change efforts either fail or seriously sub-optimize. Looking behind the numbers [he] discovered that most of the successes involved change that demanded tough decisions but little fundamental or widespread behavior shifts" (1999, 193). This is because people find it hard to change their behaviors. Change in tactics or direction are more easily accomplished because they tend not to require changes in behavior for the people involved.

Ten Principles of Behavior-Driven Change (Smith 1999)

1. Keep performance, not change, the primary objective of behavior and skill change.
2. Focus on continually increasing the number of people taking responsibility for their own performance and change.
3. Ensure that each person always knows why his or her performance and change matter to the purpose and results of the whole organization.
4. Put people in a position to learn by doing and provide them with the information and support needed to perform.
5. Embrace improvisation.
6. Use team performance to drive change whenever demanded.
7. Concentrate organization designs on the work people do, not the decision-making authority they have.
8. Create and focus energy and meaningful language.
9. Harmonize and integrate the change initiatives in your organization, including those that are decision-driven as well as behavior-driven.
10. Practice leadership based on the courage to live the changes you wish to bring about.

Smith goes on to observe that "Decisions, however, are not enough. Instead, both the decisions and the performance outcomes sought depend upon people in your organization learning specific new skills, behaviors and relationships" (1999, 202). In the end, behavior change will only occur where individuals take responsibility for themselves and the changes they are encountering.

7.4 STRATEGIC GOALS

The U.S. Department of Agriculture Risk Management Agency (RMA) describes the benefit of setting goals as follows. Goals:

- Reflect the values, interests and resource capabilities of decision makers
- Provide a basis for making decisions
- Establish priorities
- Provide a means to measure progress

Goal setting may be described as a mission-critical activity. Goals form the very heart of the management process. Following the process outlined in this chapter will help create written goal statements. In addition, it describes a way to assign costs for each goal. Knowing the goals and their associated costs provides a road map to success for the operation by measuring progress. In addition, the manager can more easily assess whether the goals are realistic by considering the available resource base.

Goals should not be only for the operation, but should also include personal dreams, ambitions, and desires. Operator goals should include way of life, hopes for family, and life mission. In addition, the operator should consider challenges, feelings of purpose, and fulfillment in life. Goals should describe outcomes or the result of the activities rather than the activities themselves.

With a mission statement in hand, the next step of drafting strategic goals can begin. These might best be viewed as a road map to follow to reach the destination set for the operation. Strategic goals are long-term goals. They are specific steps for reaching the general goal(s) described in the mission statement. Strategic goals are typically written for 10 to 20 years in the future.

7.4.1 SMART GOALS

Good goal statements should be SMART. That is, they must be a *specific* statement of what is to be accomplished; they must be *measurable* by some objective means; they must be *attainable*; they must be *related* to one another; and they must be *tractable* over time. Setting SMART strategic goals will provide the tools to manage the operation to achieve higher goals from the mission statement.

Specific: Goals should be definite, focused, and descriptive of the actions to take place. This part of the goal tells *what* must be done in precise terms.
Example: Generate enough income to allow us to maintain ownership of the operation.

***Measurable*:** Goals should be easily measured. Such goal statements provide a benchmark against which performance can be measured. This portion of the goal statement provides a means of knowing when the goal has been reached. Not all goals will have easily quantifiable outcomes; however, the expression of such goals should at least provide some notion of how they will be assessed.
Example: Generate a $3,500-per-year principal payment, allowing us to maintain ownership of the operation.

Attainable: Goals are within the reach of the operation. They can be accomplished and are realistic. Setting unrealistic goals for the operation is not helpful.

Example: This goal is realistic for the operation if it usually yields a return greater than $3,500 each year. However, if it provides only a few dollars of return over its expenses each year, this may be an unattainable goal.

Related: Goals are connected or associated with other goals set for the operation. It is clearly counterproductive to the organization to develop two goals that would require the business to move in opposite directions simultaneously for their achievement. In order for the business to realize the mission statement as quickly as possible, the goals should follow a relatively common path.

Example: Generate a $3,500-per-year principal payment, allowing us to maintain ownership of the operation.

This goal is related to other goals that move toward ownership; however, another goal might call for expansion. If this requires outside capital, it would reduce ownership of the assets pledged to the lender, making the two goals unrelated.

Tractable: Goals are manageable. These goals involve factors and resources that are controllable. They can be handled using existing resources.

Example: Generate a $3,500-per-year principal payment, allowing us to maintain ownership of the operation, while not requiring more time or new skills on the part of management.

This goal is tractable if the current management can handle the operations necessary to generate the dollars. If reaching this goal means that the manager would need to spend 5 out of 12 months a year monitoring stock reports while trying to run an agricultural operation, it would not be a very tractable goal.

Where the mission statement is the destination, strategic goals form the roadmap. Strategic goals proceed from the operation's mission statement. They are supported by the mission statement and are nurtured by the mission statement's principles. They represent the specific steps the operation must accomplish to reach its final objective. To do this, strategic goals must be prioritized. Resources available to most operations are limited; therefore, not all goals can be reached at the same time. Working on many or all goals at the same time is desirable; however, it may not be possible. Prioritizing the goals is a way of making sure the most important things are done first. If additional resources exist, they can be applied to reaching less important goals.

Strategic Goals for an Example Ranch

Market grass through the sale of livestock and wildlife products.

Market recreational experiences by utilizing the ranch resources and ambiance.
Manage all ranch resources in a profitable way, allowing all people involved to enjoy the ranching lifestyle.
Be good stewards of all ranch resources, leaving them in better condition than they were when received.
Keep all buildings, improvements, and livestock handling facilities updated and visually appealing.

Once strategic goals are written, make them visible. Hang a copy of the goals in a prominent place where they can be seen by all management team members. The refrigerator in a family operation may be the ideal location. In a corporation, the main office may be the best place. Keeping the goals where the team can see them often helps everyone keep in mind what is truly important to the operation.

7.4.2 Goal Hierarchy

The notion of goal relationship is also important. During the initial stages of the process, it is possible for the organization to set too many goals if little thought is given to the resources available, including time and energy. When discussing SMART goals, how goals are or are not related is another point easily overlooked. Richards (1986) presents the notion of goal hierarchy in an attempt to focus on this aspect of goal setting.

The goal hierarchy diagram shown in Figure 7.3 illustrates this point. Strategic Goal 1 (SG1) depends on the accomplishment of Tactical Objectives 1-3 and 1-5 (TO1-3, TO1-5). Presumably, SG1 also would depend upon Tactical Objectives 1-1, 1-2, and 1-4 as well; however, these are ignored here to simplify the presentation.

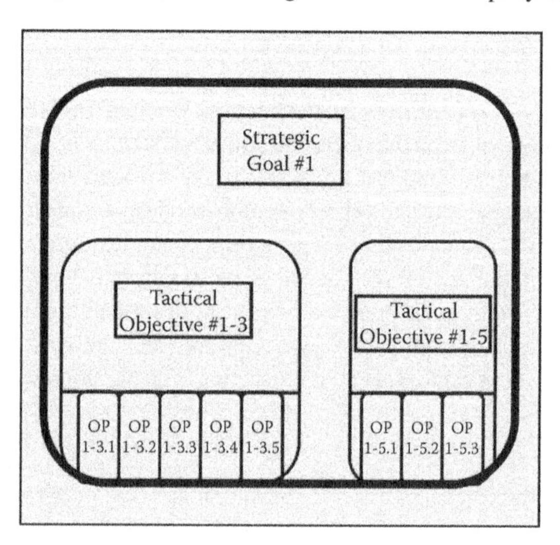

FIGURE 7.3 Goal Hierarchy. (Adapted from Richards, 1986.)

The reverse would also hold. In this case, TO1-3 is unrelated to TO1-5, except where they each support SG1 and are essential to its fulfillment.

Similarly Operational Plans 1-3.1 through 1-3.5 (OP1-3.1 to OP1-3.5) are necessary and sufficient for achieving TO1-3, but are unrelated to OP1-5.1, OP1-5.2, or OP1-5.3. OP1-5.1 through OP1-5.3 are related to each other in that they share common boundaries. As such, OP1-5.1 and OP1-5.2 must each be accomplished for either to contribute to TO1-5. The same could be said for OP1-5.2 and OP1-5.3.

Imagine a case where an operational plan supports more than one tactical objective or where a specific tactical objective supports more than one strategic goal. In such cases, developing goals, objectives, and plans is much like a house of cards—one component's failure may lead to a whole segment collapsing. In the military, skirmishes or battles may be lost, but winning the war is still a viable possibility depending on how interrelated the components remain. As such, this concept is important when constructing a resilient strategy for the future.

7.4.2.1 Tactical Objectives

Tactics refer to the techniques or solutions offered for reaching strategic goals (Robinson and Robinson, 2004). Tactical objectives provide the framework for achieving these strategic goals. They identify a sequence of events or accomplishments required on the way to reaching longer-term strategic goals. Tactical objectives generally describe *how* the strategic goals will be achieved and the order in which events need to occur. They help to outline what needs to be accomplished in the near term, as well into the future, usually the period 3 to 10 years from now.

Often tactical goals will include a description of how the performance of one resource or another must improve to allow a strategic goal to be achieved. As such, properly comprised tactical goals also include benchmarks or target levels of performance. These benchmarks provide intermediate levels of resource performance that allow the manager to measure resource improvement and movement toward strategic goals.

Tactical objectives:

- Relate directly to one or more strategic goals and the mission statement
- Are clear, concise, and understandable
- Are stated in terms of results or outcomes
- Specify a date for accomplishment
- Address no more than one major outcome per objective
- Begin with "to" and an action verb (to reduce, to increase, to replace)

Tactical Objectives for an Example Ranch

Strategic goal: Market recreational experiences by utilizing the ranch resources and ambiance.

The supporting tactical objectives are:

Tactical Objective 1-1: To market the blue-ribbon trout fishing opportunities of the ranch, available within 5 miles of the main highway, through advertisements in select fishing magazines, grossing at least $20,000 per year over the next 15 years.

Tactical Objective 1-2: To market the bird-viewing opportunities of ranch riparian areas and high-quality, overnight accommodations to members of exclusive birding clubs from the greater Chicago area, providing gross returns of at least $50,000 per year over the next 10 years.

Tactical Objective 1-3: To exploit the big-game hunting opportunities of the more remote portions of the ranch beginning next hunting season, returning net earnings of $20,000 per year over the next 10 years.

Tactical objectives also describe the steps in a transition plan. In this sense, the transition is the movement of the business from its current situation to where the mission statement describes it should be in the future. Tactical objectives then serve as the road signs along the route to reaching the strategic goals and ultimately making the mission statement a reality. Tactical objectives are necessarily specific and should include benchmark descriptions of the resource base along the way. These benchmarks need to include both quantity and quality descriptions, and an assessment of resource performance.

7.4.2.2 Operational Plans

Operational plans are concerned with describing the specific steps and timetable required for reaching the tactical objectives. Operational planning deals with the *how* and *when* of the process.

Sometimes this phase is called *action planning*, *programming*, or *implementation*. Whatever the term, it refers to the delineation of activities that must be accomplished in order to achieve the tactical objectives and the timing of these activities over the coming year. Operational plans often include a listing of the action steps, a timeline for completion, who is responsible for completion of each step, and some indicator to show the step has been completed.

Effective operational planning involves developing responses to these important questions:

- What information exists about each action?
- What resources must be included in the action steps?
- What are the interrelationships among the various resources?
- What is the ordering and timing of the various steps?
- Who should be accountable for completion of each of the action steps?

Well-developed management plans should also include an estimated cost of achievement. This is true for strategic goals, tactical objectives, or operational plans. Calculating the cost of a goal, whether it is in terms of additional annual revenue required, an increase in monthly expenses, or a one-time lump sum cost, can help the manager see what is required to make the goal a reality.

Operational Plans for an Example Ranch

Tactical Objective 1-3: To exploit the big-game hunting opportunities of the more remote portions of the ranch beginning next hunting season, returning net earnings of $20,000 per year over the next 10 years.

The supporting operational plans are:

Operational Plan 1-3.1: Consult with state Game and Fish Department personnel to set a target harvest level for the ranch that will give a 90 percent success rate at a cost of $1,500 per hunter. Person responsible: Dad. Complete by: February 15.

Operational Plan 1-3.2: The example ranch currently has 390 antelope, 120 mule deer, and about 30 whitetail deer. With $3,000, advertise hunting ventures throughout the state. Person responsible: Mom. Complete by: March 31.

Operational Plan 1-3.3: Develop 6 campsites on the ranch with camp tables, BBQ pits, and outdoor toilets. Person responsible: Dad and one hired person. Complete by: end of August.

Financing is one of many resources needed to achieve goals. Reaching a goal may require additional human resources (skills or hours of labor), added livestock performance (improved weaning weights through better genetics), or more forage resources (added rangeland or improved forage production on existing resources). Regardless of which or how much resource is required, spending the time to estimate the requirements and when the resources will be needed not only forces the manager to become serious about describing how to reach a goal, but it also increases the likelihood of achieving the goal.

7.4.2.3 Management by Objective

The management by objective (MBO) concept has been in use in business planning for several decades. The concept is a formal system for translating strategic goals into specific and relevant language for an entire organization. As such, it represents a method for describing action steps or operational plans for the entire business. Most importantly, these steps are clearly described for the people involved in carrying out those steps, as well as the measurement criteria needed to assess the level of completion of each step on a periodic basis.

Where the planning process outlined earlier in this chapter seeks to involve as many people as possible, MBO, once generally outlined for the overall organization, focuses on the supervisor/supervisee relationship. It is designed primarily for the development of work performance plans for individuals in a business. MBO is sometimes broadly applied to other segments of the work environment. The system uses standardized forms and monitoring techniques for employee management. MBO programs may provide assistance in developing useful operational plans. Further,

		Goals Achieved?	
		Yes	No
Goals Adequate?	Yes	*Satisfactory*	*Work on using goals*
	No	*Reestablish goals*	*Reestablish goals and work on implementation*

FIGURE 7.4 Goal Adequacy and Achievement Grid. (Adapted from Richards, 1986.)

it may offer additional tools for implementing operational plans after the planning process has been completed.

7.4.2.4 Adequacy and Achievement of Goals

The SRM process provides for ongoing management and evolution of the business from where it is today toward the mission statement describing its future. While strategic goals, tactical objectives, and operational plans can be assembled in an integrated fashion to link the day-to-day activities of management and labor to that future, ongoing monitoring and assessment is also needed. As shown in Figure 7.4, Richards (1986) presents a simple grid for reviewing the adequacy and achievement of goals. The adequacy and achievement grid may be a helpful means for reviewing strategic goals, tactical objectives, or operational plans on an annual, biannual, or other periodic basis. In addition, the grid can help identify what actions to take as a result.

7.5 CONTINGENCY PLANNING

Contingencies are another factor to consider in a fully integrated planning process. Monitoring is critical as plans are implemented; however, if things are not going as planned, what should happen next? Contingency planning looks at the possible outcomes to various actions planned for the business. Plans or strategies are then considered for possible outcomes that are not in the business's best interest.

Importantly, adverse events are not necessarily equal in magnitude or scope. Given this, not all contingencies should be considered with equal vigor. Greater attention is devoted to events or outcomes that threaten the business with a greater loss, whether measured in economic terms or other considerations. Further, not all businesses are equal in their willingness to accept risk. Likewise, some managers are more comfortable with risk than are others. As a result, risk and contingencies should receive serious attention in the planning process. These factors should be considered as part of planning for the future and also should be included when considering the monitoring strategies that will be used as plans are implemented. The tactical level of the SRM process includes specific steps and tools to further consider contingencies and planning for risk management.

7.6 PITFALLS TO IMPLEMENTATION

To be most effective, strategic planning must be accomplished carefully and involve the people who are critical to its success. No matter how well goals and plans are developed, no matter how much time has been invested to ensure the goals are SMART or that the process included everyone who should have been included, the strategy must be implemented. The implementation stage of the process carries its own set of risks and pitfalls. Goodstein et al. (1993) identified three general traps or pitfalls that may develop while implementing a strategy. These three categories include:

- *Failing to communicate.* This generally occurs when management follows the practice of keeping the strategic plan to itself. At the conclusion of this chapter it should be clear that strategic plans are carried out by the people in the business. In addition, no amount of window dressing will inspire others in the organization to become involved in the plan if they were not included in the process from the beginning.
- *Failing to change performance management systems.* Where a new plan that causes the business to grow in some way is to be implemented, the bar for individual performance also must be raised. When management fails to change its expectations and hold everyone accountable to those new expectations, the best laid plans will bring about only marginal change.
- *Neglecting Intangible Aspects.* Several intangible aspects can impact implementation. The first is the match of manager abilities with the strategy selected. To be successful, the manager must be able to successfully implement the strategy. Recognition that certain skills are lacking may lead to either hiring the needed expertise or to additional training. Another important factor has been called *analysis paralysis* where the organization becomes fixated on continuing to analyze the situation and cannot move to the implementation stage. This would be disastrous, as it results in no more effect than having not attempted to plan in the first place.

7.7 PART 3: APPLICATIONS AND WORKING LESSONS

Several alternative tools were presented in Part 2 above in the presentation of the Establish Risk Goals step of Strategic Risk Management. Part 3 focuses on the Risk Navigator Tools SRM developed to help accomplish this important step of managing risk.

7.7.1 TIME MANAGEMENT WORKSHEET

The Time Management Worksheet presented earlier in Figure 7.2 is a good way to assess how you and members of your management team spend the majority of your time. Set aside time in a management team meeting and ask each individual to fill in the quadrants of the Time Management Worksheet with a listing of tasks or jobs they spend their time on. The focus should be on the important things they have completed in the past month, quarter, year, or several years of a major project.

Then, spend some time discussing where each person placed the majority of his or her accomplishments in the Time Management Worksheet. Activities focused on accomplishing strategic goals and tactical objectives would usually be found in Quadrant II of the worksheet. These activities directly contribute to making the mission statement of the organization become a reality. Individuals should, where possible, shift their emphasis away from activities that are not-important-but-urgent or not-urgent-and-not-important to activities that are important-but-not-urgent to better focus the time spent in their professional and personal lives.

7.7.2 Management Team Roster

Completing a Management Team Roster (Figure 7.5) is a good way to be sure that all members of the operation are included at the appropriate level. Often, the people involved in the day-to-day management of a business forget to consider the interests of others who are not heavily involved or are geographically located at some distance from the center of business activity. Use the Management Team Roster as a tool to be sure that all individuals who may have some interest are included in business planning activities to ensure success for all those involved.

7.7.3 Mission Statement Worksheet

The Mission Statement Worksheet (Figure 7.6) can provide the means to take input from all individuals listed on the Management Team Roster regarding the business and its products. Perhaps more importantly, once completed, the Mission Statement Worksheet can also ensure that all the members of the team are reminded of what is important about being a part of the organization. Furthermore, it allows for communication of that mission to other individuals close to the business and to those outside the business when the need arises.

For the Sprague family, completing the mission statement helped all members of the team better understand what other members found important about the family business. In addition, once completed, the mission statement has helped team members to communicate those values to individuals outside the business as well. (The Sprague family mission statement may be found at the beginning of this chapter).

7.7.4 Strategic Goal Worksheet

The Strategic Goal Worksheet (Figure 7.7) included in the Risk Navigator SRM toolkit is designed to help draft complete strategic goals. To fully describe a strategic goal, a deadline for accomplishment must be set and resource demands calculated. A series of blanks is provided for this purpose. The worksheet also provides space for defining tactical objectives and operational plans that serve as steps for reaching the broader strategic goals.

Management Team Roster

Who's the coach?

Is the coach Dad , Mom, Granddad, or someone else?

Who's the quarterback?

Is the quarterback Dad , Mom, the ranch manager, an uncle or someone else?

Who are other team players?

Are the players children, other family members, hired managers or workers, parents, stockholders, or others?

Do all who are actively involved in the operation really want to be?

Have you ever asked them if they want to be involved?

Are there any players on the bench who would like to be in the game?

Are there children, other family members, hired managers/workers, parents, stockholders, or others who would like to be actively involved in the operation?

Have you ever asked them if they want to be involved?

FIGURE 7.5 Management Team Roster.

Mission Statement Worksheet

List the things you want the operation to be or become:

Describe your *vision* of the operation twenty years from now.

How do you see your business in the future?

List the things you want to do with the operation or things you want the operation to do for you or your family:

What will the operation provide for you?

What would you like the operation to *accomplish* over the next twenty years?

Describe the *focus* of the operation, what it is all about:

What does the operation produce or sell as products?

Are there products or means of marketing your resources that you see the operation being involved with over the next twenty years?

FIGURE 7.6 Mission Statement Worksheet.

<table>
<tr><th colspan="5">Strategic Goal Worksheet</th></tr>
<tr><td colspan="5">Goal Statement:</td></tr>
<tr><td colspan="5">Deadline for Goal Attainment:</td></tr>
<tr><td colspan="5">----- Goal Costs/Resources Required: -----</td></tr>
<tr><td>Basic Resources</td><td>Human Resources</td><td>Financial Resources</td><td>Livestock Resources</td><td>Wildlife Resources</td></tr>
<tr><td colspan="5">Associated Tactical Objectives:</td></tr>
<tr><td colspan="5">Associated Operational Plans:</td></tr>
</table>

The goal statement is the written description of the goal to be accomplished.

Strategic goals should have an associated deadline. This helps prioritize your goals.

Resources needed to achieve this strategic goal should be briefly described/listed in this section.

This will help in tactical planning for goal attainment.

Tactical objectives and operational plans are the means used to accomplish strategic goals.

Briefly outline here the tactical and operational steps necessary for reaching the strategic goal listed above.

FIGURE 7.7 Strategic Goal Worksheet.

The worksheet includes a section for describing supporting tactical objectives necessary for goal accomplishment. These objectives serve as the roadmap, complete with way points, for goal achievement. In addition, the worksheet includes a section for briefly describing the operational plans necessary to reach the tactical objectives. These operational plans are the actions that must occur to complete the tactical objectives along the way to reaching the strategic goals listed.

A complete planning process also includes developing an estimate of the resources needed to accomplish goals. This is true for strategic goals, tactical objectives, or operational plans. In addition to the financial costs for reaching goals, there are usually requirements from other resources as well. Reaching a goal may require additional human resources (skills or hours of labor), added livestock performance (improved weaning weights through better genetics), or more forage resources (added rangeland or improved forage production on existing resources). The Strategic Goal Worksheet contains blanks for estimating the resources needed to achieve the overarching strategic goal listed at the top of each worksheet. (The Sprague's goals are listed in the appendix at the end of this chapter.)

7.7.5 Transition Planning Worksheet

The Transition Planning Worksheet (Figure 7.8) provides workspace for developing transition plans. In general, the planning worksheet is intended to describe the changes in enterprise activities, resource use, and management proposed over the planning horizon. The overriding idea for the worksheet is to plan how changes are to be implemented across time to allow the operation to move from where it currently is toward its strategic goals. Stated another way, these changes are the tactical objectives required for strategic goal attainment.

Most tactical-level changes will not occur overnight, nor will they necessarily occur at known intervals. For this reason, the worksheet has a beginning column and an ending column, with intervening columns headed by blanks for dates. The worksheet is broken into three major sections: (1) enterprise changes, (2) resource changes, and (3) SWOT (strengths, weaknesses, opportunities, threats) analysis.

Under the Enterprise Changes section, the worksheet provides space to record the following:

- *Limiting resource:* Spaces provide a place to describe resource management issues for the overall operation that likely should be addressed in the plan. Blanks are provided across the planning horizon, as a resource that is limiting at the beginning may change over time in response to management.
- *Enterprise description:* Blanks are provided to list the various enterprises planned for the operation.
- *Changes and benchmarks:* Additional spaces to record short descriptions of the changes proposed for each enterprise activity over time. This also may include benchmarks for the overall enterprise. Information for the current situation should be recorded under the Beginning Benchmarks

Transition Planning Worksheet

		Beginning Benchmarks	Date:	Date:	Ending Benchmarks
	Limiting Resource:				
Proposed Enterprise Changes	______ Enterprise				
	______ Enterprise				
	______ Enterprise				
Resource changes	Basic Resources: Natural and Agronomic Changes to Inputs				
	Changes to Outputs Benchmarks				
Resource changes	Livestock/Wildlife Resources: Changes to Inputs				
	Changes to Outputs Benchmarks				
Resource changes	Human Resources: Changes to Inputs				
	Changes to Outputs Benchmarks				
Resource changes	Financial Resources: Changes to Inputs				
	Changes to Outputs Benchmarks				
SWOT Analysis	Strengths Weaknesses Opportunities Threats				

FIGURE 7.8 Transition Planning Worksheet.

section. Changes proposed for the future should be noted under the columns to the right, with dates noted, followed by the last column for Ending Benchmarks.

The center section of the planning worksheet is a space to describe the changes proposed in resource management. Each resource category has a separate section to describe changes to both inputs and outputs for the resource, as well as a section for benchmarks of particular interest for that resource. Again, information for the current situation should be recorded under the Beginning Benchmarks section. Changes proposed for the future should be noted under the columns to the right, with dates noted, followed by the last column for Ending Benchmarks.

At the bottom of the worksheet is the SWOT section. The SWOT concept is often used in business analysis as a technique for overall situation assessment. In that sense, it is helpful to management in deciding where it should focus additional attention when considering the business as a whole. Strengths and weaknesses are focused on internal performance. In contrast, the opportunities and threats are external to the business and analysis of these two factors usually centers on how well the business is responding to outside influences.

The intention of the SWOT section is to prompt the user to consider what strengths, weaknesses, opportunities, and threats are faced by the operation at each stage of the planning horizon. The Transition Planning Worksheet is intended to provide a unique perspective on the collective changes proposed for the operation over time. The worksheet may be used to record intended changes to enterprise management, changes in the management of particular resource outputs, or changes to attain a particular benchmark measure by a given point in time, as well as management changes for any limiting resources. When considering all these changes, the SWOT analysis can remind management of what to watch out for or what to try to capitalize on as progress is made toward various goals.

An important assumption in the discussion of the transition planning worksheet is that management will have the data necessary to assess performance as plans are being implemented. Without accurate, timely, and sufficient data, the plan implementation and associated risks cannot be assessed and implementation is more difficult, if not impossible. If adequate resource measurement processes are not already in place, they should be carefully considered, particularly as tactical objectives are outlined.

7.7.6 Action Planning Worksheet

The Action Planning Worksheet (Figure 7.9) provides assistance in planning the action steps required in greater detail for each tactical objective. This worksheet provides blanks for specific action steps, dates for completion of each step, the person or persons responsible, and the tracking or measuring system that will be used to determine if the actions have occurred as planned.

The Action Planning Worksheet might be viewed as a form for more completely describing the operational plans than can be done using the Strategic Planning Worksheet presented earlier. It is a very helpful tool in assigning specific task responsibilities to an individual or in assigning a set of activities across individuals for a week, a month, or a season.

7.7.7 Adequacy Achievement Grid

The Adequacy and Achievement Grid (Figure 7.4) may be helpful in reviewing strategic goals, tactical objectives, or operational plans on an annual, biannual, or other periodic basis. The grid may also help identify what actions to take when results are not reached as quickly as desired or where roadblocks threaten to prevent goal attainment.

7.7.8 Appendix: EWS Farms Goals

These goals were completed by the family using the Strategic Goal Worksheet described above.

Action Planning Worksheet

Tactical Objective: ________________________________

____

____

____

____

Action Steps:	Date for Completion	Person Accountable	Tracking or Measuring System
1.			
2.			
3.			
4.			
5.			

FIGURE 7.9 Action Planning Worksheet.

7.7.8.1 Strategic Goal 1

7.7.8.1.1 Goal Statement (Financial)

Ensure short- and long-term financial success by maintaining business profitability, while expanding the overall business financial resource base.

7.7.8.1.2 Tactical Objectives

Tactical Objective 1-1: Diversify the financial position of the business, as well as the business members, using all appropriate financial resources.

Tactical Objective 1-2: Allow for the inclusion of off-farm incomes, while providing enough stability from the farm to make off-farm incomes voluntary.

Tactical Objective 1-3: Restructure debt positions of the farm and the directly involved families to ensure the most efficient use of debt capital within the business.

Tactical Objective 1-4: Provide capital to the directly involved families, allowing for personal development and recreation.

7.7.8.1.3 Operational Plans

Operational Plan A-1: Increase the amount of financial investment outside the business, using the VEST model to invest in complimentary industry firms. Budget approximately $5 per acre for financial/investment activities in future cash flow projections.

Person responsible: Aaron S.

Complete by: February 11, 2006

Costs:

- Financial: $5 per acre or a total of $20,000 annually
- Human: Aaron: 10 hours per week

Operational Plan A-2: Consult with local New York Life agent and business banker to develop a plan for incorporating financial means for providing retirement funds to the directly involved families rather than relying on land ownership for retirement funds.

Person responsible: Russ and Aaron S.

Complete by: January 1, 2006

Costs:

- Human: Russ: 5 hours, Aaron: 5 hours
- Financial: To be determined at the meeting

Operational Plan B-1: Assess financial needs of each on-farm family. Identify target family living allocations and identify economic feasibility of target family living levels with current resource base.

Person responsible: Individual families followed by a team meeting

Complete by: December 20, 2005

Costs: Human: Families: 2 hours budgeting time; Team: 2-hour meeting to discuss findings

Operational Plan B-2: Identify critical farming periods, as well as periods available for off-farm work. Also, develop a schedule of off-farm work consistent with these periods to ensure that all critical farming periods are fully accounted for and all directly involved members are informed.

Person responsible: Russ, Aaron S. and Aaron M.

Complete by: January 15, 2006

Costs: Human: Russ, Aaron S., and Aaron M.: 2 hours each

Operational Plan B-3: In the event that target family living allocations are deemed not feasible with the current resource base, develop a plan and schedule to consistently work toward financial feasibility of the target family living allocations. Focus the plan on providing enough family living allocations so the wives will be able to choose to work or stay home.

Person responsible: Team

Complete by: February 20, 2006

Costs: Human: To be determined

Operational Plan C-1: Evaluate all debt positions of the farm and on-farm families and identify inefficiencies in the use of debt capital.

Person responsible: Russ, Aaron S., and Aaron M.

Complete by: December 31, 2005

Costs: Human: Russ, Aaron S., and Aaron M.: 10 hours each

Operational Plan C-2: Restructure all inefficient debt to obtain a reasonable repayment schedule. Specifically target all high-interest debt such as credit card, repair/parts, etc., and restructure these debts to be eliminated within the next five years.
Person responsible: Aaron S.
Complete by: February 1, 2006
Costs:

- Human: Aaron S.: 20 hours
- Financial: Amount to be determined from previous operational plan

Operational Plan D-1: Organize and hold yearly planning meetings to identify financial needs requested by all of the directly involved members for personal and professional development, family recreation activities, family support for travel requirements, and personal properties improvements.
Person Responsible: Kim, Desiree, and Amber
Complete by: January 1, 2006
Costs:

- Human: Individual families: 2 hours; Kim, Desiree, and Amber: 2 hours each
- Financial: To be determined at the meetings

Operational Plan D-2: Hold family meeting with all of the directly involved families to decide on acceptable proposals for personal and recreational spending.
Person Responsible: Team
Complete by: February 1, 2006
Costs: Human: Team: 2 hours

Operational Plan D-3: Budget appropriate levels of capital required for personal and recreational activities into the yearly business budgets.
Person Responsible: Russ, Aaron M., and Aaron S.
Complete by: February 15, 2006
Costs: Human: Russ, Aaron M., and Aaron S.: 10 hours each

7.7.8.2 Strategic Goal 2

7.7.8.2.1 Goal Statement (Family)

Continue to live, work, and grow our families in a rural, agricultural environment. Encourage individual development and exploration in a manner that is consistent and flexible in order to allow all individuals to reach their full potential.

7.7.8.2.2 Tactical Objectives

Tactical Objective 2-1: Effectively separate family relationships from business relationships in order to not allow the business activities and their results to undermine the family nature of the business.

Tactical Objective 2-2: Include each and every family member and their families in the business to the degree that each desires.

Tactical Objective 2-3: Allocate specific time for use in family recreation, family support, and personal development within the demanding time requirements of an agricultural production operation.

Tactical Objective 2-4: Through attainment of the financial goals previously mentioned, allow the wives in the operation to have the choice of working outside the business. Through this, provide the ability for the wives to stay at home with children to further strengthen the family base.

7.7.8.2.3 Operational Plans

Operational Plan A-1: Create a work environment that provides members mutual respect for individuality as well as business abilities. Ensure the business strength of personal diversity is not disregarded but instead used to increase the viability of the business in the long run.
Person Responsible: Team
Complete by: Immediately
Costs: Human: This goal will require a continual effort by all individuals directly involved in the operation.

Operational Plan A-2: Create a monthly business update publication to ensure the availability of all information to each business member so that all members are well informed of the business activities and circumstances.
Person Responsible: Kim, Desiree, and Amber
Complete by: February 1, 2006
Costs:

- Human: Kim, Desiree, and Amber: 10 hours per month
- Financial: $5 per publication or total of $30 per month

Operational Plan A-3: By understanding that an agricultural production operation is exposed to high levels of risk, ensure that business members realize the chance for negative business outcomes to occur. Strive as a family to not allow the negative aspects of the business to extend into the family structure of the operation and by doing so undermine the base of the entire operation.
Person Responsible: Team
Complete by: Immediately
Costs: Human: This goal will require a continual effort by all of the individuals directly involved in the operation.

Operational Plan B-1: Organize and hold a meeting to investigate the desires of family members not directly on the farm at this point in time to be a part of the business. Identify the degree to which each family member wishes to be involved with the business.
Person Responsible: Kim
Complete by: December 31, 2005
Costs: Human: Kim: 5 hours

Operational Plan B-2: Allow nondirectly involved family member desires to be incorporated into the investigation of organizational structures mentioned previously, thereby allowing for the inclusion of new family members in a way consistent with their desires.
Person Responsible: Aaron S.
Complete by: February 15, 2006
Cost: Human: Aaron S.: 10 hours

Operational Plan C-1: Plan three mandatory family gatherings throughout the year that will increase the family relationships within the business.
Person Responsible: Kim, Desiree, and Amber
Complete by: December 31, 2005
Costs:

- Human: Kim, Desiree, and Amber: 10 hours
- Financial: $3,000 to cover all of the planning, travel, and food expenses for the gatherings

Operational Plan C-2: Discuss and plan family support needs away from the farm. Allocate all details required to enable the family support needs to be met.
Person Responsible: Russ
Complete by: February 15, 2006
Costs:

- Human: Team: 2-hour meeting
- Financial: To be determined at the meeting

Operational Plan C-3: Plan summer recreation activities that will provide rejuvenation during high-demand times and provide time for the directly involved business members to evaluate and focus the business activities.
Person Responsible: Team
Complete by: March 1, 2006
Costs:

- Human: To be determined
- Financial: To be estimated

Operational Plan D-1: Identify the desires of the on-farm wives to work outside the operation.
Person Responsible: Aaron M. and Aaron S.
Complete by: December 31, 2005
Costs: Human: Aaron M. and Aaron S.: 2 hours each

Operational Plan D-2: Integrate these desires into the evaluation of financial needs. If these desires are not currently feasible, create a plan to meet the desires feasibly within the business.
Person Responsible: Team
Complete by: February 15, 2006

Costs:

- Human: Team: 2-hour meeting
- Financial: To be determined at the meeting
- Basic Resources: Addition of basic resources to the operation if needs are not currently met

7.7.8.3 Strategic Goal 3

7.7.8.3.1 Goal Statement (Organizational)

Continue to pursue organizational structures that fit the family dynamics of the operation, as well as allow for strategic goal attainment. Also, increase the business activities efficiency of the operation.

7.7.8.3.2 Tactical Objectives

Tactical Objective 2-1: Increase overall business organization.

Tactical Objective 2-2: Explore all relevant organizational structures for the business and organize business structure to meet all related goals of the operation.

Tactical Objective 2-3: Develop an effective estate plan for the operation.

7.7.8.3.3 Operational Plans

Operational Plan A-1: Remodel west side of second floor in Mom and Dad's house to create a functioning business office. Include a meeting and project work table, a communication station, a record-keeping area, and a filing and paperwork area.

Person Responsible: Russ, Aaron M., and Aaron S.

Complete by: March 1, 2006

Costs:

- Human: Mom and Dad: 5 hours planning; Russ, Aaron M., and Aaron S.: 4 weeks carpentry work each
- Financial: $5,000 to cover all materials, including building materials, furniture, and necessary office equipment

Operational Plan A-2: Organize newly purchased filing cabinet for use throughout the year. Create filing areas for current activities, monthly bills, and filed yearly paperwork.

Person Responsible: Kim, Desiree, and Amber

Complete by: January 15, 2006

Costs:

- Human: Kim, Desiree, and Amber: 10 hours
- Financial: $100 for miscellaneous needs in organizing the filing cabinet

Operational Plan A-3: Create a system for recording all field production applications on the farm. Include information for planting, fertility, herbicide, tillage, irrigation, and harvest. Organize information for day-to-day use, as well as yearly archived production records.

Person Responsible: Aaron M.

Complete by: February 15, 2006
Costs:
- Human: Aaron M.: 10 hours
- Financial: $100 to cover the cost of purchasing and creating the production application binders

Operational Plan A-4: Due to time requirements and constraints, evaluate use of hiring record-keeping help. Evaluate the feasibility (economic and time requirement) of paying Mom for record keeping in the business office. Also ensure that a minimum of one person aides in the record keeping, so the process is transferable to the future members of the business.
Person Responsible: Team
Complete by: February 1, 2006
Costs:
- Human: Mom: 20 hours per week
- Financial: $1,000 per month

Operational Plan B-1: Explore relevant organizational structures for the business through meetings with lawyer, banker, and accountant.
Person Responsible: Russ
Complete by: January 15, 2006
Costs:
- Human: Russ: 10 hours
- Financial: $200 to cover any consultation fees and related expenses for the meetings

Operational Plan B-2: Evaluate business structure alternatives in regard to other strategic goals of the business.
Person Responsible: Aaron M.
Complete by: February 1, 2006
Costs: Human: Aaron M.: 5 hours

Operational Plan B-3: Integrate the two farms into one in order to more efficiently manage the record keeping of the operation. Do this in a manner that will allow required capital to be secured and retain eligibility for all government programs.
Person Responsible: Aaron S.
Complete by: February 15, 2006
Costs: To be estimated

Operational Plan C-1: Integrate retirement plans into the financial requirements of the business through the use of the New York Life agent in town. In doing so, evaluate each of the directly involved families individually as to provide the most efficient retirement portfolios.
Person Responsible: Team
Complete by: January 31, 2006
Costs: To be estimated

Operational Plan C-2: Plan for the inclusion of any and all family members not directly involved in the business at this time. Ensure the estate plan is in agreement with the organizational structure, as well as all financial and family strategic goals. This may require meetings with professionals such as the business lawyer and/or accountant.
Person Responsible: Russ
Complete by: February 15, 2006
Costs: To be estimated

Operational Plan C-3: Ensure efficient business succession by evaluating the chosen business organizational structure in regard to the new and changing estate legislation.
Person Responsible: Russ
Complete by: February 15, 2006
Costs: To be estimated

7.7.8.4 Strategic Goal 4

7.7.8.4.1 Goal Statement (Integrated Farm Management)

Manage our farm as a co-integrated unit, while providing a step-by-step process for developing a strategic risk management plan.

7.7.8.4.2 Tactical Objectives

Tactical Objective 4-1: Manage production risk by being a low-cost producer of quality irrigated corn, dryland corn, and dryland winter wheat in northeastern Colorado and southwestern Nebraska.

Tactical Objective 4-2: Effectively expand the business, while creating a balance between meeting the related strategic goals and using a feasible business resource base.

Tactical Objective 4-3: Create an effective equipment replacement plan to update all farm machinery and vehicles in an orderly and reasonable time frame.

Tactical Objective 4-4: Increase the use of applicable marketing risk management strategies to minimize the extent to which the operation is exposed to the marketing risk present in agricultural production.

Tactical Objective 4-5: Manage the resource base of the farm so as to be good stewards of the land. Learn to manage the increasingly scarce irrigation resources of the land that are critical to the long-term success of the business.

7.7.8.4.3 Operational Plans

Operational Plan A-1: Continue to integrate cost-reducing farming practices into the operation. These practices should reduce required time, repairs, fuel, and production applications by capitalizing on new production technologies within the industry.
Person Responsible: Russ, Aaron M., and Aaron S.

Complete by: November 2006
Costs: To be estimated

Operational Plan A-2: Increase production efficiency and reduce human risk exposure through the use of appropriate production technologies.
Person Responsible: Russ, Aaron M., and Aaron S.
Complete by: November 2006
Costs: To be estimated

Operational Plan A-3: Use all farm resources fully and efficiently to increase the overall productivity and efficiency of the business resource base.
Person Responsible: Russ, Aaron M. and Aaron S.
Complete by: November 2006
Costs: To be estimated

Operational Plan B-1: Explore the possibility of adding complimentary and diversified enterprises to the business. In accordance with the financial strategic business goals, evaluate expanding the financial activities of the farm as enterprise expansion.
Person Responsible: Russ, Aaron M., and Aaron S.
Complete by: Continuously
Costs: To be estimated

Operational Plan B-2: Continue to monitor opportunities to increase productive acres farmed. Carefully evaluate any new acreage opportunities to ensure that any addition to the existing resource base is both profitable and feasible given current management requirements.
Person Responsible: Team
Complete by: As opportunities arise
Costs: To be estimated

Operational Plan B-3: In the event of an expansion opportunity, hold a team meeting to evaluate all aspects of the opportunity with respect to the mission and strategic business goals of the operation. Carefully evaluate the feasibility of the new opportunity given the current availability of management resources.
Person Responsible: Team
Complete by: As opportunities arise
Costs: To be estimated

Operational Plan C-1: Assess current equipment inventory and identify update needs. Designate needed machinery updates as critical needs, priority needs, or wants, and give justification for each.
Person Responsible: Russ and Aaron M.
Complete by: January 1, 2006
Costs: To be estimated

Operational Plan C-2: Develop a budget and schedule to address the updates identified above.
Person Responsible: Aaron S.
Complete by: February 1, 2006
Costs: To be estimated

Operational Plan C-3: Coordinate the update list, budget, and schedule for machinery and vehicle replacement with current debt requirements and cash flow feasibility.
Person Responsible: Russ, Aaron M., and Aaron S.
Complete by: March 1, 2006
Costs: To be estimated

Operational Plan D-1: Identify applicable marketing risk management strategies for irrigated corn, dryland corn, and dryland wheat production in northeastern Colorado and southwestern Nebraska.
Person Responsible: Aaron S.
Complete by: October 31, 2005
Costs: To be estimated

Operational Plan D-2: Create a marketing plan for the operation allowing for application of the marketing risk management strategies identified above.
Person Responsible: Aaron S.
Complete by: December 31, 2005
Costs: To be estimated

Operational Plan D-3: Apply the marketing plan to the crop marketing for the 2006 crop year.
Person Responsible: Russ, Aaron M., and Aaron S.
Complete by: November 2006
Costs: To be estimated

Operational Plan D-4: Evaluate the use of the marketing plan, as well as the success of the marketing risk management strategies used in the marketing plan.
Person Responsible: Russ, Aaron M., and Aaron S.
Complete by: December 2006
Costs: To be estimated

Operational Plan E-1: Use farming practices that are not detrimental to the long-term quality of the land. Ensure that the land farmed by the operation is maintained in quality, fertility, and pest and weed management, while improving these factors where needed.
Person Responsible: Russ, Aaron M., and Aaron S.
Complete by: Continuously
Costs: To be estimated

Operational Plan E-2: By having a minimum of one team member attend all applicable irrigation regulation meetings, be involved in all changes to current irrigation regulations in both northeastern Colorado and southwestern Nebraska.
Person Responsible: Russ, Aaron M., and Aaron S.
Complete by: As issues arise
Costs: To be estimated

Operational Plan E-3: Evaluate and implement all feasible water conservation strategies including, but not limited to, new crop rotations, new production technology, and new irrigation technology. In addition, modify and repair irrigation systems where needed to ensure the maximum efficiency of the systems.
Person Responsible: Russ, Aaron M., and Aaron S.
Complete by: December 2006
Costs: To be estimated

REFERENCES

Burleson, W., and C. Burleson. 1994. *Rut buster: Your visual goal setting book, that takes you step by step towards a better future.* Absarokee, MT: Sloping Acre Publishing Company.
Cairo, J. 1998. *Motivation and goal-setting: How to set and achieve goals and inspire others.* Franklin Lakes, NJ: Career Press.
Checkland, P. 1981. *Systems thinking, systems practice.* New York: John Wiley.
Covey. S. R. 1989. *The seven habits of highly effective people.* New York: Simon & Schuster.
Gable, C. 1998. *Strategic action planning NOW! A guide for setting and meeting your goals.* New York: St. Lucie Press.
Goodstein, L., T. Nolan, and J.,W. Pfeiffer. 1993. *Applied strategic planning: How to develop a plan that really works.* New York: McGraw-Hill, Inc.
Hewlett, J. P., K. Drake, G. Gade, A. Gray, F. Henderson, J. J. Jacobs, J. Jenkins, and R. Weigel. 1992. WIRE—Western Integrated Resource Education. A course in integrated, strategic management, http://agecon.uwyo.edu/wire.
Kolb, D. A. 1984. *Experiential learning: Experience as the source of learning and development.* Upper Saddle River, NJ: Prentice-Hall.
Richards, M. D. 1986. *Setting strategic goals and objectives.* St. Paul, MN: West Publishing Company.
Robinson, D. G., and J. C. Robinson. 2004. *Strategic business partner: Aligning people strategies with business goals.* San Francisco, CA: Berret-Koehler Publishers, Inc.
Smith, D. K. 1999. *Make success measurable! A mindbook-workbook for setting goals and taking action.* New York: John Wiley & Sons.
Tregoe, B. B., J. W. Zimmermann, R. A. Smith, and P. M. Tobia. 1989. *Vision in action.* New York: Simon & Schuster.

FURTHER READING

Bastian, C, and J. Hewlett. 2004. *Safety-first: A RightRisk lesson guide*. RightRisk Education Team (RR-L-1 January 04). Accessed June 2, 2009 from http://www.RightRisk.org
Hewlett, J. 2000. *Strategic planning for risk management*. Risk and Resilience in Agriculture series. Accessed June 2, 2009 from http://www.agecon.uwyo.edu/RnRinAg/

Hewlett, J P., C. Bastian., D. Kaan, and J. Tranel. 2004. Goal setting for strategic RightRisk management. RightRisk Education Team. Accessed June 2, 2009 from http://www.RightRisk.org

U.S. Department of Agriculture. *Introduction to risk management: Understanding agricultural risks: Production, marketing, financial, legal and human resource risks*. Washington, DC: U.S. Department of Agriculture Risk Management Agency (RMA3).

8 Step 4
Prioritize Risks

Dana Hoag

CONTENTS

Step 4 of the SRM process is the first task in the tactical planning section. While earlier chapters focused on discovery and goal setting to prepare for tactical planning, the purpose of this chapter is to identify, quantify, organize, and prioritize your risks. When goals are too broad, they become overwhelming. Tactical planning helps you find a more manageable problem to address. Step 4 is very important because it sets the direction of your entire management plan.

The tools in this chapter will help you develop a comprehensive but specific list of the risks facing your operation and also will prioritize which of these risks are the most important to manage later in the process. As difficult as it might seem, sometimes you will need to ignore certain risks because they don't have a big impact on you (or you simply can't have a significant influence over them). Most importantly, you will need to concentrate your energy on those risks you can impact, on those areas you can influence, and on those risks that will cause significant problems for you if they are not managed. The tools in this chapter will help you identify the risks you want to address, so you can develop a tactical action plan that specifies exactly how you will manage, ignore, retain, or ensure against these risks.

As with previous chapters, this chapter is divided into three parts. In Part 1 we look at the EWS Farms case study. The Spragues have to prioritize which risks to manage and which to ignore. The SRM process will help the Spragues identify all the risks that may be important and then show them how to choose which ones to start managing first.

The fundamental principles scientists have developed to help identify and prioritize risks are presented in Part 2. For comparison, we define and discuss common sources of risk shared by most farmers and ranchers, like low prices or bad weather. Although most managers are aware of their operational risks, it is probably insightful to review the kinds of risks other farmers, ranchers, and risk management experts have identified as common to agriculture. Finally, since not everyone likes the same approach, we describe five different methods that can be used to identify and prioritize risk.

In Part 3 we develop a Risk Navigator management tool called the Risk-Influence Calculator. This online Risk Navigator tool can help you organize and prioritize your own operational risks. You can find the Risk-Influence Calculator for this chapter on the Web site, www.RiskNavigatorSRM.com.

EWS Farms is a perfect case study as it is a traditional farm where the owner needs to evaluate risks as he contemplates how to integrate his family into the farm operation. As usual, we apply the example to the case study in order to demonstrate that Risk Navigator tools and techniques can be used in real-life settings.

8.1 PART 1: THE EWS FARMS CASE STUDY

8.1.1 Objectives

Like all farmers, the Spragues face a multitude of risks, including production, marketing, financial, institutional, and human. Managing risk is a complex process, and the Spragues do not have the resources to address all their risks at one time. No one does.

The objective of Step 4 is to obtain a list of the Sprague family's risks and to prioritize the risks to address in a Tactical Plan.

8.1.2 Strategy

We start by interviewing the Spragues to identify their risk sources, then use the Risk Navigator tools to help the Spragues prioritize which risks to address. To ensure that we don't miss any risks, we'll begin with broad, big-picture risks, and then narrow our search to prioritize the specific risks that are most important to manage. The goals the Spragues developed in Step 3 also can be used as a starting place to review broad risks.

In the EWS Farms case study, we narrow the Spragues' risk focus by using the Risk-Influence Calculator, which helps the Spragues list and prioritize their risks into the five risk categories: production, marketing, financial, human, and institutional. The Risk-Influence Calculator helps the Spragues identify and manage the risks that will have the most impact on their operation, rather than spend valuable time on controlling risks that might seem to be more convenient. A discussion about tools other than the Risk-Influence Calculator can be found in Part 2 of this chapter.

8.1.3 Implementation

We will implement Step 4 for the Spragues in three stages:

1. Review the risk goals created in Step 3 and risks cited in formal studies
2. Develop a list of the most important risks
3. Select risks to be prioritized in the Tactical Management Plan

8.1.3.1 Review Risk Goals

In Chapter 7 the Sprague family identified four risk goals for their operation:

1. Financial: Ensure short- and long-term financial success by maintaining business profitability while expanding the overall financial resource base.
2. Family: Continue to live, work, and grow with our families in a rural, agricultural environment. Encourage individual development and exploration in a manner that is consistent and flexible enough to allow all individuals to reach their full potential. Seize the opportunity to help manage business risks using the skills of additional family members, and provide support to family members who do not wish to return to the farming operation.
3. Operational Structure: Continue to pursue organizational structures that fit the family dynamics of the operation, as well as allow for strategic goal attainment. Also, increase the business activities' efficiency of the operation.
4. Integrated Farm Management: Manage our farm as a co-integrated unit while providing a step-by-step process for developing a strategic risk management plan for each of the five types of risk: production, market, human, institutional, and financial.

Obviously, the Spragues' goals center on family growth. This is important because Aaron's father is trying to make a living, while also making a place in the operation for two of his children's families. In addition, two other children are away at college and may also want to return to the farm someday.

Based strictly on the goals generated in Step 3, we would not conclude that production risks or price risks are a concern. Yet Aaron wanted more information and decided to examine literature on risk to see what other farmers were concerned about, just in case he missed anything. After reviewing the types of risks that other farmers and researchers have cited, Aaron added several more basic, production-oriented risks to his list, including price risk and yield risk. This demonstrates the importance of using multiple risk-identification approaches to assure that some risks are not overlooked. That is, if you look at something from more than one angle, you might see different needs.

8.1.3.2 Develop a List of the Most Important Risks

It is difficult for most people to pull a list of risks off the top of their heads, no matter how familiar they are with their own operations. Doing so chances missing important risks. To avoid overlooking specific risks, we developed an extensive list based on what producers identified in various surveys and studies in the literature. (We present this information in Part 2.) After reviewing his goals and reading what others said about risks, Aaron identified the following strategic risks:

Market/Price:
  Corn Price—Will my price cover my costs?
Production:
  Weather—Will rainfall support crop stand?
  Hail—Will hail destroy half my crop?
  Input (seed)—Will good corn seed be available at a reasonable price?
Financial:
  Expansion—Can the operation generate enough profit to cover new land payments?
Human:
  Family—Will my dad retire?
Institutional:
  Water—Will irrigation water be restricted?

8.1.3.3 Select Risks to be Prioritized in the Tactical Management Plan

The Spragues have little influence over some risks, such as hail; however, they can influence other risks, such as ensuring adequate water supply by updating their irrigation system. These risks can be plotted into a risk-influence chart as shown in Figure 8.1. This chart is developed by asking Aaron to assign a number from 1 (not likely) to 10 (likely) to each risk and a number from 1 (no influence) to 10 (extreme control) about how effectively each risk could be influenced or managed by the family. For example, water is an important risk for irrigated crop producers in northeastern Colorado. There is a small chance of having water restrictions, but this would have a big impact on crops; so the Spragues assigned a 5 to this risk. The Spragues, however, have a high level of influence over water since they have installed irrigation systems on much of their cropland. Therefore, Aaron assigns a 7 to their level of influence. Aaron used the Risk-Influence Calculator to create Figure 8.1.

The Probability-Impact Calculator is another tool found inside the Risk-Influence Calculator. It can help determine *the risk part* of risk and influence. Risk is a combination of probability and impact, that is, the likelihood of whether or not something will happen and how important it will be if it happens. A low risk would be unlikely and of little significance. A medium risk could be either likely to happen but of little

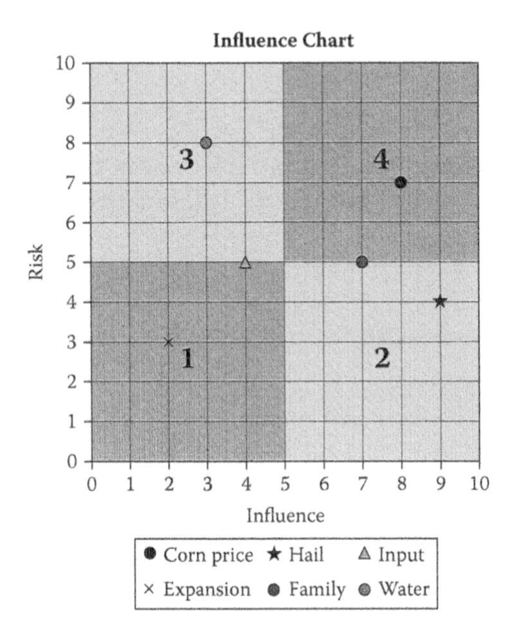

FIGURE 8.1 The Sprague's Risk-Influence Calculator.

consequence, or unlikely and significant. A high risk would be likely and have a significant impact on the farm. An example of the Probability-Impact Calculator is shown in Part 3.

The idea of the risk-influence chart is to identify needs that have a high risk and that have a reasonable chance of being influenced by the decision maker. After looking at the risk-influence chart, Aaron decides to focus on a strategy to manage corn price risk because it is the only clearly high risk that he can influence. The most important risks to focus on are in the upper-right quadrant, numbered 4 in Figure 8.1. The family also feels they have good control over most of their other high-risk options and wants to take advantage of Aaron's investment in learning about marketing in his new master's degree program in agricultural economics from Colorado State University. You can see the actual Risk-Influence Calculator that Aaron considered in Part 3. Try downloading these tools from the Web site and change some of Aaron's assumptions to see how it affects the importance of managing each of the risks.

8.2 PART 2: THE FUNDAMENTALS OF RISK IDENTIFICATION

Most farmers and ranchers are well aware of the risks they face. Nevertheless, it is always helpful to compare your experiences with others. Over time, economists and risk experts have identified the following five common sources of risk found in agriculture (Hardaker et al., 1997; Baquet et al., 1997; Harwood et al., 1999; U.S. Department of Agriculture Risk Management Agency [RMA], 1997):

1. Production risk
2. Market or price risk

3. Financial risk
4. Institutional risk
5. Human resource risk.

These risks are defined, described, and discussed in Table 8.1. The table defines each type of risk and summarizes sources and management controls. Each risk is

TABLE 8.1
Risk Management Sources and Management Controls

Risk	Defined	Sources	Management Controls
Production	Uncontrollable events such as weather, pests, or disease make yields unpredictable. Changing technology makes a manager or capital obsolescent. Inputs are unavailable or low quality.	Weather, extreme temperatures, pests, disease, technology, genetics, inputs (availability, quality, price), equipment failure, labor …	Diversification, insurance (crop, revenue), buildings, storage, vaccines, extra labor, production contracts (e.g. ensure input supply and quality), new technologies (e.g. automate watering)
Marketing and Price	Prices of inputs or outputs change after a producer commits to a plan of action. Price fluctuations stem from domestic and international supplies or substantial changes in demand.	Product quality (genetics, disease, handling, input/feed) Product price (quality, timing, global market, weather, government policy, contracts…)	Futures and options, forward contracting, retained ownership, quality controls, storage (timing), cooperatives, niche/value-added marketing…
Financial	Stems from the way a business is financed. Borrowed funds leverage business equity but increase business risks.	Market, production, legal and human risk, interest rate changes, natural disasters (drought), land market changes, foreign exchange, loan calls …	Cash reserves, equity, borrowing capacity, reducing other types of risk (production, marketing, etc.), insurance
Institutional	Government or other institutional rules, regulations, and policies effect profitability through costs or returns.	Taxes, contract disputes, regulations, government policies, lawsuits, ambiguous and/or unwritten agreements, neighbors, environmental programs …	Estate planning, tax planning, contracts, bonds (e.g. environmental liability), research and education about local laws …
Human Resources	The character, health, or behavior of people introduces risk. This could include theft, illness, death in the family, loss of an employee, or a divorce for example.	Ambiguous and/or unwritten agreements, poor planning, miscommunication, health, or other family disasters …	Family planning, including labor planning, clear contracts, training and goal setting, communication, estate planning …

described in greater detail later in this chapter and in Chapters 13 to 16. We now proceed with a more detailed discussion about these sources of risk.

8.2.1 Surveys about Risk Sources

A good place to start the risk identification process is to look at what other producers and operators across the country have reported. Table 8.2 highlights one U.S. Department of Agriculture study that looks at 6 types of risk on 15 different farms. On a four-point scale, with four being very concerned and one not being concerned at all, the average producer rated changes in government laws and regulations as their top concern and changes in technology as their lowest concern. Producers also rated price uncertainty and crop yields as somewhat concerned. An interesting observation was that producer concerns varied by crop type. For example, cotton producers rated government programs much higher than did tobacco or vegetable farmers. Price risk seemed to be lower for nurseries and vegetables than for most other producers.

We also compared and contrasted results of three risk surveys in Table 8.3. Musser and Patrick (2002) surveyed top producers in workshops at Purdue University in Indiana. The Indiana producers agreed with producers in the national study that price risk and yield uncertainty were top concerns. In contrast, these producers identified laws and regulations as their last concern. The reason for the difference may lie in the fact that risk arises where we have the least control. Perhaps top producers in the Musser and Patrick study better addressed legal and institutional problems than the USDA (U.S. Department of Agriculture) producers.

In a study of Nebraska and Texas cattle producers, Hall et al. (2003) found drought to be the most important risk. This is not surprising as this region was in the midst of a devastating drought at the time. This phenomenon serves as a warning that current events can dominate our thinking. Risk management should encompass all risks, not just those that have recently occurred. Similar to other studies, price uncertainty was also a top concern. The producers cited input costs as a serious concern, although these producers did not think labor, disease, cold weather, or the government were significant threats.

8.2.1.1 Production Risk

It might be helpful to discuss the different types of risk one at a time. We will start with specific production risks, and base our discussion on comments made by Iowa corn growers as shown in the following chart (Mickelsen and Trede, 2001). The numbers at the end of each bullet indicate the numerical rating where 0 = no concern and 4 = very concerned.

Crop Production Risk	Livestock Production Risk
Weather, wind, hail, etc.—4.22	Adequate market outlets for livestock—3.68
Disease, insects, weeds—4.00	Disease—3.51
Use of new crop varieties—3.64	Initial investment cost of facilities—3.29
Adoption of new technology—3.52	Regulations on production practices—3.16
Consolidation of input suppliers—3.42	Adoption of new technology/methods—2.97
	Obsolescence of facilities—2.65

TABLE 8.2
A National Survey of Farmer's Degree of Concern about Factors Affecting the Continued Operation of their Farms (Harwood et al., 1999)

	Mean scores[1]													
How concerned are you about each factor's effect on the continued operation of your farm?	**Other Cash Grains**	**Wheat**	**Corn**	**Soybeans**	**Tobacco**	**Cotton**	**Fruit/Nuts**	**Vegetables**	**Nursery/ Greenhouse**	**Beef**	**Hogs**	**Poultry**	**Dairy**	**All Farms**
Decrease in crop yields or livestock production	3.35	3.51	3.20	2.98	3.16	3.68	3.05	2.85	2.78	3.09	3.53	3.20	3.40	2.95
Uncertainty in commodity prices	3.41	3.83	3.40	2.93	3.15	3.75	2.88	2.82	2.63	2.96	3.31	3.09	3.54	2.91
Ability to adopt new technology	2.52	2.38	2.39	2.33	2.21	2.77	2.34	2.09	2.24	2.25	2.63	2.60	2.45	2.23
Lawsuits	2.43	2.47	2.03	2.46	1.89	2.78	2.39	2.66	2.06	2.36	2.70	2.32	2.36	2.26
Changes in consumer preferences for agricultural products	2.65	2.55	2.39	2.40	2.40	2.86	2.44	2.59	2.69	2.58	3.01	2.79	2.76	2.47
Changes in government laws and regulations	3.31	3.36	3.15	2.79	2.77	3.54	2.97	2.75	3.09	3.03	3.23	3.34	3.31	3.02

[1] 1 = Not concerned, 2 = Slightly concerned, 3 = Somewhat concerned, 4 = Very concerned

TABLE 8.3
Sources of Risk Identified in Producer Surveys

Sources of Risk[a]	Purdue University Top Farmer Crop Workshop 1999 (Musser and Patrick, 2002)	National Study-Economic Research Service (Harwood et al., 1999) Table 1, page 5)	Beef producers in Texas and Nebraska (Hall et al., 2003)
Price Uncertainty	1	3	2
Yield Uncertainty	2	2	
Business Contracts	3		
Cost of Capital Goods	4		
Government Commodity Programs	5		ns[b]
Technology	6	6	
Inputs/Costs	7		3
Injury, Illness, or Death	8		
Laws/ Regulations	9	1	ns
Legal (lawsuits)		5	
Consumer Preferences		4	
Severe Drought			1
Cold Weather			ns
Livestock Disease			ns
Labor Availability			ns
Credit/cash reserves			

[a] Sources names are adapted to integrate across studies.
[b] *ns* means rated, but neutral or of less importance in the study

Note that the participants in this study identified weather-related yield risk as the most threatening, followed closely by disease, insects, and weeds. Livestock producers fear the lack of adequate markets most, followed closely by disease. Both crop and livestock producers also report risks related to investments in technology or facilities as a significant factor. Risk also stems from the input side. Crop producers cited the consolidation of input suppliers as a threat.

8.2.1.2 Market and Price Risk

Market risk is related to the price producers receive for their crop. Price can be affected by the quality of a product, which is often within a producer's control, however, most price risk occurs off the farm and there is very little a producer can do about it. For example, corn price can be improved with proper management of moisture content. But the price is also influenced by domestic and international markets, government programs, and global events. In the Mickelsen and Trede survey (2001) shown in the following list, producers ranked narrow operating margins as one of their top concerns, largely due to price (in chart below, 0 =

no concern and 4 = very concerned.) They also cited accessibility to markets and volatility of prices as great concerns. In today's world, global markets are also on people's minds.

- Narrow operating margins—4.33
- Accessibility to markets to sell products—4.17
- Volatility in commodity prices—4.14
- Global economic conditions—3.99
- Fluctuating costs of inputs—3.94
- Trade agreements (North American Free Trade Agreement [NAFTA], etc.)—3.73

8.2.1.3 Financial Risk

In this study, supply of capital (money to borrow) was the top financial concern; however, as shown in the following list (Mickelsen and Trede, 2001), almost as important were the ability to recover from depressed times and to find lenders who were sympathetic to agriculture.

- Adequate supply of capital—4.05
- Recovery time from depressed agricultural economy—4.04
- Lenders knowledge of agriculture—4.01
- Business cycles in agriculture—3.69
- Volatility in interest rates—3.64

8.2.1.4 Institutional and Human Risk

Institutional and human risks are a bit outside the sphere of what producers generally think about when they think of risk. Institutional risk focuses on the impact institutions, such as government and the legal system, can have on producers. For example, the creation or elimination of government price and income support programs (or a special farm tax) can have a huge impact on producers. Since capital equipment has a high price tag, a change in the interest rate can have an enormous impact on cash flow. In recent years, environmental controls also have impacted many producers. As shown in the following list (Mickelsen and Trede, 2001), Iowa farmers were most concerned with changes in policies or regulations, including farm programs. Fears of increasing pressure from international groups followed closely behind.

- Changes in government policy/regulations—4.05
- Changes in government farm programs—4.04
- Foreign restrictions on products (genetically modified organisms [GMOs], etc.)—3.96
- Export trade barriers (tariffs, etc.)—3.91
- State/federal environmental regulations—3.87

Human risks stem from human behavior. For example, a divorce or death in the family can devastate a farm. The behavior of family members and employees is important, and the more critical the role of the operator, the greater the human risk in that operation. Operators told surveyors they were most concerned about the death of the operator. Injury also was a concern (Mickelsen and Trede, 2001).

- Death of owner/operator—4.18
- Injury to owner/operator—4.08
- Lawsuits—3.69
- Injury to hired help—3.53
- Divorce of owner/operator—3.53

8.2.2 Internet Resources

There is a wealth of information readily available on the Internet to help you determine which risks are most important for your situation. Some examples are shown in Box 8.1; more can be found at the RightRisk.org Web site. From the RightRisk Web site, you can link directly into the Western Risk Management Library or the National Ag Risk Education Library. Each has thousands of papers, ranging from practical to academic.

Many Web sites provide graphs, tables, and charts, and some even let you download data in Microsoft Excel. For example, the severity of price and yield risks for crops in the United States is presented in Figure 8.2. This figure shows how volatile corn prices are around the country. Price volatility is a measure of how much price varies over time. The higher the number is, the more volatile the price. (See Chapter 13 on market risk for more information about price volatility.) Dry edible beans had the highest volatility measured, and beef had the lowest. This is just one example of the information you can find about risk on the Internet. Unfortunately, there is no easy, one-stop Web site to provide you information. It takes some hard work and ingenuity to find the information you need.

BOX 8.1 FINDING INFORMATION ON THE INTERNET

- Risk Management Education (U.S. Department of Agriculture) http://www.usda.gov/rma/rme
- Weather (National Oceanic and Atmospheric Administration) http://www.noaa.gov
- Markets (Agricultural Marketing Service) http://www.ams.usda.gov/marketnews.htm (Chicago Board of Trade Market Plex) http://www.cbot.com/mplex/
- Statistical Information (National Agricultural Statistics Service) http://www.usda.gov/nass (USDA Economics and Statistics System) http://usda.manlib.cornell.edu/usda/usda/html
- Economic Information http://www.ers.usda.gov (Economic Research Service)

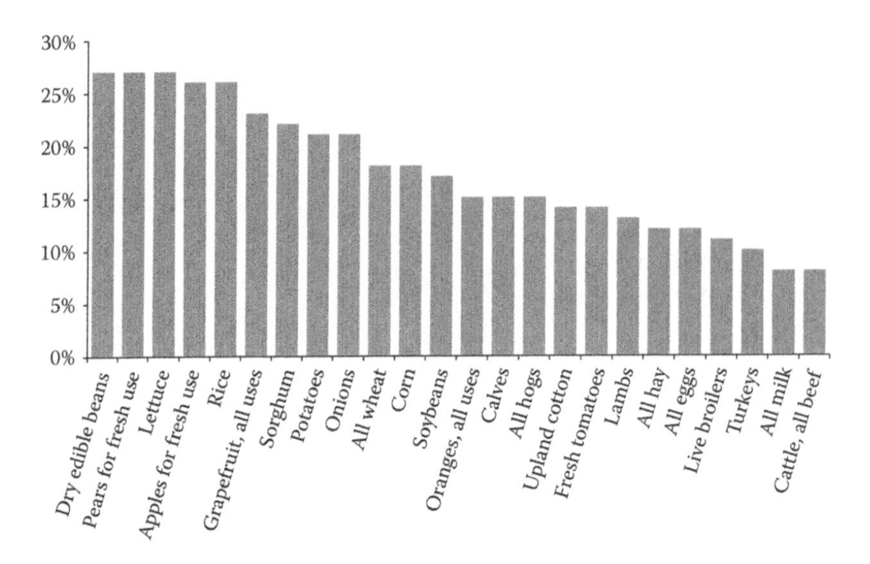

FIGURE 8.2 Price volatility of selected commodities, 1987–1996. (Harwood et al., 1999)

As we leave this section, it is important to emphasize that we cannot possibly list all of the risk resources on the World Wide Web, but spending a little time surfing could pay off big.

8.2.3 Prioritization Techniques

Most risks faced by farmers, ranchers, and agribusinesses are very complicated. Despite often extensive efforts to identify and quantify key relationships, it is frequently difficult to integrate all potential risks in a single risk management decision. As with other chapters about each step, this segment presents you with an array of tools to assist in identifying and prioritizing your risk. We recommend that you use the Risk-Influence Calculator; however, reading over the examples of the other methods is also recommended as it will help cement the concepts for listing and prioritizing decisions.

Risk management involves significant judgment on the part of the decision maker to evaluate the risks of different events and the probability that his or her actions will alter those risks (Greenspan, 2003). This process can be made easier through the use of decision analysis tools—tools that help organize information into a format that helps make better decisions. These tools often are available in electronic format. The following are four different decision analysis tools to help you identify risks:

1. Influence diagram (ID)
2. Contributing factor diagram (CFD)
3. Risk-Influence Diagram
4. Strengths, weaknesses, opportunities, and threats analysis (SWOT)

8.2.3.1 Influence Diagrams

An *influence diagram* (ID) is a graphical diagram with enough structure to organize complex and confusing relationships. IDs are discussed here because they are probably the most intuitive method available. For our purposes, a pencil and scratch paper are all that is needed to develop a useful ID for scoping out your problem in broad terms. This concept is highly recommended as a starting point; however, it is very difficult to develop a detailed ID for a complex problem. We recommend additional techniques to advance through Step 4. If you wish to develop a formal ID, there are software programs that can help.

A formal ID uses shapes, like circles, rectangles, and triangles, to represent decisions, outcomes, probabilities, and other factors. Rectangles represent decisions, ovals represent chance events, and diamonds represent payoffs (Clemen and Reilly, 2001). Arrows show how these occurrences are connected. We can demonstrate how to use an ID to evaluate the risks associated with the decision of whether to vaccinate cattle against foot and mouth disease. Cattle will be worth next to nothing if exposed to the disease. As shown in Figure 8.3, an operation's losses (profits) will depend upon the chance the herd is exposed to foot and mouth disease and whether a decision was made to vaccinate the cattle against the disease. Profit will be highest if the producer does not vaccinate and the herd is not exposed to the disease. Of course, this strategy also subjects the producer to the most risk because he stands to lose a substantial amount of money if the herd is exposed to and contracts the illness. In other words, profits will be lower if he vaccinates, but the likelihood of disease will be lower as well.

Influence diagrams can be made very complex if you draw every single box, triangle, and oval and connecting arrow. Try this simple exercise for a farm, ranch, or other business that you own or are familiar with. Put the objective "profit" in a diamond in the middle of a piece of paper. Start adding circles and rectangles to represent decisions and chance events that could have an effect on your profit. For example, add "crop yield" and "cost of production" in circles. Add "sell crop" as a decision in a rectangle. Be sure to draw every arrow connecting any two figures that are related. Remember, costs affect yield and yield affects costs, so there needs to be a two-headed arrow between them. And each of them needs an arrow pointing to profit. Then, for the second layer, add circles and rectangles that you think could have an effect on the previous circles and rectangles you drew. Continuing our example,

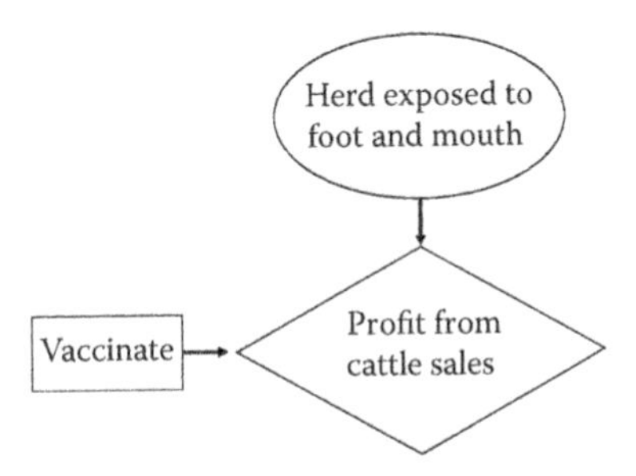

FIGURE 8.3 Influence diagram of foot and mouth vaccination decision.

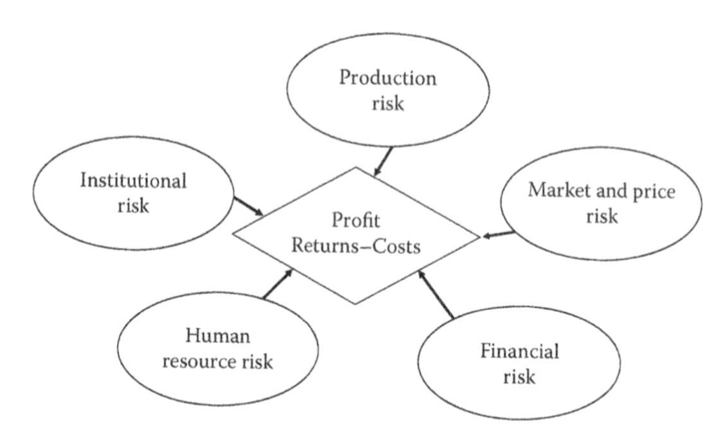

FIGURE 8.4 Contributing factor diagram of market price risks.

yield is affected by insects, water, nutrients, and weather. Insects, in turn, are affected by weather and the decision (a rectangle) to control them. Costs of production are affected by how you control pests and random events like the price of fuel. Finish drawing all the arrows between the shapes that represent related components.

Does it seem like you could go on forever? Some researchers call these "spaghetti" diagrams since a complex ID can look like a plate of noodles with round, square, and triangular meatballs. A good bit of judgment and practice may be required to find that balance between too little information to be realistic and too much to be practical. However, highly visual learners often prefer this type of decision analysis tool, and it is a popular method. Software programs, such as STELLA (iseeSystems.com), which builds models based on influence diagrams for systems thinking, are available for those who would like to use the ID method. Many public domain programs can be found free on the Web.

8.2.3.2 Contributing Factor Diagram of Foot and Mouth Vaccination Decision

To counterbalance the complex ID, we turn to a simpler diagram called a *contributing factor diagram* (CFD). The CFD focuses only on relationships between variables that contribute to a factor of interest, such as profit. Figure 8.4 is an example of a contributing factor diagram. Compared to the ID in Figure 8.3, the CFD does not include peripheral information such as how a disease is contracted, whether an exposed animal contracts the disease, the impact of temperature on infection rates, or the many factors that might have an impact on the efficacy of the vaccine.

CFDs are not flowcharts because events are not listed in sequence; they are free flowing with arrows that indicate where one item affects another. Generally, a person works backward in a CFD, starting from the problem and then working back to the issues that contribute to the problem. In this case, we will assume that most businesses start with the fundamental definition of profit below:

$$\text{Profit} = \text{Returns} - \text{Costs}$$

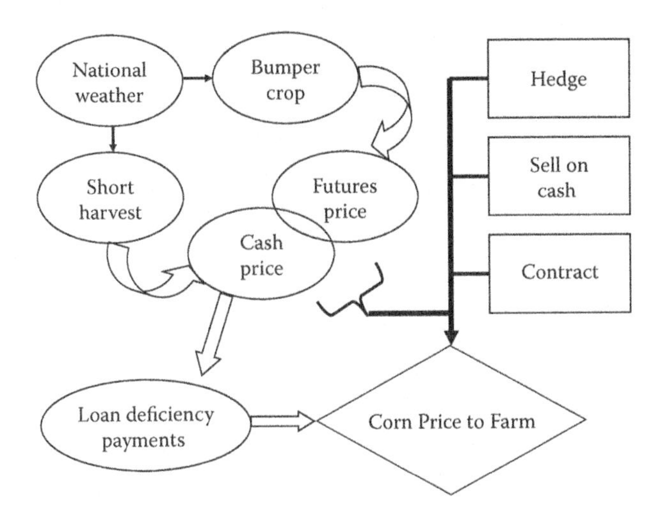

FIGURE 8.5 Contributing factor diagram of the EWS Farms corn marketing problem.

Working backward as shown in Figure 8.4, we focus only on how profit is affected by our five types of risk. We can then get more specific about where each risk comes from and prioritize how to affect that risk. A decision maker could choose price risk and move on to define it. Let's continue this example, since that's exactly what Aaron chose when he prioritized price risk for corn production. The CFD is expanded to illustrate the price risk decision on our case farm as illustrated in Figure 8.5. The net price a farmer receives is a function of how weather affects national supply, how the government sets price support program loan deficiency payments, and three decisions:

- Whether or not to sell on the cash market
- Whether or not to forward contract
- Whether or not to hedge on the futures market

The decisions shown in Figure 8.5 are typical for producers facing price risk, including EWS Farms. The Spragues would like to receive the highest price possible for their corn crop. They can sell on the cash market at harvest or store the crop, hoping for a better price later. They also could forward contract with the local elevator or their other option is to hedge on the futures market. (Note that storage is not represented in our simple illustration above.) The marketing plan developed by the Spragues, described in Chapter 13, allows them to choose any of these options, including a combination that includes selling one-third of the crop in each option.

8.2.3.3 The Risk-Influence Matrix

Up to this point we have discussed some rather elaborate methods for identifying risk. One of the most straightforward methods is called a *risk-influence matrix* (Clark and Timms, 1999). The concept of a risk-influence matrix is to rank each risk by how much impact it has on you and how much influence you have on it, as demonstrated in Figure 8.6.

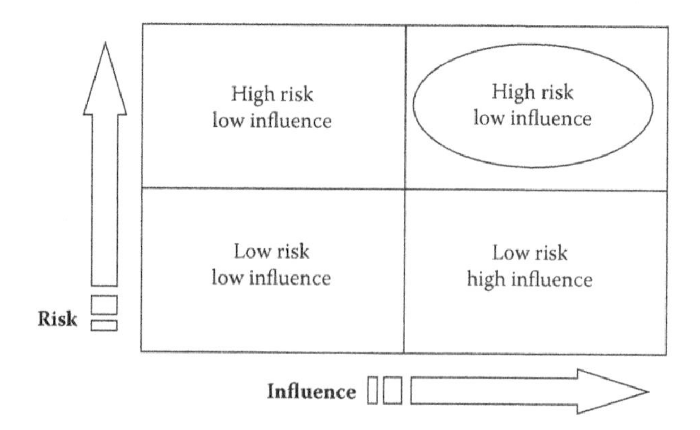

FIGURE 8.6 The Risk-Influence Matrix.

As can be seen in this simple illustration, some risks have a bigger impact on producers than others. For example, a hail storm might be considered a bigger risk than an equipment breakdown. This could be because the probability of the storm is greater than the probability of a breakdown, because the impact on profit is greater for a hail storm, or a combination of both. A manager or producer also has a varying degree of influence over different types of risk. There is little that can be done to reduce hail damage, but most producers are skilled at keeping equipment going. The purpose of the matrix is to identify where risks are high and where the operator has the most influence to address the risk. This area is indicated by the circle in Figure 8.6. Generally, you would prioritize managing the risks in the upper right-hand corner and setting aside those in the lower left-hand corner.

You can build your own customized matrix with the Risk-Influence management tool in Part 3. The tool allows you to compare impact and probability, which determines risk, and then to compare risk against influence. A detailed example can be found in Part 3.

8.2.3.4 SWOT Analysis

The last tool we will look at is a SWOT analysis. SWOT stands for *strengths, weaknesses, opportunities*, and *threats*. It involves the strengths and weaknesses internal to your business. For example, strengths on your farm or ranch could include good weather, well-working equipment or facilities, solid training, or a strong genetic line in a livestock herd. Weaknesses could include the opposite of any of the strengths, such as being in an area prone to hail damage. Strengths and weaknesses are internal to the business. Opportunities and threats, however, can come from inside the business or from external sources. An on-farm threat would be a recurrence of disease or weeds. An off-farm threat might include policy changes, rezoning, or your landlord terminating your lease.

The purpose of the SWOT analysis is a little different from the previous tools. SWOT helps you think about what your risks are by looking at where your SWOTs are, rather than looking at where your risks are. For example, suppose you were considering what career you would most like for yourself. You might fancy becoming an NBA basketball player, with becoming a polo star as second choice. However, after

On Farm	Off Farm (External)
Strengths Grew up farming in the area, father available for consultation and other help, family nearby, masters degree in agricultural economics	
Weaknesses New farmer, limited access to capital, rented land	
Opportunities On-farm storage, family help	**Opportunities** Crop insurance, rented land and water, government price support programs
Threats Estate claims by siblings	**Threats** Possibility of losing irrigation rights, hail

FIGURE 8.7 SWOT analysis on EWS Farms.

considering your SWOTs, you might scrap being in the NBA since you are only 5 feet 10 inches tall and can't even touch the hoop when you jump. You might give up on your polo dreams, too, if you drive a 1998 Dodge Neon and have trouble finding gas money. Your strength is your education in computer science, and you have an opportunity to get a job at Dell.

Consider Figure 8.7. At age 26, Aaron Sprague is a young farmer and he has a wife and three kids to support. This poses a serious weakness given how much capital and experience it takes to farm. Nevertheless, Aaron has a supportive family and is farming near where he was raised. He can take advantage of his family's extensive experience and capital reserves, including land, equipment, and even finances. On the farm, Aaron has some opportunities to use his family's help and crop storage facilities. But sooner or later the family will have to decide how to split the home farm and assets. Aaron's sister and her husband also farm, while his brother and other sister have not yet decided what they wish to do. Off the farm, Aaron has an opportunity to rent land and water, to insure his crops, and to use government programs. He is constantly threatened by hail and his source of groundwater is subject to a regional lawsuit with neighboring states.

8.3 PART 3: RIGHTRISK NAVIGATOR MANAGEMENT TOOLS

While we discussed several evaluation methods in the previous section, we now focus on the Risk Navigator SRM tool, the Risk-Influence Calculator. The purpose of this tool is to draw a visual map of risks in a way that helps you prioritize which risks need to be managed most. You will be asked to list up to ten risks and then indicate the probability that each will occur, the impact it will have on your operation if it occurs, and the amount of influence you may have to reduce the risk.

8.3.1 Risk-Influence Calculator

This tool is divided into three steps. In the first step, you are asked to enter the risks you wish to consider for prioritization. You are provided two slots in each of five risk

Risk Navigator
STRATEGIC RISK MANAGEMENT

By: Dana Hoag and Eihab Fathelrahman
Contact: dana.hoag@colostate.edu

Risk-Influence Calculator

Data Matrix | Probability - Impact | Risk - Influence | Help | Save, Load, and Delete

Risk Category	Risk Type	Description	Probability	Impact	Risk	Influence
Market/Price:	Corn Price	Will my price cover my costs?	4	10	7	8
					#VALUE!	0
Production:	Hail	Will hail destroy half my crop?	1	7	4	9
	Input	Can quality corn seed be found?	6	4	5	4
Financial:	Expansion	Can expansion costs be paid?	3	4	3	2
					#VALUE!	0
Human:	Family	Will dad retire?	2	8	5	7
					#VALUE!	0
Institutional:	Water	Will irrigation restrictions apply?	7	9	8	3
					#VALUE!	0

FIGURE 8.8 Risk-Influence Calculator.

categories: market, production, financial, human, and institutional. You may ignore the risk category when you enter risk types if you like. That is, you can enter more than two risks in any of the risk types by entering extras in other risk categories and ignoring the risk category label. The objective is to assign a level of risk and influence to each risk. Start by naming each risk and writing a brief description. Then enter the level of probability and impact for each risk type—risk will be determined automatically. Risk is equal to the (probability score + impact score)/2, rounded downward to the nearest integer. Then enter your influence level.

Figure 8.8 shows what we entered for Aaron from Part 1. Notice that when you assign a 1, there is little chance of a bad outcome and when you enter a 10, there is a high chance of a bad outcome. Aaron asks, "Will my price cover my costs?" Enter a 1 to mean that the costs will be covered and 10 to indicate that costs will not be covered. Likewise, enter 1 when the impact of the risk is very low and 10 when it is very high. Finally, we want to know how much influence you have on the risk. Enter 1 if you have no effect and 10 if you can control it perfectly.

The scores that Aaron Sprague used come up automatically in the tool if you download the EWS Farms example; otherwise the tool comes up empty and ready for you to use. Aaron assigned a relatively high probability of not being able to find good corn seed and to getting his irrigation water cut off. Aaron is probably concerned that corn seed might be short due to the increased plantings in recent years, and in eastern Colorado, farmers are embroiled in several battles that have resulted in lost water rights. He felt that a severe hail storm was unlikely and that his dad would probably not retire any time soon. Only a low market price and loss of irrigation water landed in Quadrant 4, where both impact and probability are high. Aaron entered his influence last. He felt that he had a lot of influence over the price of corn because he just completed his master's degree in agricultural economics at Colorado State University where he learned advanced marketing techniques. He assigned a low influence score to being able to do much about the loss of irrigation water. The

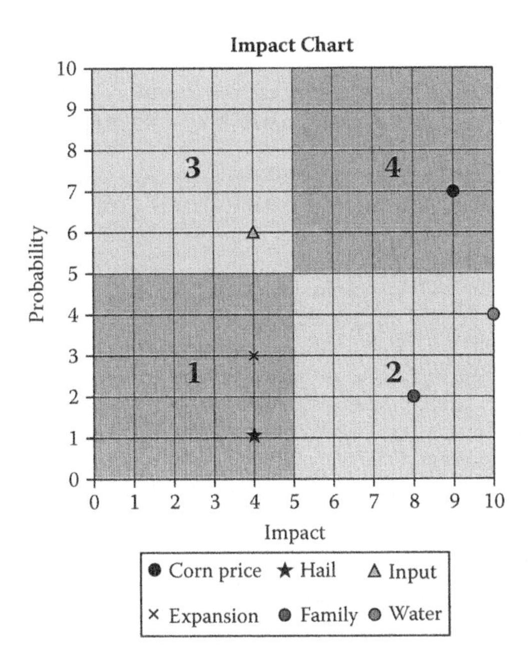

FIGURE 8.9 The Risk-Impact Diagram.

reason Aaron provided a high influence score for hail damage was not because he could control the weather, but because he could buy crop insurance to cover hail losses.

The second step is to examine and prioritize your list of risks. Start by clicking on the Probability-Impact tab. Risk is a combination of probability and impact. Use the probability-impact chart to examine where your risks are coming from (Figure 8.9). Risks in the upper-right quadrant (Quadrant 4; quadrants are numbered in the book) are a priority because they have a high chance of happening and will have a big impact if they happen. Risks on the lower left (Quadrant 1) are not a priority. Risks in Quadrants 2 and 3 are of concern. For example, a risk with a probability score of 7 and an impact score of 6 would have a risk score of 6, which will plot into Quadrant 4 in the chart ((7 + 6)/2, rounded down).

Finally, you can determine your risk priorities. Click on the Risk-Influence tab. Referring back to Figure 8.1, you can see the risk level of corn price is 7, as described previously. Aaron assigned an influence score of 8 to this risk, so it lands in Quadrant 4 as our top risk to prioritize. Hail has a high enough influence score to be a priority, but is only a 4 on the risk level, so it lands in Quadrant 2. Even though the probability of Aaron's father retiring is low, the risk lands on the border in Quadrant 4 because the impact would be high if he did retire; so it should be taken seriously.

The risk tool allows you to print or save your results. It is also a good idea to readjust your numerical estimates to make sure the charts represent what you really think.

REFERENCES

Baquet, A., R. Hambelton, and D. Jose. 1997. *Understanding agricultural risks: Production, marketing, financial, legal, and human resources.* Washington, DC: U.S. Department of Agriculture Risk Management Agency.

Clark, R., and J. Timms. 1999. *The better practices process: Focused action for impact on performance.* Gatton, Australia: The Rural Extension Centre.

Clemen, R., and T. Reilly. 2001. *Making hard decisions with DecisionTools®*, 2nd ed. Pacific Grove, CA: Duxbury-Thomson Learning.

Greenspan, A. August 29, 2003. Monetary policy under uncertainty. Remarks at a symposium sponsored by the Federal Reserve Bank of Kansas City, Jackson Hole, WY.

Hall, D. C., T. O. Knight, K. H. Coble, A. E. Baquet, and G. F. Patrick. 2003. Analysis of beef producers' risk management perceptions and desire for further risk management education. *Review of Agricultural Economics* 25: 430–448.

Hardaker, J. B., B. M. Hurine, and J. R. Anderson. 1997. *Coping with risk in agriculture.* New York: CAB International Publishing Company.

Harwood, J., R. Heifner, K. Coble, J. Perry, and A. Somwaru. 1999. *Managing risk in farming: Concepts, research, and analysis.* Washington, DC: U.S. Department of Agriculture, Economic Research Service, Market and Trade Economics Division and Resource Economics Division, Report 774.

Mickelsen, S., and L. D. Trede. 2001. Identifying and applying learning modes to risk management education to Iowa farmers. Iowa State University, 28th Annual National Agricultural Education Research Conference, December 12.

Musser, W., and G. Patrick. 2002. How much does risk really matter to farmers? In *A comprehensive assessment of the role of risk in U.S. agriculture*, edited by Richard Just and Rulon Pope, 537–556. Boston, MA: Kluwer Academic Publishers.

U.S. Department of Agriculture, Risk Management Agency (RMA). 1997. *Introduction to risk management: Understanding agricultural risks: Production, marketing, financial, legal and human resources.* Washington, DC: U.S. Department of Agriculture.

9 Step 5
Identify Risk Management Alternatives

Jay Parsons and Dana Hoag

CONTENTS

Like it or not, every business operator is a risk manager. Accounting for risk, even when there is a great deal of subjectivity, makes you think more deeply about how to manage it (Hardaker and Lien, 2005). Being lucky will help, but it's best to be prepared. Fortunately there are many ways to manage or control risks. In this chapter, we discuss several risk management tools and techniques and explain how to go about identifying which risk management alternatives you want to consider when making your final risk management plans (Chapter 11).

9.1 PART 1: RISK MANAGEMENT ALTERNATIVES FOR EWS FARMS

9.1.1 Objective

Many risk problems have numerous management alternatives. In the previous chapter, Aaron identified a number of risks to be concerned about and his ability to address them. That led him to the conclusion that he wanted to focus on strategies to manage corn price risk.

The objective of Step 5 is to identify risk management alternatives for EWS Farms that can be used to address corn price risk.

9.1.2 Strategy

The strategy in this step is to identify a set of useful management techniques, then organize them so that probability can be added in Step 6 (Chapter 10) and one can be selected in Step 7 (Chapter 11). We start by collecting information about possible risk management alternatives that can be used to address corn price risks. It is important to keep an open mind in the gathering phase before whittling down the choices to a reasonable set of alternatives. A technique that tends to work well in this regard is to do your own thinking first before gathering opinions from others (Hammond, Keeney, and Raiffa, 1999). Of course, this should be followed up by a comprehensive search for other input through the Internet, books, trade journals, and peers.

By doing your own thinking first, you are not allowing others to limit your thought processes. Relying too much on your first thoughts is a form of psychological anchoring. You do not want your first thoughts to include a conglomeration of opinions from others or you'll often never get to some of the more interesting alternatives in the back of your mind. Do your own independent brainstorming first and then talk to others. When you do talk to others, let them offer their thoughts and ideas before mentioning any of yours. This adds to the independent thought processes and enriches the list of alternatives.

Once the strategies you want to consider have been selected, they need to be organized in some fashion that combines the information into a format that can be used for risk management. We chose to manage price risk in Step 4 (Chapter 8) and will pick which management techniques we want to consider in Step 5, this chapter. The information from this chapter will be combined with the information about probabilities in Step 6 (Chapter 10) for making a final management choice in Step 7 (Chapter 11). We will use a decision tree to organize our information.

9.1.3 Implementation

9.1.3.1 Identifying Alternatives

Aaron's recent education in agricultural economics allowed him to independently form a fairly comprehensive list of risk management alternatives to address corn price risk. This list included:

- Cash market sales
- Forward contracting to the local elevator
- Hedging on the futures market
- Spreading out crop sales across the year
- Maintaining flexibility in regards to the timing of sales

By talking to family members, friends, neighbors, marketing experts that he knows at the University, and local market representatives, he was able to add to this list. Family members and other farmers provided ideas on what has worked in the past for them, current land leasing trends that could have an impact on market risk exposure, and their own sources of market outlook information. His university professors offered thoughts on how to put together a comprehensive marketing plan and their own opinions on where to find the best information.

In the end, Aaron concluded that formulating a comprehensive marketing plan was the way to go. In this plan, he decided to include three alternatives: selling grain on the cash market, using futures contracts to hedge on the national market, and forward contracting with the local elevator. This step made combining these alternatives into a marketing plan that maintains flexibility while spreading out crop sales a priority for addressing risk.

9.1.3.2 Decision Trees

Decision trees are an important tool that can provide a clear graphical representation of decision problems involving risk. The main value of decision trees lies in their construction (Rae, 1994), which requires the decision maker to identify all relevant courses of action, events, and payoffs while being clear on the order of time in which they occur. As we saw in Chapter 8 with influence diagrams, decision trees use a standard set of symbols to represent various elements of the decision problem. Branches or forks in the tree represent alternative actions or events. A rectangle is used to represent a decision node, circles are used to represent a chance event node,

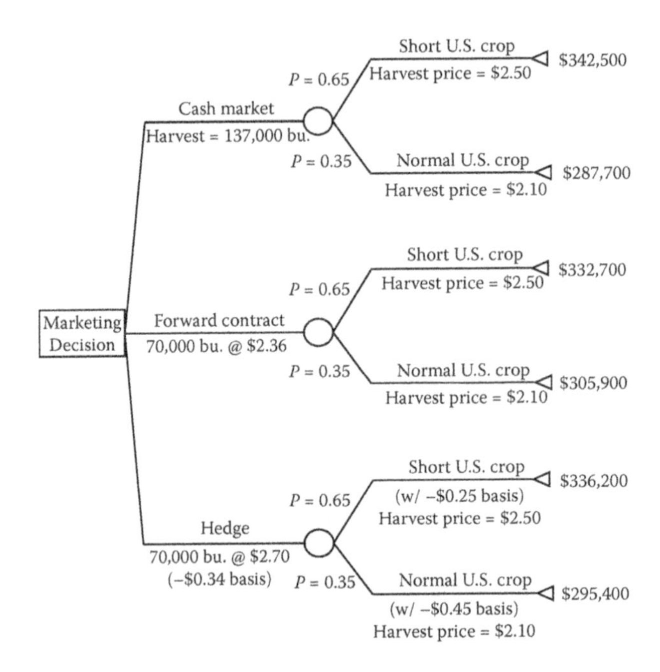

FIGURE 9.1 A simple decision tree of the EWS Farms marketing decision.

and triangles represent payoffs. In general, action forks and event forks appear in the tree from left to right in the order they occur over time.

Consider Aaron's decision problem about how to market the Sprague's corn crop. He's narrowed it down to three alternatives including selling grain on the cash market, using futures contracts to hedge on the national market, and forward contracting with the local elevator. In Figure 9.1, a decision tree is displayed representing three possible marketing alternatives for 137,000 bushels of corn produced on EWS Farms. For simplicity, the model only considers the following three marketing alternatives: (1) market the entire 137,000 bushels on the cash market at harvest, (2) forward contract 70,000 bushels at $2.36 per bushel and market the remaining 67,000 bushels on the cash market at harvest, (3) hedge 70,000 bushels at $2.70 (a –$0.34 basis) now and market the crop on the cash market at harvest. In reality, there is a very large number of marketing combinations that could be used involving the cash market, forward pricing, and/or hedging. The decision tree forces the decision maker to define specific choices, thus clarifying the analysis. Also, notice that the figure depicts these choices as mutually exclusive alternatives, meaning that only one can be chosen. This is not always the case and the decision tree is an important tool that can be utilized to distinguish whether or not this is true.

In terms of the market risk, or chance event, Figure 9.1 shows there are two possible outcomes. Aaron speculates that there is a 65% probability of a short U.S. crop resulting in a harvest cash price of $2.50 and a local basis of –$0.25.* There is a 35% probability of a normal U.S. crop resulting in a harvest cash price of $2.10 and

* We discuss how to estimate probabilities in the Chapter 10.

Risk Outcomes	Harvest Cash Price	Harvest Basis	Probability	Decision Alternatives #1 Cash Market 137,000 bu.	#2 Forward Contract 70,000 bu. @ $2.36	#3 Hedge 70,000 bu. @ $2.70 (–$0.34 basis)
Normal U.S. Crop	$2.10	–$0.45	35%	$287,700	$305,900	$298,200
Short U.S. Crop	$2.50	–$0.25	65%	$342,500	$332,700	$339,000
			Expected Value	$323,320	$323,320	$324,720
			Standard Deviation	$38,749	$18,950	$28,850

FIGURE 9.2 A payoff matrix for three marketing plan alternatives.

a local basis of –$0.45. The six payoffs for each decision alternative/event outcome combination are shown in terms of total marketing revenue. Notice that none of the decision alternatives dominates another alternative. That is, when comparing the payoffs, no alternative is better than any other alternative 100% of the time.

The decision problem in Figure 9.1 can also be summarized in a payoff matrix. The advantage of a payoff matrix is that it is more compact, less graphically intense, and it naturally fits into a spreadsheet tabular form where summary statistical calculations can be presented with little additional work. Figure 9.2 shows the payoff matrix for Aaron's decision problem using the six possible payoffs displayed down the right-hand side of the decision tree in Figure 9.1. Notice that the last two lines display the expected value and standard deviation for each of the three alternatives being considered. This information, and more, can be extracted from the payoff matrix to help rank best risk management actions. That is, the evolutionary process of identifying a set of management practices, narrowing them down to the ones you want to consider, and then building a decision tree sets the information up for constructing the payoff matrix, which is what we use in Step 7 to rank best management practices. A more complete description of the payoff matrix is provided in Chapter 11 where Step 7 is described.

We learned in Chapter 2 how the probability distribution, or PDF, could be used to describe risks. The PDF contains a lot of information about how a variable is distributed, including the mean, mode, variance, and confidence intervals. Using a PDF, in histogram form, a risk profile for each of the three marketing alternatives can also be developed from the decision tree or the payoff matrix. The risk profiles for the EWS farm decision tree are presented in Figure 9.3. The figure labeled Decision Alternative 1 represents the cash marketing strategy for all 137,000 bushels produced by the Spragues. Alternative 2 represents the strategy alternative that involves forward pricing 70,000 bushels and selling the remainder on the cash market. And Alternative 3 represents the strategy that involves hedging 70,000 bushels.

At first glance, the risk profiles do not look that different. This is partly due to the scale drawn in the graphs. A closer look at the numerical summary statistics in Figure 9.2 shows the differences more clearly. Alternatives 1 and 2 have identical average returns, but alternative 2 has less risk, since the standard deviation is lower. The third alternative has the highest average return and medium risk. Methods for

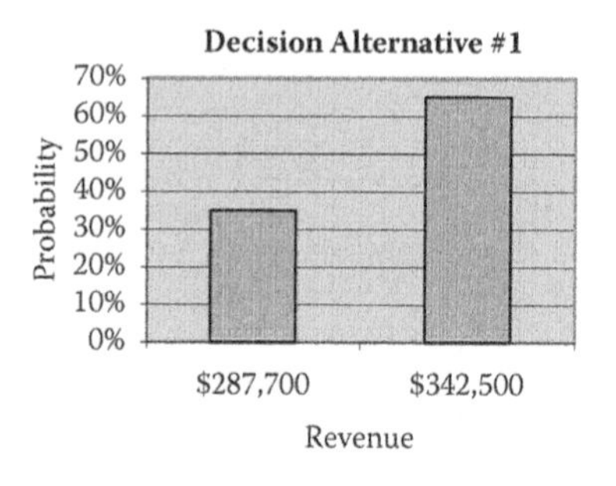

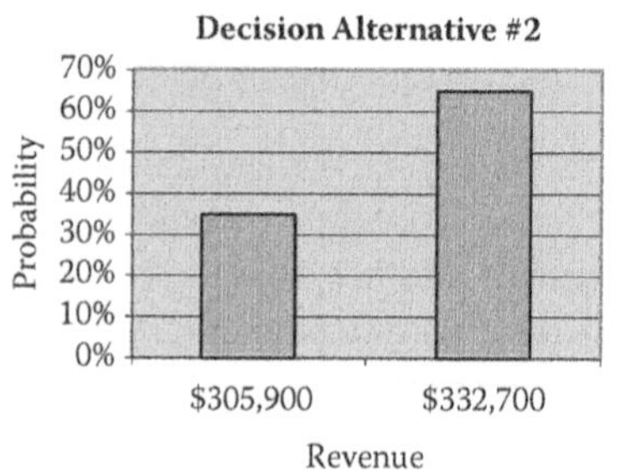

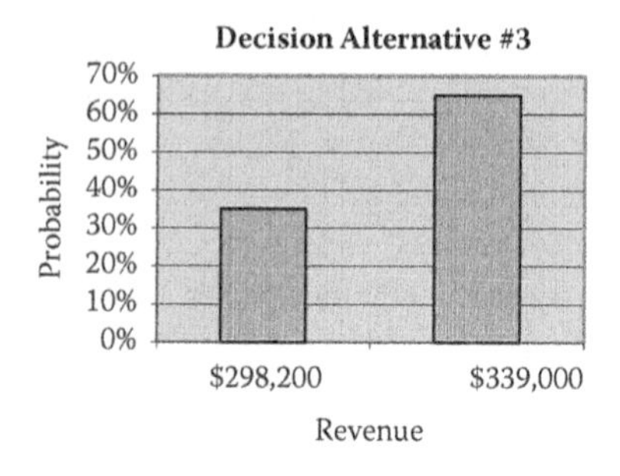

FIGURE 9.3 Risk profiles for three marketing plan alternatives: cash, forward price, or hedge. (Based on information in Figures 9.1 and 9.2.)

ranking these alternatives will be addressed more thoroughly in Step 7: Choose the "Best" Risk Management Alternative. For now, think about whether the cash alternative is better for any reason. If not, then the Spragues should put together other alternative marketing strategies that compete better with alternatives 2 and 3. This is the type of thought process that Aaron will be undertaking as he puts together his marketing plans.

9.2 PART 2: THE FUNDAMENTALS OF IDENTIFYING RISK MANAGEMENT ALTERNATIVES

The previous section showed how the Spragues chose the management alternatives that they wanted to consider and how to represent them in a decision tree, payoff matrix, and risk profile. In this section, we present a more thorough and expansive examination of the different strategic approaches for managing risk and the specific tools and techniques that are available to address risk through one of these approaches.

9.2.1 Four Strategic Approaches to Managing Risk

There are four basic ways to manage or control risks:

- Avoid risks
- Transfer risks
- Assume or retain risk
- Reduce risk

These strategies show generically where risk is managed. We will discuss each one then move into a discussion about the tools and techniques that can be used to address risk in a later section.

One way to manage risk is to *avoid* it wherever possible. Metaphorically, people who keep their money under their mattress are avoiding the risks of letting others hold their savings. While it is understandable that people want to avoid risks, extreme risk avoidance can have extreme impacts, like significant losses in income potential and even introducing the decision maker to new and potentially greater risks. Again, metaphorically, putting your money under a mattress eliminates any earning potential and exposes you to the risk of fire or flood. A business example of avoiding risks would be to avoid crops like onions or potatoes, which net wilds swings in earnings, in order to pick relatively safer crops like hay.

Someone who does not like risk might want to *transfer* his or her risks to someone else. This is often a better option than risk avoidance if there is an appropriate market to which to transfer the risk. There are many formal modes for transferring risks. In the case of insurance, risk is transferred from an individual to a corporation that can tolerate more risk. A producer pays a firm more than the expected indemnity to avoid the risk. The company earns a living from the risk premiums. It can afford to pay for accidents and catastrophes because it is pooling risks over many people, or types of coverage. Even insurance companies are required by law to reinsure so that they maintain diversified pools. A company that specializes in hurricane insurance, for example, swaps some of its coverage with a company that covers automobile accidents so that neither company is caught short should a crisis occur that is too large for them to handle, like the devastation that insurance companies had to pay for in New Orleans when Hurricane Katrina came through in August 2005.

The other important mechanism for transferring risk in agriculture is the futures market. Producers swap risks with speculators by hedging and with options. The market for swapping is large enough to distribute the risks across many people.

Those people who do not mind dealing with risk may want to *assume* or *retain* their risk. The motivation for putting up with risk is that there is usually a positive correlation between risk and return. Those people who take on more risks, though they have more ups and downs in their lives, make more money in the end. That is, they make more money *if* the ups and downs don't put them out of business. A study at Colorado State University (Jianakoplos and Bernasek, 1998) showed that because women are more risk averse than men, on average, they end up with less savings at older ages. People that assume risks can take actions to make them able to bear them, like good access to capital, as is discussed later.

The fourth and final way to manage risk is to *reduce* it. Avoiding risk or transferring all of the risk to others can be thought of as one extreme of reducing risk to zero. The other extreme is retaining risk but doing nothing about it. However, whenever risk is present, people will usually want to take action to reduce it to whatever extent possible in an economically feasible manner. For example, instead of totally avoiding risky crops or assuming the most risk I can by growing the riskiest crop I can find, I could reduce my risk by diversifying my crop selection. I could grow some onions and some wheat, for example. This form of self-protection reduces but doesn't eliminate the probability that a loss will occur. Other examples of risk-reducing behavior include planting a drought-resistant variety of seed in a drought-prone area, preventative maintenance performed on equipment, or spreading out sales like EWS Farms is looking to do through market planning to reduce the probability of receiving a low price.

9.2.2 Four Ways Managers Influence a Risk Profile

Using PDFs, we can show that there are four basic ways that managers can influence risks (e.g., Fleisher, 1990). As shown in Figure 9.4, any starting risk profile can be represented by a solid-lined PDF. For example, think of this starting risk profile as the price distribution for corn. A risk manager can influence risks by taking an action that shifts the distribution, as shown by the dotted lines. These shifts could be favorable, unfavorable, or ambiguous.

The first way that a producer can shift risk (shown in panel 1 of Figure 9.4) is to squeeze down the dispersion, while maintaining the mean. Any alternative practice that has the same mean but a lower variance than other alternatives will be preferred by all managers. On average, the outcome is the same and it is better off with respect to risk. Our price distribution might be narrowed by hedging the crop, for example. The average price might be about the same, but the benefits of high prices and losses from low prices would be passed on to the futures market, shrinking the distribution. Other ways to shrink the distribution include new technology such as better fertilizer or new crop varieties.

The second way a manager might influence risk is to raise the mean without affecting the dispersion, as shown in panel 2. Again, this would be universally preferred since a producer is no worse off on risk, but has a better average. A producer might raise his average corn price without reducing his variance by storing the crop. As another example, buying specially bred heifers instead of raising your own replacements might raise mean weaning weight without having much impact on the dispersion or variance around the mean weaning weight.

The third way risk could be influenced, as shown in panel 3, is to skew the distribution. Skewing a distribution means to push the probability to one end or the other of a distribution. The example shown depicts more probability of higher prices. The storage example above might skew the distribution rather than shift it. Another

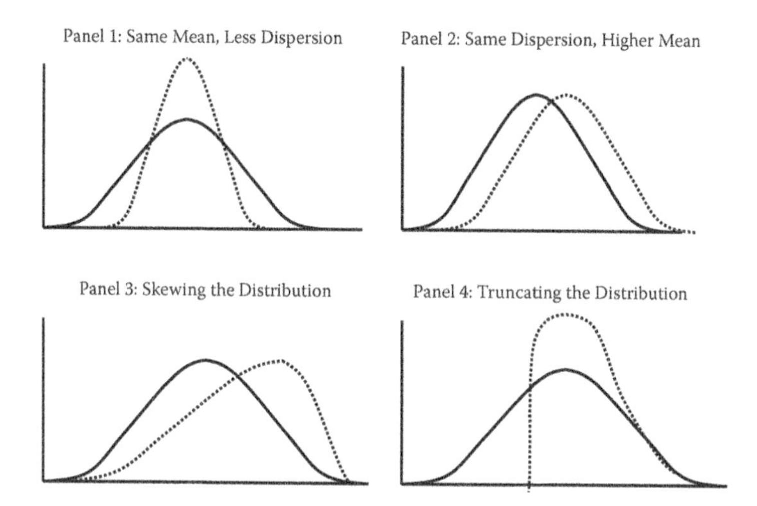

FIGURE 9.4 Four ways that risk management strategies can alter probability distributions.

example is an irrigation system that reduces the probability of getting low yields and increases the probability of getting higher yields.

The fourth way is to truncate the distribution. This essentially cuts off the chance of any occurrence beyond a certain level. The example shown in panel 4 depicts a case where low outcomes are not possible. Insurance is the most common way people truncate a distribution because it protects us on the downside but does not limit us on the upside. However, there are many other ways to truncate a distribution. For example, buying options on the futures market can protect a producer on the downside without limiting the upside. Hedging, on the other hand, limits both the up- and downside (see Chapter 13).

Finally, it is important to point out that there is a "fifth" way to influence risk, which is an adaptation of the first. A producer who is successful at totally eliminating risk creates a degenerate PDF. This is the same thing as shrinking the variance in panel 1 until it is infinitesimally small, forming a vertical line over the value. In more practical terms, no risk implies a single value. A producer who forwards prices might consider himself risk free.

9.3 PART 3: TOOLS AND TECHNIQUES FOR MANAGING RISKS

We are now ready to discuss specific tools to manage risk, including avoiding, retaining, reducing, or transferring it. As in Chapter 8, a good place to start is to look at what other producers and managers have said in surveys. Table 9.1 shows responses to manage risk mentioned in three different surveys. It is interesting to note what people ranked the same and equally interesting to see what they ranked differently in the separate surveys.

Two of the surveys indicated that being a low-cost producer as the best way to manage risks. The third indicated that having cash reserves was the most important thing to do. Forward pricing and using the futures markets were also important, but

TABLE 9.1
Ranking of Responses to Risk

Risk Management Response	Relative Ranking by Study (1 = top): Top Crop Manager Workshop 1999 (Musser and Patrick, 2002)	Beef Producers in Texas and Nebraska (Hall et al., 2003)	National Study-ERS-Agricultural Resource Management Study, 1996 Data (Harwood et al., 1999)[a]
Being a low-cost producer	1	1	
Liability insurance	2		
Government program participation	3		
Forward contracting or pricing	4	ns[b]	3
Flexible production technologies	5 (tie)		
Futures—hedging, options	5 (tie)	ns	4
Life insurance	6		
Debt-leverage management	7		
Maintain animal health		1	
Cash/credit reserves		3	1
Off-farm investments		4	
Specialization in management		5	
Diversify enterprises		ns	2

Source: Harwood et al., 1999.
[a] ERS survey based on actual use rather than on preferences.
[b] *ns* means rated, but neutral or less importance in the study.

recall that these same surveys in Chapter 8 indicated price risk as the number one source of risk. The top manager survey indicated that liability insurance was the second most important way to manage risk. This is probably a sign of a more litigious time and of the subjects of the study. Beef producers had a very different list from other producers. These surveys indicate different priorities because they are taken in different regions, at different times, among different types of producers. This shows the importance of maintaining your own independence when evaluating what others have said.

Now we will work our way through a list of many different responses to handling risk management. Each of these alternatives will be presented with a short description and examples that tie them back to the four ways to manage risks. As examples of some of the most commonly used responses by producers, the first four alternatives on our list will be given a more thorough treatment than the rest.

9.3.1 Crop and Revenue Insurance

The use of insurance for farmers dates back to the fifteenth century. The Monte dei Paschi, which became one of the largest banks in Italy, was established in Siena in 1473 to serve as an intermediary for agricultural cooperatives set up to insure one another against bad weather. As a member of the cooperative, farmers in areas with a good growing season would agree to compensate those whose weather had been less favorable (Bernstein, 1996).

Insurance is generally a way to transfer a risk to others. An insurance program is called *actuarially fair* when the total premiums collected by the insurer over time exactly match the indemnities paid out. One of the first conditions that will determine if an insurance program is going to exist in the marketplace is whether or not enough information exists to determine the probability distributions and magnitudes of expected losses in order to calculate the actuarially fair premium rate.

Multiperil crop insurance policies have been around in the United States for over one hundred years protecting farmers from potential yield losses. During that time, some private crop insurance companies failed because of their inability to establish a satisfactory actuarial foundation for the premiums. Today, a quasi-public institution called the Federal Crop Insurance Corporation (FCIC) makes multiperil crop insurance policies available to producers either through direct sales or by acting as a pooling agent for private companies. As a pooling agent, the FCIC provides reinsurance for the private companies along with the actuarial structure they need for setting premiums.

A multiperil crop insurance policy is based on four to ten years worth of yield data or Actual Production History (APH). A producer insures a percentage of the APH average yield to determine the guaranteed yield. If the actual yield falls below the guaranteed yield, the producer is entitled to an indemnity payment. Of course, the higher the percentage of the APH insured, the greater the premium will be for the policy based upon the actuarial calculations that determine the expected indemnity payouts.

Because premium rates are set according to some average level of risk such as county-level data, crop insurance leads to a classic adverse selection problem. Adverse selection occurs when the producer has more information about their individual risk of loss than the insurer. Two producers signing up for crop insurance for the first time who have established the same APH average and chosen the same coverage level will have similar premiums even though one producer may have a much more volatile production history and (implied) probability of a loss occurring. This may lead to higher participation rates among producers with more variable yields and a riskier pool of participants overall. Goodwin (1993) verified this using county-level data for Iowa corn by determining that the elasticity of demand for insurance was significantly smaller in riskier counties.

In addition to adverse selection, a moral hazard problem also affects the viability of crop insurance. A moral hazard occurs when producers have the ability to influence their expected indemnity by undertaking certain actions. Hardaker et al. (2004) describe moral hazard as "a form of post-contractual opportunism." It is an asymmetric information problem where the insured producer has the ability to do things that increase the probability or the magnitude of the indemnity payments. The presence of insurance can lead the producer to take on more risk than he would if insurance were not available. For example, we've heard producers justify planting a crop into marginal or poor growing conditions based on the fact that it was insured. Smith and Goodwin (1996) confirmed that insured Kansas wheat farms spent an average of $4.23 per acre less on fertilizer and chemical inputs than did uninsured farms.

Adverse selection and moral hazard are two of the market failures often cited as reasons justifying the government's involvement in the administration of the multiperil crop insurance program (Harwood et al., 1999). In short, the existence of market failures leads to a riskier pool of participants, which results in an actuarially fair premium rate that is higher than many producers are willing to pay. To encourage participation, the government provides a premium subsidy that lowers the cost of insurance to producers. This has varied over the years and across policy options, but has grown significantly since the Federal Crop Insurance Reform Act of 1994 and is now as high as 59% for some policies. Government subsidies also cover some delivery and administrative expenses that a private insurance company would normally add onto the actuarially fair rate to determine the total premium.

So, farmers have a couple of major incentives for participating in a multiperil insurance program. In addition to the risk protection they provide to reduce the negative impact of low yields, government subsidies lower the premium costs below the actuarially fair rate, resulting in a positive expected return to the producer.

Two other market conditions also exist that are regularly mentioned as reasons for participation. One of these is that many producers have the purchase of crop insurance placed on them by a lender as a condition for receiving short-term production loans or long-term investment loans. Another reason is that producers involved in forward contracting their crop can use crop insurance to avoid yield risk and the implied threat that poses to their ability to fulfill the contract.

Multiperil policies are yield guarantee policies only, and generally do not provide market protection. However, there have been a number of insurance products developed in the last twenty years that do provide some market protection. Crop Revenue Coverage (CRC), Revenue Assurance (RA), and Income Protection (IP) are examples of policies that establish revenue guarantees through price and yield coverage level selections. Of course these types of coverage are more expensive on a cost/dollar of coverage basis than a multiperil policy because they protect against downside risk affecting either one of the factors in the revenue equation rather than just yield. Recently, some new whole-farm revenue products such as Adjusted Gross Revenue (AGR) and AGR-Lite have also been introduced. These whole-farm policies can be combined with and serve as a pseudo-umbrella policy for individual commodity policies in a risk management plan.

All of the insurance products mentioned to this point are, at this time, government supported through premium subsidies, expense reimbursements to the insurance companies, and/or FCIC reinsurance policies. There are also a few private-market insurance products available. One of these is hail insurance. Hail occurs frequently enough in some parts of the country, like the eastern plains of Colorado, to encourage a large number of producers to purchase hail insurance. Unlike multiperil and most other forms of crop insurance, producers can wait to purchase hail insurance until after the growing season has started. For wheat producers like EWS Farms, it is typical to wait until well into the growing season before making the decision about whether or not to purchase hail insurance. This decision will largely be based upon the perceived yield potential of the crop still in the field and the implied risk hail imposes. Producers pay a premium that covers the insurance company's expected indemnity, administrative costs, and profit. In return, if their crop is damaged by hail, they receive an indemnity that covers their losses subject to their selection of coverage level.

There are a number of reasons why hail insurance is a viable insurance product in the private market. These reasons include the fact that losses from hail are not correlated over time. Hail losses generally do not cover a wide area like drought and some other hazards. Hail losses also cannot be directly influenced by managers' decisions. However, hail losses can easily be measured using well-established and standardized methods. This leads to an accurate calculation of an actuarially fair premium and a market incentive for the producer to pay a premium equal to the actuarially fair rate plus a risk premium that covers expenses and profits accruing to the private insurance company. The producer is willing to pay the risk premium in exchange for the downside risk protection the policy provides.

9.3.2 Forward Contracting

For well-established commodities like corn, wheat, cattle, etc., numerous marketing alternatives are available to manage price risk. A forward contract is an agreement between a buyer and a seller (producer) that spells out an agreed-upon price, quantity, and delivery date for a commodity. Many producers use forward contracts with a local buyer to lock in a delivery price for their grain at a later date. Two of the most common reasons for doing this are to lock in a harvest delivery price for crops that are still growing or to make adjustments in their delivery date for tax purposes.

Forward contracting transfers the market price risk from the seller to the buyer. The contract *bid* price offered by the buyer is determined by the appropriate futures price as coordinated with the delivery date minus a *basis* as determined by the buyer. The basis represents the marketing costs for the local buyer with the national exchange.

Forward contracting does not transfer production risk. The producer retains the risks associated with yield quality and quantity. If the commodity does not meet the standards of the buyer, the price received may be severely discounted or the delivery may be rejected altogether. If the producer is unable to deliver the quantity and quality called for in the contract, they could be forced to purchase the commodity on the cash market to fulfill the contract. Referring to Figure 9.2, forward pricing squeezes the size of the PDF in return for reducing the mean.

9.3.3 Futures and Options

Unlike forward contracting, futures and options contracts do not involve the producer physically delivering the commodity. They both represent positions taken on the national exchange and are transacted through a commodities broker. As such, they involve brokerage fees and standard contract quantities. When a producer uses these tools to offset a local cash market it is called *hedging*. All of this is explained in more detail in Chapter 13.

These numerous marketing alternatives require a significant amount of effort to understand. The goal of transferring risk to others is consistent, but differences in how they work have an impact on how they manage risk. For example, some marketing alternatives like hedging do nothing to manage *basis* risk. Aaron has studied all this in college and has identified it as an area where he has a comparative advantage in adding value to the EWS Farms operation. It is easy to understand why he would like to concentrate on formulating a risk management plan as the risk management alternative for EWS Farms.

9.3.4 Diversification

Diversification can be used to reduce risk in a farming operation. It can take many forms including enterprise selection, market timing, geographic locations, business structures, investments, and vertical integration. Aaron's plans to address risk through the formulation of a comprehensive marketing plan will likely involve diversification. Locking in the price or selling your crop at multiple points throughout the year is commonly called *spreading out your sales*. While spreading out your sales guarantees that not all of your crop will receive the highest price, it also guarantees that at least some of it will not receive the lowest price either. Marketing plans usually involve the use of multiple marketing alternatives like combining cash sales with forward contracting, for example. This too is a form of diversification and reduces risk to the operation because not all of your eggs are in one basket.

Diversifying your operation by producing a number of different crops and/or livestock is designed to reduce risk by lowering the probability of having a complete production failure. For example, an operation that raises cattle and wheat will have less exposure to the risk associated with summer hail storms than a producer who raises nothing but wheat. However, you should be careful to not add an inefficient enterprise to your operation just for the sake of diversification. An enterprise that loses money year in and year out generally cannot help you enough via diversification to justify including it in your operation. Also, care should be taken not to add enterprises that are too much alike thinking that they are going to add diversification. For example, two crops that are planted and harvested at similar times may expose you to more production risk than if you just picked one and focused on being a very efficient producer of that crop.

Diversifying your operation by producing a number of different crops and/or livestock can also lower the probability of having a complete marketing failure. For example, livestock and feed grain prices tend to move in opposite directions. When feed grain prices are high or moving higher, feeder cattle prices tend to be low or moving lower. Having both enterprises in your operation will lower the risk associated with poor market conditions.

Disbursing the farming operation over a wide geographic area is another form of diversification. This practice reduces the risk of adverse weather conditions affecting production. For example, hail storms tend not to cover a wide geographic area. So a producer with land spread out over a vast geographic area is unlikely to suffer a complete crop loss from one hail storm. Transportation costs have to be considered when using this strategy because even though the geographic disbursement might be appealing, it comes with the cost associated with increased travel.

Diversification is one of the most commonly used risk management strategies. In fact, many of the alternative responses to risk mentioned in this chapter meet the definition of diversification. The idea behind diversification is to lower your exposure to risk by selecting a mixture of activities that are low or negatively correlated with one another (Hardaker et al., 2004). If done well, the effect is to stabilize your net returns.

However, is diversification always good? Duchene (2007) states that diversity has its dark side because by definition, diversification adds complexity. Too much complexity can lead to chaos, added expenses, poor management, and an overall mess. Magnusson (1969) describes a "range of diversification" inside which diversification makes sense, but outside of which specialization becomes optimal.

Magnusson's range of diversification is defined by the ratio of the difference between the means of the activities compared to their correlation coefficient. Diversification from one activity to two activities makes sense if the difference between the means is not too big and the correlation coefficient is not too high. If the means are too far apart relative to the variance reduction realized, specialization in the activity with the higher mean makes more sense. A producer with two activities whose returns are not perfectly correlated but have the exact same mean will, of course, achieve the same average return with a lower variance by engaging in both activities rather than just one. However, most activities will not have the same expected return so it becomes a standard risk–return trade-off as the producer grapples with the decision on using the lower return activities in order to stabilize returns.

Finally, it is important to understand that diversification is a firm-level decision with a lot more to consider than just means and correlations between various activities. There is a lot that goes into an agricultural operation and it is no accident that most operations exhibit some diversification behavior. The benefits of diversification are sometimes hidden within the operation but discernable nonetheless. Musser and Patrick (2002) describe one study that reported a correlation coefficient between returns for soybeans and corn in the South as 0.68, which would imply that it would not be a good combination for diversification since these two crops tend to move in common directions. However, because of joint production relationships (pesticides reduced by eliminating corn rootworm problems; soybeans leaving nitrogen for corn; machinery and labor inputs that are highly complementary) this rotation increases yields, while reducing insecticides, nitrogen fertilizer, machinery, and labor. There is no risk–return trade-off. Nevertheless, a producer is obviously better off with a diversified corn–soybean rotation.

9.3.5 Technology

Investment in technology is a producer activity that reduces risk in a number of ways. A producer who buys a new piece of equipment is likely to have significantly decreased the probability that equipment problems will severely decrease production. Utilizing the latest technology can increase the quality of the product harvested and decrease the time it takes to harvest it. It can also decrease the probability of production losses during the growing season. For example, investing in good irrigation equipment will lower the probability of water problems decreasing crop yield. A sheep producer who invests in a really good guard dog is also using technology to reduce risk because it lowers the probability that they will lose a lot of animals to predators. Biotechnology is getting a lot of attention in recent years. A crop producer who grows the latest varieties will likely be reducing risk in some important respect. Drought-resistant varieties, herbicide-tolerant varieties, and disease-resistant varieties are just a few examples of genetic improvements in seed targeted at reducing risk. Where in the past, much of the focus may have been on increasing yield, recently genetic research has focused much more on stabilizing yield by reducing downside production risk.

9.3.6 Flexibility

Maintaining flexibility means you have the ability to make adjustments to meet changing conditions. Coordinating crop rotations, machinery investments, labor, etc., to allow flexibility in decision making can not only improve the bottom line, but also avoid financial disasters brought on by being in the wrong place at the wrong time with no good alternatives to turn to. An example would be a cow-calf producer who maxes out their feed resources on the cow-calf enterprise only to be faced with a long-term drought. With no flexibility in finding reasonably priced additional feed resources, the producer could be faced with the unfavorable task of selling off part of the cow herd in a down market. Selling cows has long-term implications that could take a while to overcome. Whereas, if the producer had maintained a little flexibility either in feed resource alternatives or by running some yearlings with a smaller cow herd, more desirable responses to the drought situation may have been available. Depending upon how it is done, maintaining flexibility will either reduce risk or avoid it altogether. One of the most common and simple ways to maintain flexibility is to rent rather than own.

9.3.7 Excess Capacity

Excess capacity simply means to have greater ability to address a problem than you think you will need. Excess capacity can reduce or eliminate risk in many different situations. As alluded to above, excess feed capacity can reduce the risk associated with drought conditions. Excess machinery capacity can reduce the risk associated with machinery breakdowns. Of course, potential losses should always be weighed against the cost of owning and maintaining the extra capacity. This cost includes the opportunity cost associated with having the dollar value of the excess capacity invested elsewhere. Maintaining adequate liquidity by having cash reserves or having assets that can easily be converted to cash is another example of excess capacity as is maintaining a credit reserve that can be accessed in a time of need.

9.3.8 Off-Farm Employment

Off-farm employment can help a producer avoid financial risk. The cash flow associated with off-farm employment has kept a number of producers in the farming business over the years. While it may not be the most desirable response to risk, it can be the most practical one in some cases. This is especially true when the producer or a family member possesses skills that make him or her very employable off the farm and/or the farm isn't big enough to justify some particular risk-reducing alternatives.

9.3.9 Information

Obtaining additional information reduces risk or can help you avoid risk with your decision making. Crop production information and animal performance information can help you reduce the risks involved with production decisions. Obtaining additional market information can help you reduce the risks involved with marketing decisions. Up-to-date financial information helps you reduce the risk of making a bad financial decision. In general, more information is always better than less when it comes to risk. More information will not necessarily guarantee things turn out the way you want but it will increase the odds.

9.3.10 Low-Risk Enterprises

Studying the crop and livestock enterprise alternatives available to you and selecting the ones that have a low variation in returns from year to year can reduce risk considerably. These low-risk enterprises may earn their low-risk status by exhibiting relatively stable yields, prices, or a combination of both. For example, irrigated alfalfa hay is viewed as a low production risk enterprise compared to irrigated corn. Hay is harvested multiple times throughout the year, and therefore exhibits a less variable yield from year to year than corn.

On the market side, government programs limit the downside market risk associated with commodities like wheat and corn. Proso millet, for example, has a very volatile market compared to traditional grain markets. Commodities that have had government policies in place over the years that stabilize prices are much more widely grown than most commodities that have not had government intervention. Another example would be a livestock producer who chooses to raise cattle rather than buffalo because of the more established market for cattle where he can forward contract much of his output if he desires and reduce market risk.

9.3.11 Low-Risk Production Practices

Using low-risk production practices, such as irrigation, can help stabilize your income by stabilizing yields. Fallow rotations and conservation tillage practices on dryland cropland are other examples of low-risk production practices. In livestock production, vaccinations are considered a low-risk production practice. Some low-risk production practices may not result in maximum yields or maximum profits. However, if the benefits they provided from a risk reduction standpoint outweigh the cost associated with lower potential yields, producers will often adopt the low-risk practices for their own long-term benefit.

9.3.12 Land Lease Arrangements

Negotiating land lease arrangements can help reduce risk in a number of ways. Producers like crop share agreements from a risk reduction standpoint because the land owner shares in the yield variability. The percentage shares stay constant from year to year so the producer's lease costs go up and down with yield variability. In poor production years, lease costs are inexpensive because the actual quantity given to the land owner is reduced. For example, in the extreme case, if a wheat producer with a crop share agreement is hailed out one year, the rent due to the land owner is zero. However, in good years, the lease costs go up, of course, but the producer is often happy to give up a little on the upside in exchange for being able to share downside risk. Crop share lease agreements also reduce market risk. If market prices are low, the land owner as well as the producer has to deal with it. So, lease payments are essentially reduced with the lower market conditions. Some producers have started to negotiate flexible cash rent agreements. In these agreements, the land owner shares in the market risk but the producer retains all of the production risk. The producer agrees to pay rent based on a fixed quantity of production times whatever the harvest price turns out to be.

Land leasing strategies can also be used to reduce financial risk. This occurs as a by-product of shared production and/or market risk with the land owner. However, a producer can also diversify by leasing from several different land owners in order to reduce the risk that one land owner can severely impact the operation by not renewing a lease.

9.3.13 Renting or Leasing Equipment

Renting or leasing equipment can help a producer avoid the financial risk associated with equipment loans. It can also reduce production risk if the leased equipment represents more reliability than the alternative equipment that would be purchased. The producer must be careful not to enter into an unfavorable rental agreement, but generally, leasing equipment provides more flexibility and an overall reduction in risk to counteract the extra operating cost.

9.3.14 Forward Pricing Inputs

Use of forward pricing inputs is a response to risk that can stabilize costs. For example, recent years have seen a continuous increase in fuel and fertilizer costs. Grain producers who forward price these inputs can better establish their production costs. This reduces the uncertainty in some future decisions and stabilizes income overall.

9.3.15 Spreading Out Sales

Spreading out sales is a form of diversification and it can take a number of different forms. A marketing plan like the one Aaron is putting together will usually involve a strategy that spreads out sales over time to some degree in an effort to stabilize the average price received. A livestock producer who retains part of their weaned animals is spreading out sales from that year's crop. The fact that the retained animals will be sold into a different market also provides enterprise diversification, even though that may not be foremost on the producer's mind. Retaining animals obviously involves retaining risk, but spreading out sales provides a counteracting reduction in risk as long as some of the weaned animals are sold on the cash market at weaning. Whether it is livestock or crops, the main advantage of spreading out sales is the ability to hit different parts of the annual price cycles.

9.3.16 Business Structure

Incorporating all or part of your operation can financially isolate high-risk activities and limit your risk exposure. Your liability is limited to what you put into the business entity. For example, you could choose to keep your land out of your farming corporation. By leasing the land to the corporation, business continues as usual but the land is sheltered from the risks of the farming enterprise. These types of arrangements can help you avoid the risk of foreclosure or someone coming after the land value as a claim in a lawsuit versus the farming corporation.

9.3.17 Property Insurance

Maintaining adequate property insurance helps transfer the risk associated with casualty losses to an insurance company. This includes things like homeowner's insurance, insurance on farm buildings, and loss or damage insurance on vehicles and equipment. It is important to consider the size of the loss represented by what is being insured and the premium cost associated with the policy. Insurance companies will often provide very good coverage for a small annual premium because of the low probability that a loss will occur.

9.3.18 Liability Insurance

In today's litigious environment, liability insurance to protect against legal and social risks is becoming increasingly important. Employees or other parties can sue for large sums of money claiming injury or property damage. Neighbors can sue claiming their rights have been violated. Labor, especially young family labor, can pose a significant liability risk to an operation. It is important that should something bad happen, it is not compounded by the fact that you have inadequate liability insurance. Transferring some of that risk to others could be important to your long-term sustainability.

9.3.19 Health Insurance

You don't have to be a farmer or a rancher to understand the importance of health insurance. Health costs continue to rise. As new health procedures, medicines, treatments, and so forth continue to be discovered, it is almost a certainty that prices will continue to rise. Much of your health care cost risk can be transferred to an insurance company through a good health insurance policy. However, as with all insurance, the premium costs need to be weighed against the protection the policy provides.

9.3.20 Life Insurance

Life insurance can be used as a risk management tool in a number of creative ways. Most importantly, life insurance can be used to ensure the proper transfer of the farm to the surviving spouse or the next generation in the event of death. Like all insurance, life insurance is transferring risk. In this case, it is transferring the risk of death to an insurance company. However, life insurance can be managed much more along the lines of an asset than your typical insurance policy. Some life insurance policies, like whole life policies, actually gain in value over time. When a death occurs in a family, it is always a traumatic time. Adequate life insurance can be pivotal in ensuring that cash exists for the survivors to smoothly continue the operation.

9.3.21 Estate Planning

Life insurance is one aspect of good estate planning. Proper estate planning reduces the risk that the death of a key member of the operation will cause irreparable harm to its viability. Passing down the operation can be a sensitive subject, but it is best to address it while everybody is healthy and before people have entered the panic mode. Entire books and numerous resources have been dedicated to proper estate planning. We'll not go into any more detail here other than to say that establishing an estate plan, keeping yourself informed, and keeping it up to date can provide tremendous piece of mind.

9.3.22 Management and Key Personnel Backup Plans

Losing a key employee or a manager can be devastating. Having a backup management plan in place can reduce the dependency that the operation has on key personnel for its continued success. This is a little more comprehensive than just having insurance or an estate plan. Losing a critical employee or owner through death, illness, injury, or resignation poses a significant risk to most businesses. Plans should be made that allow for a smooth transition and continued operations. This can be done by training backup managers to take over, cross-training employees on different jobs, planning to bring in temporary help, or hiring custom work done. Involving others in the management of the business helps reduce the risk that losing key personnel can pose.

9.4 SUMMARY

There are many risk management alternatives that can be used to address most situations. We have listed several of the common responses to risk that managers have used over the years. This list is fairly comprehensive but not all encompassing. The tools available to manage risk are growing every day as are the creative ways to use them. A manager's number one priority in this step is to identify all of the alternatives to manage the risk in question so as not to eliminate the best alternative from consideration. This may seem obvious but it is easier said than done. A decision maker must approach this task with an open mind and not get locked in on a small set of common responses. A comprehensive list of alternatives is one of the big keys to making good decisions.

REFERENCES

Bernstein, P. L. 1996. *Against the gods: The remarkable story of risk.* New York: John Wiley & Sons, Inc.

Duchene, L. 2007. Are all your eggs in one basket? Diversity in the field and in your business plan is a key strategy for making your farm truly sustainable. The New Farm, Rodelle Institute, http://www.newfarm.org/features/2007/0507/diversity/duchene.shtml (accessed May 2008).

Fleisher, B. 1990. *Agricultural risk management.* Boulder, CO: Lynne Rienner Publishers.

Goodwin, B. K. 1993. An empirical analysis of the demand for multiple peril crop insurance. *American Journal of Agricultural Economics* 75(May): 425–434.

Hall, D. C., T. O. Knight, K. H. Coble, A. E. Baquet, and G. F. Patrick. 2003. Analysis of beef producers' risk management perceptions and desire for further risk management education. *Review of Agricultural Economics* 25: 430–448.

Hammond, J. S., R. L. Keeney, and H. Raiffa. 1999. *Smart choices: A practical guide to making better life decisions.* New York: Broadway Books.

Hardaker, B., and G. Lien. 2005. Towards some principles of good practice for decision analysis in agriculture. Working paper 2005-1, Norwegian Agricultural Economics Research Institute, Oslo, Norway.

Hardaker, J. B., R. B. M. Huirne, J. R. Anderson, and G. Lien. 2004. *Coping with risk in agriculture*, 2nd edition. Oxfordshire, U.K.: CABI Publishing.

Harwood, J., R. Heifner, K. Coble, J. Perry, and A. Somwaru. 1999. *Managing risk in farming: Concepts, research, and analysis.* Washington, DC: U.S. Department of Agriculture, Economic Research Service, Market and Trade Economics Division and Resource Economics Division, Report No. 774.

Jianakoplos, N. A., and A. Bernasek. 1998. Are women more risk averse? *Economic Inquiry* 36: 620–630.

Magnusson, G. 1969. *Production under risk: A theoretical study.* Uppsala, Sweden: Almqvist & Wiksells.

Musser, W., and G. Patrick. 2002. How much does risk really matter to farmers? In *A comprehensive assessment of the role of risk in U.S. agriculture*, edited by Richard Just and Rulon Pope, 537–556. Boston, MA: Kluwer Academic Publishers.

Rae, A. N. 1994. *Agricultural management economics: Activity analysis and decision making.* Wallingford, Oxon, U.K.: CABI Publishing.

Smith, V. H., and B. K. Goodwin. 1996. Crop insurance, moral hazard, and agricultural chemical use. *American Journal of Agricultural Economics* 78(May): 428–438.

FURTHER READING

U.S. Department of Agriculture, Risk Management Agency (RMA). 1998. *Building a risk management plan: Risk-reducing ideas that work.* Washington, DC: U.S. Department of Agriculture.

10 Step 6
Determining the Likelihood of Outcomes

Catherine Keske, Dana Hoag, and Eihab Fathelrahman

CONTENTS

In Chapter 2 we outlined the basic concept of a risk profile. There are many variables to consider in a risk profile such as weather, inputs, prices, and yields. Identifying the variables that can affect your operation and determining the likelihood that these events may occur can be daunting. As in other chapters, Chapter 10 is divided into three parts to provide hands-on experience in developing your own risk profile. In the first part, EWS Farms develops a risk profile for an operational decision. In this case, EWS Farms is looking at price and basis risk for marketing its corn. In the second part, the statistical terms and concepts behind risk profile development are defined and explained. The final third of the chapter provides the Risk Navigator tools, which can be applied to develop your own risk profile.

10.1 PART 1: DEVELOPING A RISK PROFILE FOR EWS FARMS

10.1.1 Objective

In Chapter 8, Step 4, Aaron's priority risk was corn price. It is well known that fluctuations in agricultural prices are a main source of risk to producers. In order to minimize or manage this risk, it is worthwhile to understand the range of prices the producer may face during the next season and the likelihood that each of these prices may occur. This is the process of developing a probability density function as we introduced in Chapter 2, commonly known as a PDF, which coincides with Step 6. The PDF is the core concept of a risk profile and it embodies the statistical attributes of a risk.

The objective of Step 6 is to develop a risk profile using the likelihood of various corn prices that EWS Farms will face in the next year.

10.1.2 Strategy

The strategy in this step is to collect historical information on the potential values for corn prices, use the information to develop a PDF, and then develop a risk profile from the PDF. As part of his strategy, Aaron used information from the futures market and the cash basis to define cash price: cash price equals future price plus the basis. (See Chapter 13 for more information.) Price and yield data were collected from both futures contracts and the local basis information from northeastern Colorado. Data for futures contracts were obtained from historical records for the December No. 2 Yellow Corn Contracts, from January 1995 to December 2004. The price basis information was collected from the historical records of the Holyoke Cooperative Association in Holyoke, Colorado. Farm-level yield information was also collected from annual EWS Farms and insurance reporting records.

10.1.3 Implementation

We will implement Step 6 for the Spragues in two steps:

1. Collect data on local, historical corn prices
2. Build a risk profile from the data

TABLE 10.1
December Corn Futures from the Chicago Board of Trade

Year	December Corn Futures by Week			
Week	11F	20F	29F	48F
1995	$2.62	$2.69	$2.91	$2.72
1996	$3.16	$3.58	$3.55	$2.71
1997	$2.93	$2.62	$2.48	$2.18
1998	$2.82	$2.59	$2.32	$1.89
1999	$2.46	$2.33	$2.15	$2.05
2000	$2.60	$2.54	$1.94	$1.99
2001	$2.45	$2.20	$2.29	$2.41
2002	$2.28	$2.32	$2.42	$2.43
2003	$2.39	$2.47	$2.12	$1.92
2004	$3.00	$2.90	$2.35	$2.06

10.1.3.1 Collect Data

A PDF of cash prices for corn is created by simulating weekly price variations in the futures market and basis prices in order to construct a set of probabilities. Cash price is essentially the end price the producer receives for his or her crops. First, we use the futures data on price per bushel of corn, which was obtained from the Chicago Board of Trade, and presented in Table 10.1. The four data columns show the average closing price per bushel at different weeks of the year (weeks 11, 20, 29, and 48, annotated with an "F" for futures price) from 1995 to 2004.

Table 10.2 presents basis prices for corn during the same years and weeks, annotated with a "B" for basis price. This table shows that the basis for a bushel of corn is usually negative, due to transportation costs from Holyoke to the Chicago Board

TABLE 10.2
Basis Bids, Holyoke Cooperative

Year/Week	11B	20B	29B	48B
1995	–$0.21	–$0.17	–$0.05	–$0.17
1996	–$0.03	$0.02	$0.78	–$0.24
1997	–$0.35	–$0.24	–$0.10	–$0.27
1998	–$0.28	–$0.18	–$0.17	–$0.41
1999	–$0.39	–$0.44	–$0.41	–$0.49
2000	–$0.45	–$0.39	–$0.10	–$0.15
2001	–$0.18	–$0.18	–$0.19	–$0.29
2002	–$0.19	–$0.19	–$0.13	$0.10
2003	–$0.06	–$0.10	–$0.06	–$0.22
2004	–$0.25	–$0.01	$0.20	–$0.29

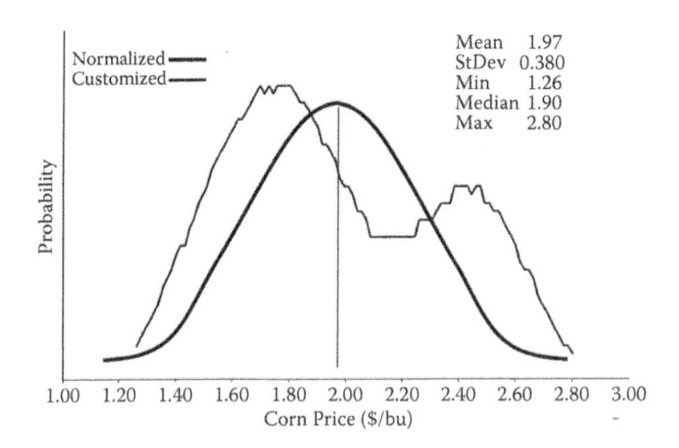

FIGURE 10.1 Normalized and customized PDF for corn price at EWS Farms.

of Trade; however, the basis can be positive, which means that there is more demand for the corn locally than there is in Chicago. The cash price will be a combination of the probability of a futures price and the probability for a basis price that week. In other words, the cash price is the futures price plus the basis.

10.1.3.2 Build a Risk Profile

By combining the information in Tables 10.1 and 10.2 we can produce the probabilities associated with cash prices, which we normalize and summarize in a PDF for EWS Farms. Two PDFs are shown in Figure 10.1. One PDF is a normalized, bell-shaped curve. The other PDF is a customized curve, or histogram, that plots the actual shape of the PDF. On the horizontal axis, we see the different historical cash price levels faced by EWS Farms. The vertical axis shows the probabilities that are associated with the cash price levels. For those familiar with the concept of a bell curve, average cash prices are shown in the middle peak of the curve, illustrated with a bold line. The price-level data for EWS Farms with a normal distribution generates an average price of $1.97 per bushel of corn. From the graph, we can develop more information for our risk profile. From the statistical summary, we can see there is a 5% probability of a low price level of $1.26 per bushel, and on the high side there is a 5% probability of $2.80 per bushel. Again, on average, EWS Farms will earn $1.97 per bushel of corn, shown at the highest point of the normalized curve.

When we say we are *normalizing* the data, we are forcing the probability data to fit a normal PDF distribution (bell-shaped) as closely as possible, even if the data is not really bell shaped. While normalized data can make a risk profile easy to understand, it can be misleading. To see this, we can customize a risk profile by plotting values one by one in a histogram. The customized risk profile illustrated in Figure 10.1 turns out to have two humps, rather than taking on a bell shape. Typically, a histogram is shown as vertical bars, but to illustrate our point here we can trace a continuous line along the tops of the bars. Customization can create a more accurate risk profile because it gives the producer a more accurate view of the potential prices he or she may face. The customized risk profile for EWS Farms uses annual price data for individual yields. This is exactly the same data used in the normalized curve, but it looks different because the data are customized. Notice that contrary to the normal PDF distribution, the *mean* of $1.97 in the EWS Farms profile is one of the lower-probability outcomes. We describe the custom curve as a *bimodal distribution* since it has two humps, so to speak, which represent common outcomes. This bimodal distribution has several risk management implications for EWS Farms. That is, it is more likely that EWS Farms will get either a lower ($1.75) or higher ($2.45) price than the mean.

In other words, if you were the operator of EWS Farms, you would have a higher likelihood of receiving a price less than average than you would of having the average price where the hump occurs for a normal distribution; you would also have a relatively higher likelihood of achieving a higher than average price. This occurs because the two humps for the histogram occur on the right and left side of the average. This illustration shows the importance of adding a graphical picture to the risk profile. Given this customized risk profile, your decisions regarding how to manage your risk will likely depend upon your risk preferences, which were discussed in Chapter 6.

In the second part of this chapter, we discuss more of the technical information about how these probabilities are determined. We also provide you with a tool in the last section of the chapter to create your own customized and normalized risk profile. The Risk Profiler Tool in Risk Navigator includes a graph of the PDF, the CDF, a histogram, and a summary of statistical information like mean, minimum, maximum, and standard deviation.

10.1.4 Relying on History to Define Probability

It may seem that the past is the best predictor of the future. But this statistical analysis for EWS farms confirms that's not always the case. Using the past ten years of data, we estimated that the mean price would be about $2.00 bushel. The very next year prices skyrocketed, earning corn producers over $6 per bushel. So the past did not do so well on predicting the future. But the high prices did not last either. After two years, prices returned to the $2 range. The important lesson here is to use statistical tools carefully, supplementing historical information with current information, and perhaps even your intuition. Over the long run a structured, systematic process, even with its flaws, will likely beat trying to predict the future through other means.

10.2 PART 2: FUNDAMENTAL MEASURES OF PROBABILITY

Using EWS Farms as an example, we have discussed how average price data and a customized PDF can be used to convey information about annual production expectations. The purpose of this section is to provide a tutorial on statistical concepts that will enable you to build your own PDF and CDF. You don't have to have a strong understanding of these concepts to use the Risk Navigator SRM tool to build a risk profile, but it helps to understand as much as you can. These concepts are critical to risk management. There have been many books written about statistics, but we focus mainly on the topics that will lead you to fulfilling the goal of creating your own PDF and CDF.

The main idea of this tutorial is to expose you to the many different ways to determine the frequency of an event. This is a very brief tutorial about other statistical methods that can be used to evaluate performance, both at the individual and at the market level. We provide several examples of these methods and we utilize these techniques in the online Risk Navigator SRM Tools section. (If you need more practice or would like more in-depth information about statistical procedures, you may want to consult a statistics textbook or look at one of the many Web sites devoted to explaining statistics.)

The first topic we discuss is a graphical depiction of risk through the PDF and CDF. Next, we discuss how to calculate measures of central tendency. Measures of central tendency convey how the data are distributed or, in other words, how the data are concentrated. For example, consider the fulcrum of a teeter-totter, where the highest concentration of pressure is exerted on the teeter-totter, but where the teeter totter is most balanced. This image is somewhat analogous to measures of central tendency because it reflects where the data are most balanced. Common measures of central tendency for sample data are the mean, the median, and the mode.

10.2.1 Probability and the Risk Profile

There are many ways to measure how frequently events occur and what may be considered a "typical" event. This leads us to perhaps the most important statistical concept in this part of the tutorial: probability. As we move through this section, we will learn about probability and how to use it to build a risk profile. The risk profile displays the likelihood of outcomes in various ways so a decision maker can better understand the potential consequences of different management strategies.

The risk profile shows risk from different directions to broaden your understanding. At the core of the risk profile is the PDF. In Chapter 9 we showed that the shape of the PDF, its location (right to left), and its width all have implications for risk. For example, a PDF that has the same mean as another but that is wider is less desirable because it has more risk without any extra return. In this chapter, we explain the PDF in detail to give the reader a more complete understanding of the concept. The Risk Navigator SRM Risk Profiler displays four basic parts: (1) a continuous PDF, (2) a discrete histogram of the PDF, (3) a CDF, and (4) summary statistics. We discuss probability terms, the PDF and CDF, measures of central tendency, and measures of spread before moving on to how to elicit and produce your own risk profile.

10.2.2 Terms Related to Probability

The probability of an event is the chance that the event will happen. In its most simple form, if an experiment has n number of outcomes and j number of these outcomes belong to or are favorable to event I, the probability that I occurs, $P(I)$, is the ratio j/n. This is expressed mathematically as:

$$P(I) = \frac{\text{number of outcomes belonging to } j}{\text{total number of outcomes } n}$$

For example, when playing "heads or tails," the probability of flipping a coin and seeing a head is 0.5 because there is one outcome in j (heads) and n is equal to two total outcomes.

At this point, we are only considering situations where outcomes must each have an equal chance of occurring and must be mutually exclusive, meaning they cannot occur at the same time. Later we will discuss the situation where the outcomes do not have an equal chance of occurring.

Before moving on, it is important to define a few variables:

- ***Random or statistical experiment***: This term refers generally to a measurement or process of observation whose outcome cannot be predicted with certainty. Usually, there is more than one possible outcome.
- ***Population or sample space***: This is the set of all possible outcomes of a statistical experiment. For example, the total number of possible outcomes after rolling a pair of dice is 36.
- ***Event***: This is a particular collection of outcomes when an experiment is performed. It is a subset of the sample space. For example, the total number of doubles after rolling a pair of dice {(1,1), (2,2), (3,3), (4,4), (5,5), (6,6)} is an event. The probability of an event is always between 0 and 1, inclusive. If an event cannot occur, its probability is 0. If it is certain to occur, its probability is 1. So the probability of obtaining a double after rolling a pair of dice is 6/36 or 1/6.
- ***Mutually exclusive***: Two events are said to be mutually exclusive if they have no outcomes in common. Three events are mutually exclusive if no two of them have common outcomes.
- ***Sample point***: This term refers to each member or outcome of the sample space.
- ***Random variable***: This is a quantity or characteristic that can assume different numerical values determined by chance or the outcome of an experiment. Examples of random variables include the sum of a pair of dice after they are rolled, GDP, prices, wages, and the lifetime of a flashlight. A random variable may be either discrete or continuous. A *discrete random variable* is a random variable that takes on only a finite (or countably infinite) set of numbers. An example is the number of heads in a toss of two coins (0, 1 or 2). On the other hand, a *continuous random variable* is a random variable whose possible values can take on any value in some interval of values. Examples include temperature and weight.

10.2.3 The Probability and Cumulative Density Functions

Probability is presented graphically in a PDF. These functions are a little bit difficult to interpret at first, but investing in understanding them pays off by making it easier to understand a risk profile. We discussed how the shape of the PDF conveys information about risk in the previous chapter. To complete your understanding, we now show you precisely how the PDF is built.

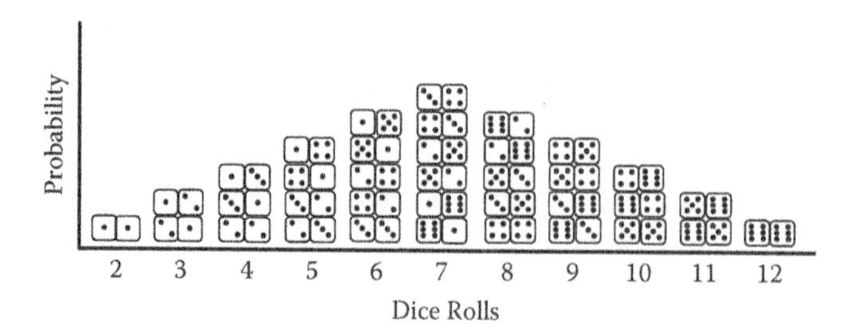

FIGURE 10.2 Probability density function (PDF) for two dice.

Consider an example of a probability distribution obtained from a gambler playing craps. The gambler has two dice. His payout is a function of what he rolls. If he rolls a 7 on the first roll, he wins; if he rolls a 5, he wins only if he rolls another 5 before rolling a 7 again. His luck will depend upon the probability of rolling a 7 compared to rolling a 5. This is a case where understanding the probability of a given roll could be very helpful. As shown in Figure 10.2, there is only one roll that will result in a sum of 2. There are two rolls that will produce a value of 3, 3 rolls for 4, and so on until you reach 7. There are six combinations that produce a value of 7. Then, the probability starts to decline from five combinations that make an 8, four that yield a sum of 9, and so on until there is only one combination that yields a 12. Graphing the height of the probability for each roll forms the PDF.

If a curve were fitted along the top of the dice distribution, it would form a smooth and continuous PDF in the standard bell-shaped form. Try to remember this example when looking at a PDF; those outcomes that are most likely will have the highest height above them. The shape of the function is very important. The symmetrical bell-shaped PDF is a *normal* distribution. It has several helpful properties; for example, as shown in Figure 10.3, the mean, median, and mode (described in the following section) are equal and are right in the middle of the distribution. In addition, we know that about two-thirds (68.27%) of the time a random draw from a normal distribution will yield a value within one standard deviation (σ) away from the mean. About 95% of the time the draw will be within two standard deviations, and it will be within three standard deviations over 99% of the time. This allows us to form reasonable expectations about the future and to perform helpful sensitivity analysis for likely events.

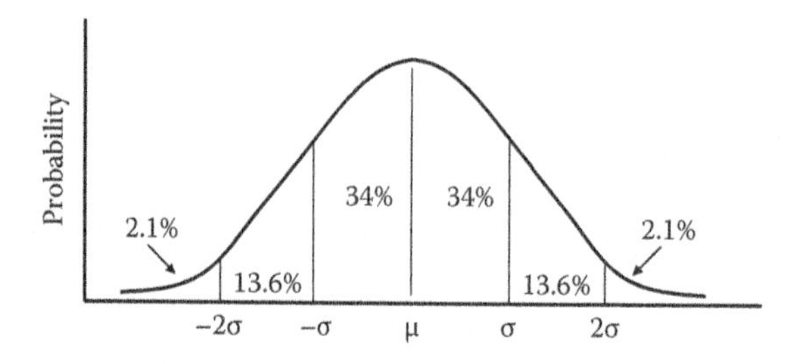

FIGURE 10.3 Properties of a probability density function (PDF).

TABLE 10.3
Probability Density Function (PDF) and Cumulative Density Function (CDF) for Rolling Two Dice

Outcome	Frequency	PDF	CDF
1	0	0	0
2	1	1/36	1/36
3	2	2/36	3/36
4	3	3/36	6/36
5	4	4/36	10/36
6	5	5/36	15/36
7	6	6/36	21/36
8	5	5/36	26/36
9	4	4/36	30/36
10	3	3/36	33/36
11	2	2/36	35/36
12	1	1/36	1

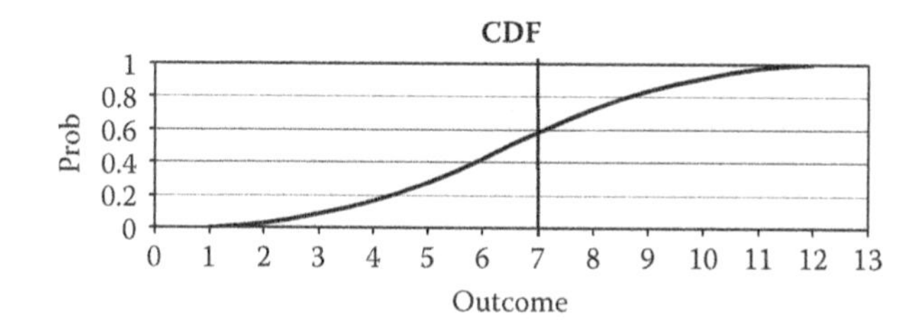

FIGURE 10.4 Cumulative probability density (CDF) for rolling two dice.

A very important counterpart to a PDF is the CDF. The CDF is calculated by adding the probabilities together. The probabilities, including the cumulative probability, are shown in Table 10.3.

When looking at a CDF, a person can very quickly see the probability of exceeding or falling below any specific outcome by simply drawing a vertical line over the value of interest. For example, as shown in Figure 10.4, the probability of getting an outcome below 7 is approximately 0.58, or 58%, and the probability of obtaining a number above 7 is 1 – 0.58, which is 42%. The roll will be below 2 0% of the time and 12 or below 100% of the time.

10.2.4 Measures of Central Tendency

10.2.4.1 The Mean

Up to this point, we have primarily discussed how to find the mean, or average, of a data set. The mean of a data set may be taken of an entire population, such as all of the beef growers in the state of Colorado, in order to determine the average number of head of cattle owned on Colorado ranches. However, finding the population mean can be challenging—and expensive. Therefore, the mean of a *sample population* may be found as an estimate. For example, if done properly, we can obtain a sample of cattle producers in the state of Colorado, and from that sample mean, we can project the correct value of the population mean of cattle owned on Colorado ranches.

Sometimes it is difficult for beginning students to understand whether the data set is a population or a sample. Take heart—even experienced researchers may have difficulty distinguishing the difference in some cases. Likewise, there are situations when the population is so small that one should determine the population mean rather than the sample mean. Typically, the population mean is designated by the Greek symbol μ pronounced "me-you." The sample population is represented by the symbol often referred to as $\bar{x}$. To find the mean, add the values and divide by the number of values in the sample, even if the values are "0" or negative. We illustrate this in the example below:

Example 10.1: Find the Population Mean

DATA:

4 7 −12 22 0 −8 12 −12 12 25

CALCULATION:

$$\mu = \frac{4+7+(-12)+22+0+(-8)+12+(-12)+12+25}{10} = 5$$

If we consider the values in this example as thousands of dollars in profit for ten years in an agricultural operation, then we have determined that the average amount of profit is $5,000. We also can take a sample of the last five years and determine the sample mean:

Example 10.2: Find the Sample Mean

$$\bar{x} = \frac{(-8)+12+12+(-12)+25}{5} = 5.8$$

In this case, the sample mean of the last five years of production is $5,800. Whether or not a sample is representative of the population mean (equal to $5,000) is always debatable and subject to more in-depth testing. However, given the variation in the population mean (which we discuss momentarily), we could still assert that we have achieved a sample mean that does a good job of reflecting the population mean. When there are a lot of data, researchers take just a sample of the population with the hope that it represents the entire group. This is not always the case, and researchers take great care to collect representative samples.

Sometimes there are a large number of values that must be calculated in order to find the mean. Doing this requires some basic mathematical operations. Multiply each value by its frequency of occurrence. Next, add all of the summed values and divide by the number in the population. For example:

Example 10.3: Find the Mean of a Large Population

DATA:

Values	Frequency
40	4
44	6
48	9
52	4
64	7

CALCULATION:

$$\mu = \frac{(40\times 4)+(44\times 6)+(48\times 9)+(52\times 4)+(64\times 7)}{4+6+9+4+7} = 50.4$$

$$\mu = \frac{1512}{30} = 50.4$$

This is easily accomplished in Excel or another spreadsheet program.

10.2.4.2 The Median

The median is the middle value of the data set when all of the values are listed in either ascending or descending numerical order. To obtain the median for a data set with an odd number of values, simply find the middle value in the ordered array. To find the median for a data set with an even number of values, we average the two middle numbers or values in the ordered array.

Example 10.4: Find the Median for an Odd Number of Values

DATA:

4 6 10 −4 −8 −5 2 −9 12 7 1

First, arrange the values in either an ascending or descending order. In this case, we order the numbers from lowest to highest:

−9 −8 −5 −4 1 2 4 6 7 10 12

Because there are an odd number of values in this data set, the median will be the middle value. This means that the median of the ordered set above is 2.

Example 10.5: Find the Median for an Even Number of Values

2 6 8 −10 4 5 −12 7

Again, arrange the sample values, in this case from lowest to highest:

−12 −10 2 4 5 6 7 8

The median will be the average of the middle values, 4 and 5. The median is therefore (4 + 5)/2 = 4.5.

10.2.4.3 The Mode

The mode is the value in the data set that occurs most frequently. If only one value has the highest frequency of occurrence, the data set is said to be unimodal. The data set is bimodal when two values have the highest, but equal, frequency. It is important to note that when all the values in the data set have the same frequency of occurrence (as is the case of a single occurrence) there is no mode.

Example 10.6: Find the Mode

Values	Frequency
40	4
44	6
48	9
52	4
64	7

The value with the highest frequency is 48, with a frequency of 9. The mode of the data set is 48; therefore, the data set is said to be unimodal. To return to the concept of a histogram, a unimodal data set is distributed as seen in Figure 10.5.

Notice that the most common value is "V," shown by the highest bar in the histogram.

To reinforce the concept discussed above, when two values occur at the same frequency, it creates a bimodal distribution as illustrated in the example below and in Figure 10.6.

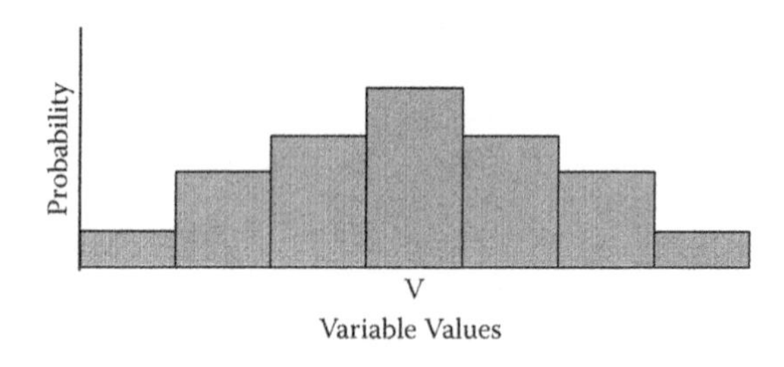

FIGURE 10.5 Unimodel frequency distribution.

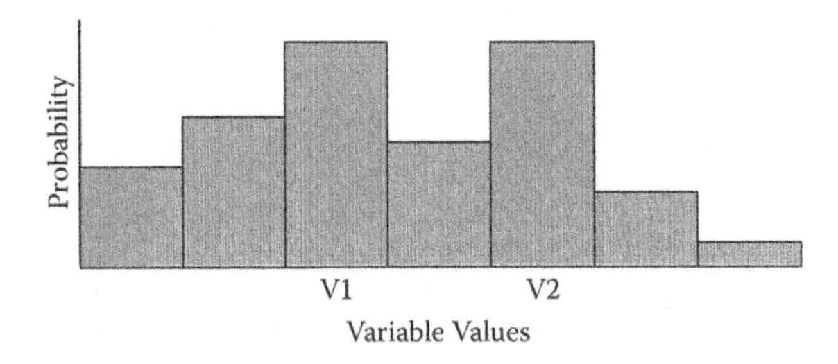

FIGURE 10.6 Bimodal frequency distribution.

Example 10.7: Bimodal Frequency Distribution

Values	Frequency
40	4
44	6
48	9
52	4
60	9
64	7

The values that occur at the highest rate are 48 and 60. Each occurs with a frequency of 9 observations, as shown by V1 and V2 in Figure 10.6.

10.2.4.4 Skew

When we look at the frequency distribution of the mode, it provides a picture of how evenly the values occur. The distribution of the mode is referred to as *skew.* Recall that a normal density function has a symmetrical shape, with an equally likely chance of events occurring to the right of the mean and to the left.

While there are a number of skew scenarios, this section focuses on three of the general types:

- Negatively skewed distributions
- Positively skewed distributions
- Symmetrical distributions

Negatively skewed distribution: The label *negatively skewed distribution* may first appear contradictory to its histogram and frequency distribution as shown in Figure 10.7. A negatively skewed distribution is described as such because the mean is closer to the negative (or left side) of the histogram than the other measures of central tendency. The location of the mean implies the distribution is *negatively skewed.* The following is a description of a negatively skewed distribution:

- The other measures of central tendency (median, mode) fall to the *right* (or on the positive side) of the mean.

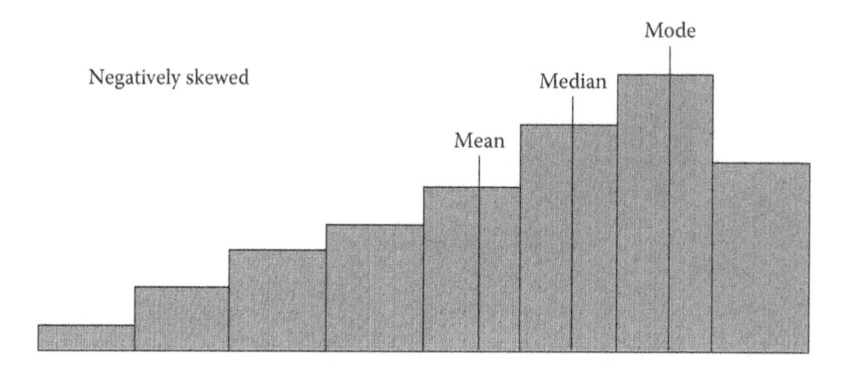

FIGURE 10.7 Negatively skewed distribution.

- The tail of the distribution (values that occur at lower frequencies) is to the *left* (the negative side) of the mean. The mode occurs farthest to the right. That is, this is the largest positive value. The median is between the mode and the mean.

Positively skewed distributions: A *positively skewed distribution* can be described as the mirror image of a negatively skewed distribution. The mean is closer to the positive (or right side) of the histogram than the other measures of central tendency; hence the data have a *positive skew*. A visual example of a positively skewed distribution is shown in Figure 10.8. The following is a description of a positively skewed distribution:

- The other measures of central tendency (median, mode) fall to the *left* (or on the negative side) of the mean.
- The tail of the distribution (values that occur at lower frequencies) is to the *right* (the positive side) of the mean. The mode occurs farthest to the left. That is, this is the smallest positive value. The median is between the mode and the mean.

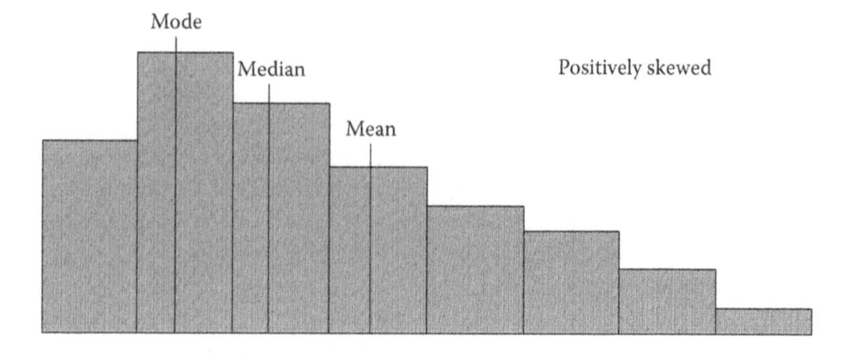

FIGURE 10.8 Positively skewed distribution.

Symmetrical distribution: A symmetrically skewed distribution looks much as the description implies: the measures of central tendency are concentrated in the middle of the data set and the data are evenly distributed. A symmetrical distribution looks very similar to the normalized bell-shaped curve shown by the bold line in Figure 10.1. The following is a summary of the characteristics of a symmetrical distribution:

- When the distribution is unimodal, the mean, the median, and the mode are located at the center of the distribution and they are equal to one another.
- The data values are equally distributed on both sides of the mean.

10.2.5 Measures of Spread

Measures of spread describe how the data observations are distributed (or spread) around the mean. As is the case with the mean, measures of spread can be used to assess either sample or population data. Measures of spread provide additional insights into how the data are distributed, and this information is complementary to the other measures of central tendency. In this section we will discuss two ways in which we can measure spread:

- Variance (using a single variable such as weight)
- Standard deviation (the standard deviation of a single variable)

We then discuss two related measures, covariance and conditional probability, in the following sections.

Why is it important to measure spread? If a hobby farmer has 5 horses that all weigh exactly 1,100 pounds, the mean, median, and mode are all 1,100 pounds. This is a unimodal frequency distribution with no variation in weight. However, if these 5 horses weighed 900, 900, 1,100, 1,300, and 1,300 pounds, respectively, the mean and the median are the same (1,100 pounds), and the distribution is bimodal; however, the variation in weight provides insight into different dynamics. Even though the mean is the same as the first scenario, the second scenario reflects that there are both small and large horses. This implies different equipment, feeding, and exercise schedules are probably necessary. Therefore, dietary management conclusions may not be the same if only information about the mean were considered. Measures of spread tell you something about the range of weights around the mean.

In the next section we discuss three different measures of spread and how to calculate these measures. Some of the formulas are complex, and typically calculations like variance and covariance are performed with computer software or a spreadsheet. We simply provide a short discussion with a few examples so that you appreciate and understand how the data observations and the spread can reveal more information than just the measures of central tendency.

10.2.5.1 Variance

Like population and sample mean, the difference between sample variance and population variance is that the population variance considers all observations. Variance measures the spread of the distribution of individual observations about its mean. Variance can be applied to either population or sample data. As will be discussed later in the chapter, variance also can be used with randomized data. Sometimes it can be difficult to determine whether one is using an entire population or a sample, particularly when the number of observations is large. For example, to find the variance for a hobby farmer who owns 20 horses, we would consider the weight of *all* 20 horses.

The variance for the population data is denoted by σ^2 (read as sigma squared) and is expressed mathematically as:

$$\sigma^2 = \frac{\sum (x-\mu)^2}{N}$$

where

x = the single observation
μ = the population mean
N = the number of observations in the population
Σ is a symbol for "sum"
σ^2 means population variance

As shown in the example, the population variance contains all of the farmer's horses. For example, the weight of each horse on a horse farm with 20 horses is recorded in the first column of Table 10.4. The average weight for that column, 1,087.35 pounds, is subtracted from each individual horse weight and the result is squared as shown in the numerator of the variance formula. The squared differences are summed in the last column. Plugging these values into the population variance formula yields a variance of 26,778.93 pounds. It is difficult to make intuitive sense of a variance, but we will make this link when we discuss standard deviation, which is the square root of the variance.

The following shows a sample variance, which is the same equation but divided by $n - 1$, where n is the size of the sample instead of the population, N. The term $\bar{x}$ replaces μ only to reflect that it is the sample mean rather than population mean that is being subtracted. The sample variance is denoted S^2 instead of σ^2. The term S^2 can be confusing because we don't actually square the term on the right.

$$S^2 = \frac{\sum (x-\bar{x})^2}{n-1}$$

Using the horse example, let's say there are 5 horses that weigh 900, 900, 1,100, 1,300, and 1,300 pounds, respectively. If the owner has 20 horses, these 5 horses are part of a sample of 20. We can calculate the sample variance in the example in Table 10.5. The resulting variance is 40,000 pounds.

Note that the population mean (1,087.35 pounds) is not too far off of the sample mean (1,100 pounds). This is one indication that the sample is representative of the population. The variance of the sample is high, however, and it reflects that fact that there is a large difference in horse size.

TABLE 10.4
Population Variance for Horse Farm

Horse Weight (x)	Mean	Difference	Squared Difference
850	1087.35	–237.35	56335.02
850	1087.35	–237.35	56335.02
900	1087.35	–187.35	35100.02
900	1087.35	–187.35	35100.02
900	1087.35	–187.35	35100.02
902	1087.35	–185.35	34354.62
905	1087.35	–182.35	33251.52
1100	1087.35	12.65	160.0225
1100	1087.35	12.65	160.0225
1100	1087.35	12.65	160.0225
1105	1087.35	17.65	311.5225
1125	1087.35	37.65	1417.523
1130	1087.35	42.65	1819.023
1200	1087.35	112.65	12690.02
1200	1087.35	112.65	12690.02
1250	1087.35	162.65	26455.02
1300	1087.35	212.65	45220.02
1300	1087.35	212.65	45220.02
1310	1087.35	222.65	49573.02
1320	1087.35	232.65	54126.02
Mean			
Sum of Squared Differences			**535578.6**
Divided *N* (20)			**26778.93**
Variance			**26778.93**

TABLE 10.5
Sample Variance for Horse Farm

Horse Weight (x)	Mean	Difference	Squared Difference
900	1100	–200	40000
900	1100	–200	40000
1100	1100	0	0
1300	1100	200	40000
1300	1100	200	40000
Sum of Squared Differences			**160000**
Divide by *n* – 1(4)			**40000**
Variance			**40000**

10.2.5.2 Standard Deviation

The standard deviation is an easier value to use than variance; and it is defined as the square root of the variance:

$$\sigma = \sqrt{\frac{\sum (x-\mu)^2}{N}}$$

The standard deviation of all twenty horses is therefore approximately 164 pounds. Likewise, the sample standard deviation is the square root of the sample variance, 200 pounds. Notice that this makes the sample and population measures of spread look closer than they did for variance (because variance squares the differences).

The standard deviation is clearly a much smaller value than the variance, which makes it easier to work with. The standard deviation tells us how the horses' weights vary around the mean of 1,100 pounds. If this example were assumed to be a normal distribution, two thirds of the horses will either be within 200 pounds (more or less) of the mean of 1,100 pounds, and about 99% will be within 600 pounds more or less.

10.2.6 Covariance

Up to this point we have discussed measures of spread around one variable, like variations around the average horse's weight; however, it is not uncommon for a second variable to change when the first variable changes. To make this easier, we can return back to the horse example where we can measure how both the weight and height of a horse (measured in hands, which is equal to 4 inches) vary together.

Covariance measures the degree to which two variables vary together. The covariance is defined as:

$$Cov(X,Y) = E\left[(X-\mu_x)(Y-\mu_y)\right]$$

where:

$Cov(X,Y)$ means Covariance
E means expected value, or the average
μ_x = the mean population of X
μ_y = the mean population of Y

Essentially, covariance multiplies the differences between variable X and variable Y and their respective means, which is multiplied by the average product. Computers generate covariance at lightning fast speed, and sometimes this formula can be cumbersome to compute manually. Through clever mathematical manipulation, we can rearrange the formula to calculate the covariance more easily:

$$Cov(X,Y) = (E\{X,Y\} - E\{X\}E\{Y\}$$

where

$Cov\ (X,Y)$ means Covariance

$E\{X,Y\}$ = the mean of $X*Y$

$E\{X\}$ = Mean of X

$E\{Y\}$ = Mean of Y

10.2.7 Conditional Probability

Frequently, the probability of one event depends upon another. This is known as *conditional probability*. The conditional probability of event G is the probability that this event occurs provided event F has occurred. The conditional probability of event G, conditional on event F occurring, is expressed mathematically as:

$$P(G/F) = \frac{P(G \cap F)}{P(F)}$$

This means the conditional probability of G, given F, is equal to the ratio of their joint probability to the marginal probability of F. Suppose there are 100 cows in a herd, divided by breed and by production category as follows:

Cows by Breed and Production

	Milk Production	Beef Production
Holstein Friesian	45	10
Jersey	10	35

You are asked to randomly pick a cow from the herd:

- p(Holstein Friesian) = 55/100
- p(Jersey) = 45/100
- p(milk production) = 55/100
- p(beef production) = 45/100

This is the basic idea of conditional probability based on frequency. Then, you are asked to pick a cow in milk production given that it is a Holstein Friesian. The probability of doing that is:

$$p\ \text{(cow in milk production given that is a Holstein Friesian)} = \frac{45}{55} \approx 75$$

This also can be expressed as a ratio of probabilities as follows:

$$p\ \text{(cow in milk production given that is a Holstein Friesian)} = \frac{45/100}{55/100} \approx 75$$

10.2.8 Estimating Probability

The last concept we need to discuss is how to elicit a PDF that will be useful enough to help you make decisions. Estimating probabilities is a tricky business and difficult to master; however, many new techniques have been developed and weaknesses can be minimized by factoring them in when you build the PDF. We will discuss some of these techniques below and help you develop your own PDF with our Risk Navigator SRM tool, the Risk Profiler.

At one time there was debate about *objective* and *subjective* probability. A PDF built from data was thought to be objective, but a PDF built from logic, advice, or experience was said to be subjective. The more contemporary view, however, is that all probabilities are subjective at some level, even those built purely from data. For example, estimating the probability of corn price in 2008 from historical data would have been way off of the record-breaking prices reported that year. Ironically, a guess based on a newspaper report could have been much closer. So, which estimate was objective: the data, the press, or neither?

10.2.8.1 Building a Normalized PDF

With these caveats said, the first and most straightforward way to elicit probability and build a PDF is by using data. We built a normalized PDF in Part 1 for Aaron's corn prices in Figure 10.1. But how did we go from data to drawing the PDF? There are many commercial programs to build PDFs, and we will do it for you in the Risk Navigator SRM Tools. But for the sake of example, let's build a PDF and CDF in Excel. In the first column of Figure 10.9 with data, we enter ten years of Aaron's actual corn price data. In the next column, we used his actual corn price data to generate a range of prices from a low and high price between plus and minus three standard deviations from the mean. In the next column, we use the Microsoft Excel NORMDIST function to determine the cumulative density of the price ranges generated for the fourth column and the PDF. We used this function to estimate the PDF value in the last column. The CDF value is computed by the function = *normdist(x, mean, standard deviation, true),* where x is the price of corn, and mean and variance are displayed in the graph. The command "True" tells Excel to compute CDF. The command "False" would tell Excel to compute a PDF. The PDF cannot be graphed from the NORMDIST function; therefore, we estimate probability density by subtracting the current and previous CDF values. That is, the probability of a price of $1.30 is the difference in the CDF at 1.30, 0.04, and the CDF value at 1.15, 0.01. (The probability reads 0.02 instead of .03 due to rounding.) Technically, this is the probability of the interval from 1.15 to 1.30.

10.2.8.2 Eliciting Probabilities

We have to rely on reason, experience, or pure guesswork if we have no data or if we decide the data are not meaningful. There are at least two major ways to elicit probability information (Clemens and Reilly, 2001; Frey, 1998): asking the expert about his or her degree of belief about an outcome or presenting the expert with a lottery that reveals their probability values.

	Formula	Corn Price ($/bu)	Corn Price ($/bu)	Cumulative Probability	Probability Density
		2.55	1.15	0.01	0.01
		2.47	1.30	0.04	0.02
		1.91	1.45	0.08	0.05
		1.48	1.60	0.16	0.08
		1.56	1.75	0.28	0.12
		1.84	1.90	0.42	0.15
		2.12	2.05	0.58	0.16
		2.33	2.20	0.73	0.15
		1.70	2.35	0.84	0.11
		1.77	2.50	0.92	0.08
Mean	= average(C3:C12)	1.97	2.65	0.96	0.04
Variance	= var(C3:C12)	0.14	2.80	0.99	0.02
Std Dev.	= stdeva(C3:C12)	0.38	2.95	1.00	0.01
Min	= max(C3:C12)	1.48	3.10		
Max	= min(C3:C12)	2.55	3.10		

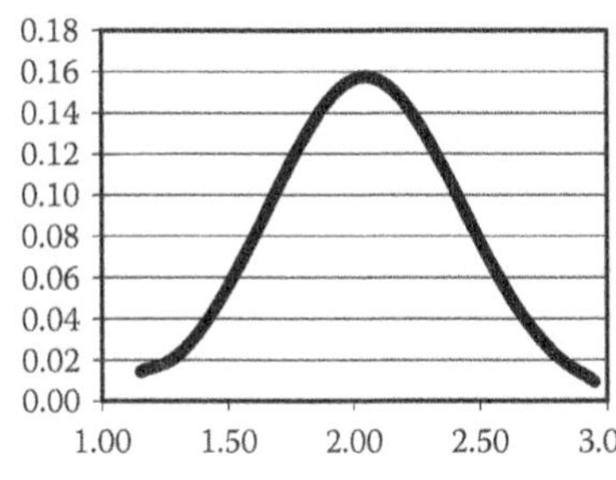

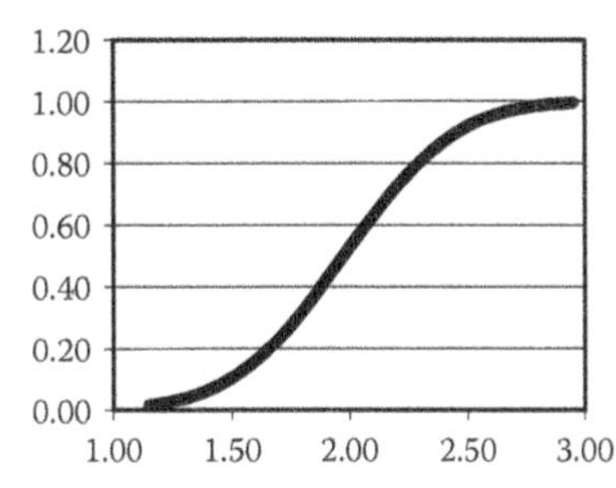

Probability and Cumulative Density Function, Corn Price

FIGURE 10.9 Estimating a CDF in Excel.

While it seems straightforward, just asking a person to tell you his belief about probability is rather difficult to do. The first method we will discuss is to build a representative PDF. Imagine having 100 pennies to represent 100% probability. You can lay them out in stacks to make a histogram like the dice example in Figure 10.2. You would have to choose intervals of the variable you are interested in first, such as 0–50, 50–100, 100–150. and 150–200, for corn yield. In the dice example, the values are 2, 3, 4, 5, 6, 7, 8, 9, 10, 11, and 12. Now, stack your coins over any value until the heights represent probability. If you were representing the probability of throwing any particular value of two dice, the stack of coins would look like the PDF in Figure 10.2. The tallest stack would be over the number 7.

A person could simply plot lines for the PDF or CDF on a piece of paper or in a spreadsheet. Clemens and Reilly (2001) call this the *fractile method*, where probability (the CDF is preferred) is plotted for a segment of the horizontal axis. Two methods favored among experts in decision analysis are the *fixed value method* and the *fixed probability method* (Frey, 1998).

The fixed value method asks the expert about the probability that actual value is higher or lower than some arbitrary number for each value on the horizontal axis.

Keep in mind that an "expert" can be an individual familiar with the commodity—from a farmer/rancher to a commodity broker or academic. One could go along the interval at the bottom of the PDF and ask what the probability would be for each value, making sure that the total probability sums to 1. For the dice example, the expert is asked, "What is the probability that one would roll a total score of 4, for example, in any singe roll?" Researchers have found it best to ask whether the probability would be higher or lower than some value and then to hone in on the actual value. Therefore, the second question is likely, "Is the probability higher or lower than 0.3 that I will roll a 4?" A third, follow-up question is whether the value is higher or lower than 0.4.

The fixed probability method reverses the process and asks the expert to determine the appropriate range of the random variable that corresponds to a given probability. For the dice example, the question is what value would we expect the dice to be when the probability is 0.2?

We use a method that combines these two approaches in Navigator. We ask you to enter the value or midpoint for five ranges of the random variable (such as $1 to $3, $3 to $5, $5 to $8, etc.) and then to enter the corresponding probability. This avoids having to enter the size of the intervals on the horizontal axis.

10.2.8.2.1 Eliciting Probability with a Lottery

Another way to elicit the probability is to offer the user a lottery. In this method, an expert plays games or places bets that reveal their beliefs about probabilities. For example, a person might be given a choice between the following two bets (Clemens and Reilly, 2001):

Bet 1: Win $X if the Lakers win
Lose $Y if the Lakers lose
Bet 2: Lose $X if the Lakers win
Win $Y if the Lakers lose

Bets 1 and 2 are opposite sides of the same bet. Which bet the expert picks indicates his or her beliefs about probability. Since there are countless ways to conduct reference lotteries, the details about how to calculate probability can be found in Clemens and Reilly (2001). We rely on a version of lottery in one of our Risk Navigator tools.

10.2.8.2.2 Estimating Probability Based on Experience

Another way to elicit probability, and one that we also use in our Risk Navigator tool, does not involve trying to visualize and copy the PDF or CDF. This method asks the expert to estimate the low, mode, and high values of the random variable. This can be less arduous than trying to visualize the PDF in some situations. In Navigator, we can use these three points to estimate a PERT (Program Evaluation and Review Technique) distribution. Think of the PERT as a special type of distribution that is similar to the normal distribution in how it looks and is interpreted, but it has special properties that make it the best function to use with only three data points.

A PERT distribution is a parametric distribution that has been adapted to allow experts to estimate it based only on the minimum, most likely, and maximum values. The PERT gets its name from its cousin concept in PERT networks. It is defined as:

$$\text{PERT}(a,b,c) = \text{Beta}(\alpha_1, \alpha_2) * (c - a) + a$$

where a is the minimum value, b is the most likely value, c is the maximum value, and Beta is the beta distribution (Vose, 2000). Furthermore:

$$\alpha_1 = \frac{(\mu - a) * (2b - a - c)}{(b - \mu) * (c - a)}$$

$$\alpha_2 = \frac{\alpha_1 * (c - \mu)}{(\mu - a)}$$

and

$$\text{Mean}(\mu) = \frac{a + 4 * b + c}{6}$$

In the Risk Navigator Tools, we build a PERT distribution based upon information that you provide on the minimum, maximum, and most likely values. This is shown later in Part III.

10.2.9 Challenges in Estimating a PDF and CDF

There are many problems involved in asking a person to estimate a PDF or CDF (Hardaker and Lein, 2005). We review some of the most common ones because knowing about these challenges can help you avoid making similar mistakes, and therefore make your estimates more realistic.

Overconfidence in judgment is a common problem. One solution is to simply gather more information. Another solution is to involve more people, perhaps a spouse or other family member, in your estimations. Avoidance of uncertainty is another problem. Hardaker and Lien (2005) found evidence that many people don't even want data about the probability of extreme events because the events are viewed as uncommon. Of course, the solution is to understand that the pain of having to think about risks up front might be less than dealing with the consequences of ignoring them. Misconception about chance leads people to consistently overestimate small probabilities and underestimate large ones (Buschena, 2002).

Anchoring makes people have a hard time estimating probability because they are focused on something from the past. For example, it would would have been hard for wheat producers to estimate prices in 2008, when they were as high as $15 per bushel or more, if their minds were anchored in the past when prices were about $3 per bushel. Motivation can force decision makers to have trouble estimating probabilities, particularly when part of the motivation is to avoid discomfort. For example, you might want to avoid the embarrassment of reporting those years when you were unable to harvest anything.

Based on these observations about how estimating probabilities can go wrong, we recommend involving more than one person, to promote honesty; using historical records, to avoid anchoring and inflating or deflating high and low values; and forcing yourself to be as factual as possible, no matter how uncomfortable the facts may be.

10.2.10 Finding Data to Estimate Probabilities

We hope that we've made it clear that estimating the probabilities associated with different agricultural outcomes is a worthwhile activity. In fact, the value of your risk management profile, which you will develop in Part 3, greatly depends upon the accuracy of the probabilities and data used to build the PDF and CDF. Of course, the real challenge lies in being able to collect all of this information.

When it comes to finding data and sources of information about probabilities, many students and agricultural producers have a hard time getting started. Once they begin the hunt for data and gain momentum, however, most people are surprised to find just how much agricultural information is available for historical commodity prices (at the national, state, and regional level), weather, land values, and water consumption, just to name a few. To find general data related to agricultural issues, a good place to begin is the U.S. Department of Agriculture (USDA) Web site (www.usda.org). At this site, you'll find information about various agencies, including The Agricultural Research Service, Economic Research Service, Animal and Plant Health Inspection Service, Natural Resource Conservation Service, National Agricultural Statistics Service, and Risk Management Agency. They all have special sections where you can download data on everything from A to Z.

Local elevators and the Chicago Board of Trade, both used in Aaron's example, also present a wealth of information about prices. You can start looking for price data at the USDA Agricultural Marketing Service (http://www.ams.usda.gov). Many local organizations collect and provide price data to their members. As shown in Chapter 2, the Kansas Farm Bureau keeps records of prices and posts them on the Web. In addition to commodity data, Yahoo! Finance (http://finance.yahoo.com/) provides current and historical data about the financial markets. Information is also available for publicly traded agricultural companies like Caterpillar, Inc. The National Drought Mitigation Center has a lot of weather-related data and can direct you down to the level of data for the closest weather station near you (http://www.drought.unl.edu/index.htm).

Records from your farm or ranch's operation yield one of the most accurate portrayals of the probabilities associated with certain events. If your operation does not keep detailed records of price, yield, or input expenses, you are not alone, but it is never too late to start. The better information you retain about past operations, the more accurate picture you'll be able to portray for the future using the tools in Risk Navigator. Local data are preferred to regional data. If you don't have information about your farm, you can see if there is county data available as many agencies keep data at the county level. Yield data may be available at the Natural Resource

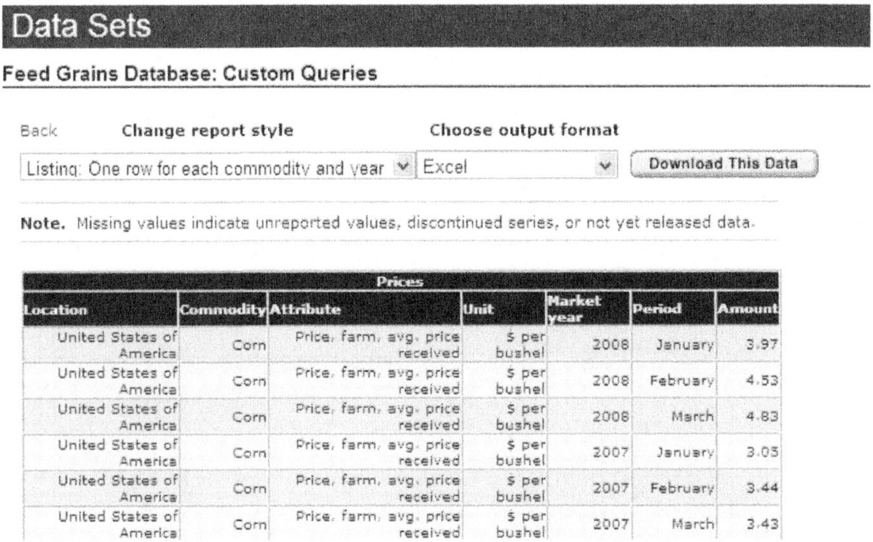

Data Sets

Feed Grains Database: Custom Queries

Back Change report style Choose output format

Listing: One row for each commodity and year | Excel | Download This Data

Note. Missing values indicate unreported values, discontinued series, or not yet released data.

Prices						
Location	Commodity	Attribute	Unit	Market year	Period	Amount
United States of America	Corn	Price, farm, avg. price received	$ per bushel	2008	January	3.97
United States of America	Corn	Price, farm, avg. price received	$ per bushel	2008	February	4.53
United States of America	Corn	Price, farm, avg. price received	$ per bushel	2008	March	4.83
United States of America	Corn	Price, farm, avg. price received	$ per bushel	2007	January	3.05
United States of America	Corn	Price, farm, avg. price received	$ per bushel	2007	February	3.44
United States of America	Corn	Price, farm, avg. price received	$ per bushel	2007	March	3.43

FIGURE 10.10 Corn price data download from the USDA Economic Research Service.

Conservation Service or Agricultural Statistics Service. Local weather data can be found at the National Oceanic and Atmospheric Administration (NOAA).

It is also worth mentioning that forecasts from experts, from the national level down to the county extension agent, can provide guidance into the probability of events. Finding the right source is simply a matter of being persistent on your Web searches. Try searching directly for data and then try searching for forecasts or articles.

The following is an example of how to find data on the Web: Let's say I want to download data about corn prices, like Aaron did for EWS Farms. I started searching at the National Agricultural Statistics Service and the Agricultural Marketing Service. It took a few minutes, but I download data from the Economic Research Service (www.ers.usda.gov). After clicking Data Sets at the top, I click Feed Grains Database. From there, I click Custom Queries. Next, I answer questions about the data I want, (monthly corn prices for the United States from 1987 to 2008). What I find is shown in Figure 10.10. Note that it is all set to download into Excel. Follow the directions to download the data and then do with it as you wish. You could make a PDF and CDF, for example, using the data shown in Figure 10.10.

10.3 PART 3: RISK NAVIGATOR MANAGEMENT TOOLS

10.3.1 Risk Profiler

Risk Profiler helps you build your own risk profile, using one of three methods. The program is set to open on the first tab, or method, which is to describe your PDF by features. The page, as it appears, is shown in Figure 10.11. Enter your information in the upper left quadrant. This first method is the easiest of the three. Enter the minimum, maximum, and most likely values to create a PERT distribution.

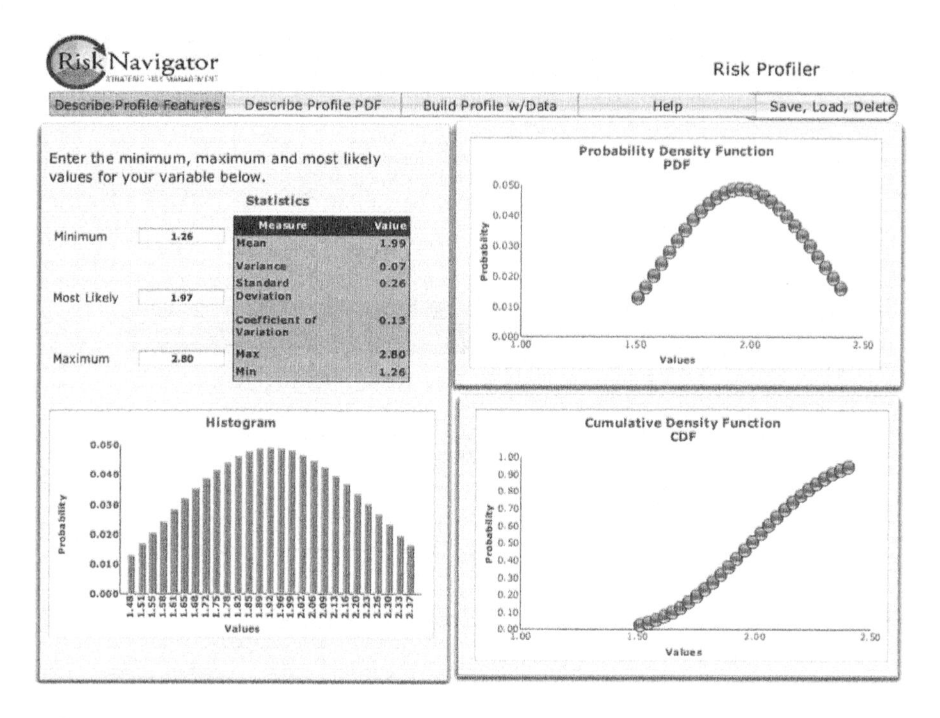

FIGURE 10.11 Risk Profiler home page.

The risk profile is created instantly, showing a PDF, CDF, histogram, and summary statistics. Recall from Part 1 that Aaron's PDF is bimodal when graphed as a histogram. The PDF and CDF in Navigator force the data into normal distributions for the graphs on the right side. The histogram on the lower left, however, can reveal whether the PDF is something other than normal; that is, it might not be bell shaped. This first estimation method does not have enough information to compute anything but a normal distribution for the histogram; however, the other two methods will reveal differences in the histogram when they exist.

You can print, save, or load a previously saved profile, or click on a different tab to estimate probability using a different method.

The second tab lets the decision maker build a PDF by describing the outcomes and probabilities. An example is presented in Figure 10.12. Enter five values for the random variable. In this case, Aaron entered $1.26, $1.45, $1.97, $1.67, and $2.80 per bushel. Next, enter the probability that the random variable occurs. Navigator applies a symmetrical and even interval around each point estimate, which makes each bar the same width. Notice that in this case the histogram shows a different shape than the normal PDF. Try entering these values on your own to see how this PDF differs from the first method.

The final method is to enter data about the random variable. The screen for this method is shown in Figure 10.13. In this example, we have entered the cash price Aaron received (shown in Figure 10.1 in Part 1). Again, this produces a risk profile by a different method, one that shows the bimodal distribution in the histogram more accurately than the other methods.

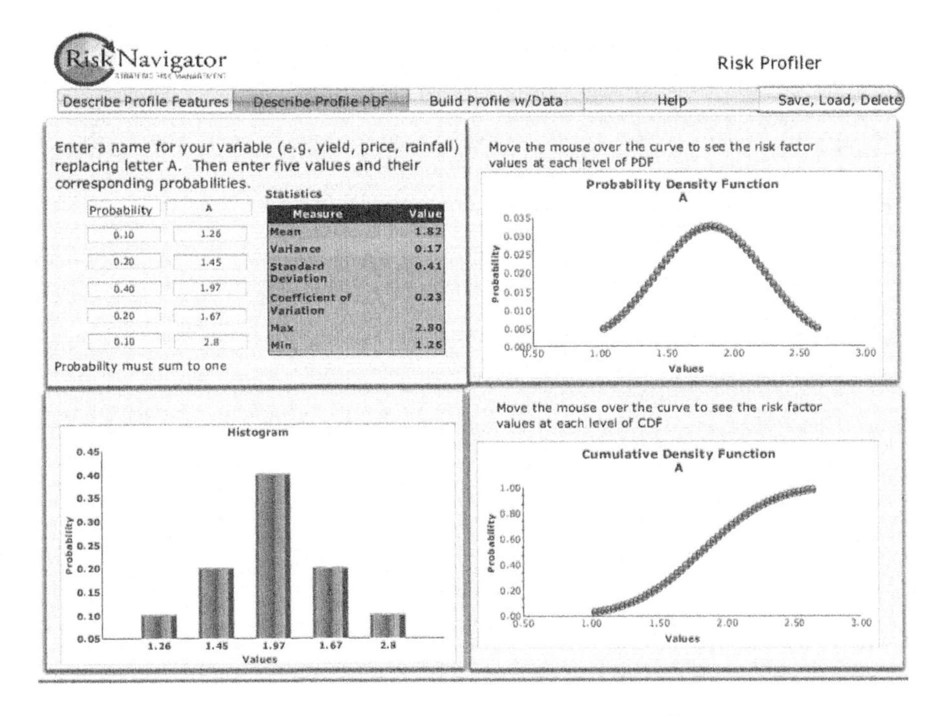

FIGURE 10.12 Build a PDF by describing outcomes and probabilities.

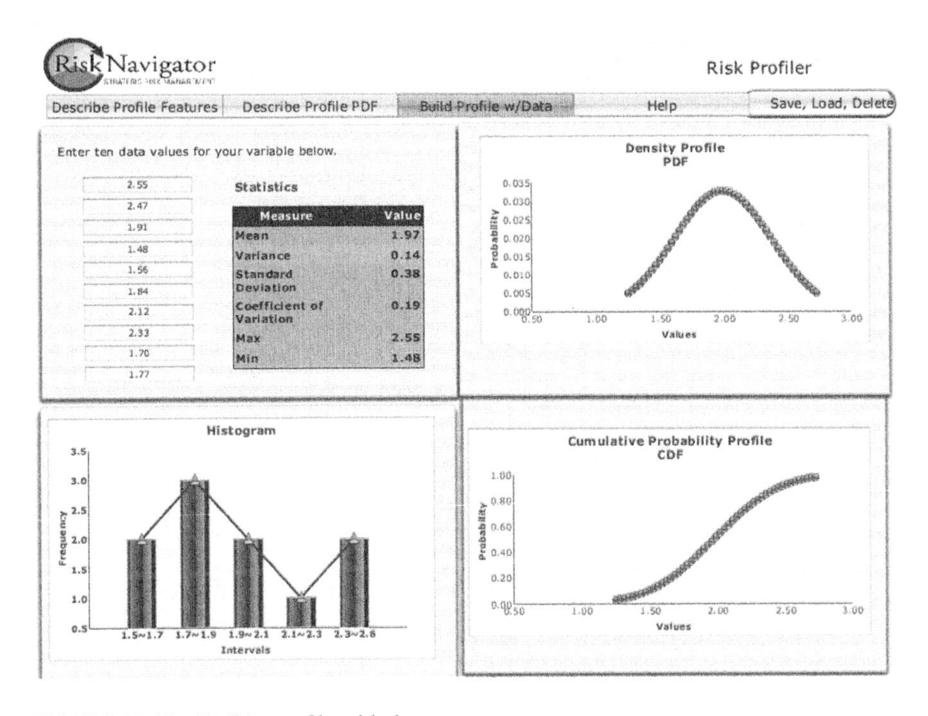

FIGURE 10.13 Build a profile with data.

REFERENCES

Buschena, D. 2002. Non-expected utility: What do the anomalies mean for risk in agriculture? In *A comprehensive assessment of the role of risk in U.S. agriculture*, edited by R. Just and R. Pope. Natural Resource Management and Policy Series. Boston: Kluwer Academic Publishers.

Clemen, R., and T. Reilly. 2001. *Making hard decisions with Decisions Tools*[R]. Pacific Grove, CA: Duxbury, Thomson Learning.

Frey, C. 1998. Briefing Paper Part 1: Introduction to uncertainty analysis. Department of Civil Engineering, North Carolina State University. Accessed June 9, 2009 from http://www.ncsu.edu/~frey/freypub.html

Hardaker, B., and G. Lien. 2005. Towards some principles of good practice for decision analysis in agriculture. Working paper, Norwegian Agricultural Economics Research Institute, Oslo, Norway.

Vose, D. 2000. *Risk analysis: A quantitative guide,* 2nd ed. New York: John Wiley & Sons.

11 Step 7

Choose the "Best" Risk Management Alternative

Dana Hoag, Eihab Fathelrahman, and Jim Ascough

CONTENTS

The purpose of the Strategic Risk Management (SRM) tactical phase in Risk Navigator is to finalize our tactical plan in preparation to take action in the next phase. The tools in this chapter take advantage of a payoff matrix and an individual's risk personality to rank management actions where there are trade-offs between risk and returns. For example, you could compare selling on the cash market to hedging or forward contracting. Perhaps the cash market makes the most money, followed by hedging, and then forward contracting. But the risks might be exactly reversed; for example, forward contracting might provide the most stable income followed by hedging and then the cash market. How would you choose which of these strategies is best for you? Every person is unique—you will balance risk and reward trade-offs differently than someone else. Choosing a strategy depends upon your risk personality, which in turn depends on your strategic resources such as financial health, risk attitude, and goals (Steps 1 through 3 in the strategic phase).

Part 1 of this chapter illustrates the personal side of risk management, using EWS Farms as a case study. As we determined in Chapter 9, Aaron has three options for marketing his corn: cash market sale, hedging in the futures market, and forward pricing. In Part 2 we explain all the gory details about how to rank these risk management options to determine which is best suited for Aaron and his family. Over ten different ways to rank risk are described; seven are included in the Risk Navigator tool called Risk Ranker. Others are included in two other Navigator tools, Value at Risk and the Risk Efficiency Tool. Each of these methods takes advantage of a different way of thinking about risk. Lastly, we discuss when it is necessary to use information about your risk preferences. Many of the methods presented do not require risk preference information, but sometimes it is absolutely necessary. We describe all three tools in Part 3. As in the previous chapters, we demonstrate these tools using EWS Farms as an example.

11.1 PART 1: THE EWS FARMS CASE STUDY

11.1.1 OBJECTIVES

In Step 4 Aaron decided that he wanted to look at corn price risk. In Step 6 he determined three management actions to consider: cash marketing, hedging, and forward pricing.

The objective of Step 7 is to develop a payoff matrix for EWS Farms and rank management actions for pricing corn by using a ranking criterion best suited to Aaron's risk personality.

11.1.2 STRATEGY

We rely on a payoff matrix to help Aaron organize all the information about his risk management choices and teach him about methods to help him choose a risk management strategy. As discussed in earlier chapters, a payoff matrix is a table that acts as a decision tool for ranking risk management actions, given potential payoffs and their likelihood of occurrence. Aaron's payoff matrix for selling corn, which he prioritized in Step 4 (Chapter 8), is built using the management actions he has chosen to consider from Step 5 (Chapter 9) and the risk profiles from Step 6 (Chapter 10). Step

7 is to choose which management action to use. The payoff matrix consolidates risk information into one place in a format that makes it easy to read and compare. We describe several methods to rank risks using the information in the payoff matrix. Finally, we enter the payoff matrix into Risk Ranker, which ranks each management action in the payoff matrix using seven different methods. From there, it's up to the Spragues to choose the management action that best suits them.

To make his decision, Aaron can draw from his family's risk personality. He knows his financial health from Step 1 and his personal risk preference score from Step 2. This information may be the determining factor to rank management alternatives when there are trade-offs in risks and returns. That is, your risk personality will dictate exactly how much return you would be willing to give up, if any, to avoid a given level of risk. However, sometimes your risk personality doesn't even change which management action would be best. For example, an irrigation system that reduces risk and increases profit will be an improvement no matter how risk averse you are. Your only risk personality consideration would be your financial health. In other cases, risk matters, but not very much. It is for this reason that Risk Ranker contains a special tool, called *risk indexing*, which shows whether your risk preferences have an impact on management rankings.

11.1.3 Implementation

We will implement Step 7 on EWS farms in three stages:

1. Develop a payoff matrix
2. Rank management actions by criterion
3. Determine if risk should be considered.

11.1.3.1 Develop a Payoff Matrix

We developed a payoff matrix from Aaron's decision tree in Chapter 9 (see also Figure 11.2). The payoff matrix arranges information in a handy format for making comparisons, so that we can rank risks based on your risk personality. The payoff matrix lists Aaron's management actions, potential scenarios that describe what might happen to prices, their probabilities, and the profit that Aaron would make for each scenario and risk management action. A more detailed description of how to build and interpret the payoff matrix is provided in Part 2 of this chapter.

11.1.3.2 Rank Management Actions by Criterion

Once Aaron develops his payoff matrix, he can use it to rank the desirability of each of his corn marketing alternatives. The ranking will consider risk and return simultaneously. Several tools are available to complete the ranking, as explained in the next section, Part 2, and demonstrated in Part 3 under the Risk Ranking tool.

The results of different ranking strategies for EWS farms are presented in Figure 11.1. Aaron developed five risk management alternatives, the three original actions described previously; plus a second, modified, hedging strategy (Hedge 2) and a second, modified, forward contracting scenario (Forward Price 2). The rankings,

	Management Actions				
Risk Index	Cash	Hedge 1	Hedge 2	Forward Price 1	Forward Price 2
Maximum Expected Value	1	5	2	3	4
Maximax Index	1	2	3	5	4
Most Likely	3	1	2	5	4
Minimax Regret	1	5	2	4	3
Hurwicz Index	1	3	2	5	4
Maximin Index	4	5	2	1	3
Laplace Insufficient Reason	1	5	3	2	4
Mode	1	5	2	5	4

FIGURE 11.1 Risk management rankings for EWS Farms.

from 1 to 5, are listed for each of the criteria considered for each risk management alternative. For example, the cash system ranked first when using the maximum expected value criteria and third using the most likely criteria. The Hedge 1 scenario ranked fifth for the maximum expected value criteria but first for the most likely criteria.

So, which is the best system? To answer this question a decision maker would need to understand which rule fits her risk personality best; that is, the "best" management actions will depend on the criteria used. At that point, the decision maker can complete Step 7 by choosing which management action to use and move on to implementing her plan in Step 8. Of course, it can add comfort if more than one rule supports your choice.

11.1.3.3 Should Risk Be Considered?

Risk does not always matter; and sometimes it matters but not enough to make it worth the effort required to manage it. This might be especially true in this case since Aaron is not very risk averse. Risk Ranker and the Risk Efficiency Tool (RET) provide information about how much an individual's personal aversion to risk will affect the ranking of management actions. It is safe to say that Aaron's risk aversion level is so low that he could probably concentrate on maximizing profits and ignoring risk. We will prove this later when we discuss risk indexing in Part 2.

11.2 PART 2: THE FUNDAMENTALS OF RANKING RISKS

11.2.1 The Payoff Matrix

In Chapter 2, we introduced you to a payoff matrix and promised to show you how to use it to rank your risk management alternatives. We are now ready to do that. Focusing on the risk chosen in Chapter 8, the price of corn, we will build a payoff matrix using the risk profiles from Chapter 10 and the management actions selected in Chapter 9. In this simplified example, Aaron narrowed the probability states down to just two possibilities, a normal crop and a short crop. His outcomes are reported as total revenue rather than price per bushel. The resulting payoff matrix is shown in Figure 11.2. If there is a normal market, Aaron predicts the gross revenue will be \$287,700, which is about \$2.10 per bushel. But he feels the price will increase to

Probability		Management Actions		
States	Probability	Cash Market	Forward Contract	Hedge
		Outcomes (Total Revenue)		
Short U.S. Crop	0.65	**$342,500**	$332,700	$336,200
Normal U.S. crop	0.35	$287,700	**$305,900**	$295,400

FIGURE 11.2 Simplified payoff matrix example for EWS Farms.

$2.50 if there is a short crop, yielding him $342,500 in total revenue. He estimates a 65% chance of a short crop and a 35% chance of getting a normal crop. He has three management actions to consider. He can sell all his grain on the cash market or, alternatively, he could forward price or hedge about half of his crop. As shown in Figure 11.2, Aaron enters the outcomes (revenues) for the remaining management options for each probability state, a normal crop and a short crop.

Aaron enters this information into the standard payoff matrix format (Figure 11.2). On the left, he enters the states of nature that could occur: in this case, a short or normal crop. Think of states like scenarios. You might list weather scenarios like low rainfall, average rainfall, and high rainfall. Or if you were looking at agricultural pests, your list might read no infestation, light infestation, medium infestation, or high infestation. You can make your list as long as you wish; however, you will have to provide probability and outcome information for each item on your list, so practicality might dictate keeping your list brief.

The next part of the matrix is the probability that each state will occur. In Chapter 10 we discussed many methods to estimate probability, but it could be as simple as entering your gut feeling. Aaron picked two states of nature, a normal crop and short crop. He feels there is a 65% chance there will be a short crop, leaving a 35% chance for a normal crop. The other part of the matrix focuses on the management actions. Aaron entered three, moving from a cash market on the left to hedging on the far right, though this list also can be as long as you desire. Finally, the matrix outcomes have to be filled in. Each cell represents what would occur given the state (to the left) and action (above). For example, Aaron predicts he will receive $342,500 in total revenue if he sells on the cash market and there is a short U.S. crop. He estimates his revenue will fall to $336,200 if he hedges and there is a short crop. The core of the matrix is the outcomes section; it lists all of Aaron's potential revenues for each combination of states and management actions.

If you had a crystal ball and knew that there would be a short crop, you would choose to sell on the cash market (outcome bolded in the short crop row of Figure 11.2) because it makes the most money when there is a short crop. Likewise, if you knew the crop would be normal you would forward contract (bolded). But you don't know whether there will be a short crop, which is why this is a risk problem. Take a moment and try to decide which management action you would choose based on this risk payoff matrix. Write down how you made your choice. That is, what was it that made one management action better for you than another? Trying to rank the options in this example will help you understand better when we show you some ways to rank actions later in this chapter.

Including risk can make decisions much more difficult to make. But before we talk about how to manage risks, let's talk about when it is appropriate to include it.

11.2.2 When to Include Risk

Generally speaking, the more risk a producer is willing to take on the more profit she can make. Therefore, it stands to reason that transferring your risk to someone else should reduce profits. For example, crop insurance can be purchased to reduce risk, but your income will be reduced by the cost of the premium the insurance company charges you. Forward contracting also reduces risk, but the person buying your crop will certainly offer a lower price than you could expect, on average, if you sold the crop yourself. Therefore, a big part of risk management is to determine if the benefit of managing risk yourself is worth the extra money, compared to the cost that someone would charge you to transfer the risk to them. But sometimes managing risks does not produce enough benefit to make it worthwhile to even bother with looking at the costs to manage it, whether it is you or someone else doing so.

There is no question that for many decisions, producers gain a great deal from managing risk. Kastens (2001) looked at records from 2,000 growers in a Kansas Farm Management Analysis and Research Program. He wanted to know how the top third of producers differed from average producers. The advantage of the top third of producers is shown in Figure 11.3.

Kastens found that the top third earning producers, across all types of farms, had a 31.9% lower production cost per acre than an average farm. The value of this advantage averaged $48.61 per acre. It is clear from this study that managing cost is perhaps the best strategy for making money. Interestingly, variability in farm income ranked second. Top producers had two-thirds less variability, which earned them a $28.50 per acre higher return. Considering this lesson from top producers, risk does matter to someone.

On the other hand, Musser and Patrick (2002) concluded that farmers are not very willing to accept a loss in order to take on a risk. Producers tend to be more

Management Strategy	Change from Average (%)	Overall Value ($/ac)
Decrease cost	31.9	48.61
Increase in yield	14.4	7.56
Increase in price	8.7	0.28
Increase in government payments	70.3	9.60
Increase in size	80.0	11.95
Increase in farm income variability	66.8	28.5

FIGURE 11.3 Value of management strategies for top third of producers in a Kansas study. (Data from Terry Kastens, "Risk and Reward: How Do Farm Returns Stack UP? Should Farm Managers Invest in the Stock Market?" Kansas State University Extension Web site, http://www.agmanager.info/farmingmgt/finance/investment/LandvsStocks.PDF, 2001.)

risk neutral than risk averse, contrary to conventional wisdom that says producers are primarily risk averse. Musser and Patrick's producer survey found that 33.6% of cotton and 36.8% of corn producers would not be willing to reduce their average yield in trade for a technology that gave the same yield every year (no yield risk). Furthermore, 51.3% did not agree with the statement: "My primary marketing goal is to reduce risks rather than raise my net sales price." Only 34.3% were willing to take a lower price to reduce risk. That is, only about one third of the producers claimed they were risk averse.

Musser and Patrick's study raises the concern that not everyone is risk averse. But perhaps it's the size and type of the risk that determines whether it matters. For example, it is generally true that the greater the debt/asset ratio in a farm business, the more a manager will worry about having a poor operating profit; after all, capital is needed to service debt. It takes a lot of effort to include risk in the decision-making process, and sometimes considering risk does not alter your management choices anyway. In these cases, including risk in the decision might not be worth the effort. Therefore, consider risk when it is worthwhile. Two of the Navigator tools discussed in this chapter will help you know when risk is important to consider.

Deciding not to consider risk does not necessarily mean that you are risk neutral. Instead, it may be that the consequences of failing to manage the risk are small or very unlikely. Likewise, the benefits of an action might overwhelm any concern for the potential cost. For example, would you buy a lottery ticket that had an 80% probability of paying $10,000? Most of us would say yes if the ticket cost $1. Therefore, a risk study is not warranted, i.e., the benefit is so large compared to the cost that virtually everyone would still buy a ticket. Another situation where you might choose to ignore risk is when the benefit is too small to make it worth the effort.

Risk Navigator SRM gives you the skills and tools you need to manage risk; however, it also provides you with the tools needed to decide when you can ignore it. Risk Ranker includes a special risk indexer that can help. As shown in Figure 11.4, we develop a certainty equivalent index (CEI) and a risk premium index (RPI). Recall that the certainty equivalent (CE) is the amount that would make you indifferent to a risky prospect. For example, you might be indifferent between an investment in the stock market that is expected to earn $10,000 and putting your money in CDs to earn a guaranteed $8,000. Your CE would be $8,000 and your risk premium (RP) would be $2,000. The CEI is simply the CE divided by the expected value (EV). For example,

	Risk Preference (0.5 = somewhat risk averse, 4.0 = highly risk averse)				
	0.5	1.0	2.0	3.0	4.0
Expected Value (EV)	$12,000	$11,000	$10,000	$9,000	$8,000
Certainty Equivalent (CE)	$10,400	$9,200	$8,000	$6,800	$5,600
Risk Premium (RP)	$1,600	$1,800	$2,000	$2,200	$2,400
CE Index (CEI, % of EV)	86.7%	83.6%	80.0%	75.6%	70.0%
RP Index (RPI, % of EV)	13.3%	16.7%	20.0%	24.4%	30.0%

FIGURE 11.4 Risk indexing to determine impact of risk management efforts.

if your risk preference was measured in Chapter 6 as 20, the CEI is 80%, and the RPI equals the 1-CEI, or 20%. If this was your index, your risk personality would be willing to sacrifice up to 20% of the expected value to trade away the risk for certainty. Therefore, risk would be important to you. The basic idea is that it takes effort to manage risk; so don't manage it unless it is important to your decision. Everyone is different, but many of us might not want to factor in risk when the CEI is very high, say over 95%, because the gains don't typically make it worth the cost to manage.

The Risk Navigator tool reports your CEI for the full spectrum of risk aversion levels, one risk management action at a time (e.g., Figure 11.4 shows the CEI for cash sales). You simply read off the CEI or RPI for the risk aversion level that best represents you.* Then you decide if the gain from managing risk makes it worth the effort. In Aaron's case, for example, he knows his risk preference lies somewhere between 0.5 and 1.0, which means he has an RPI of about 1 to 2%. That's about $3,000 to $6,000 for the investments he is considering. The question then becomes, will Aaron spend more time and effort than $3,000 to manage risk? If so, he loses more than he gains when he manages risk.

Another tool that can be used to determine when risk is important to consider is the Risk Efficiency Tool (RET). RET uses stochastic efficiency with respect to a function (SERF) analysis to rank outcomes for all levels of risk preference. SERF works by identifying utility-efficient alternatives for a risk preference and ordering preferred alternatives in terms of certainty equivalents (CEs) as the degree of risk aversion increases. The CEs are readily interpreted because, unlike utility values, they can be expressed in monetary terms. For a rational decision maker who is risk averse (the typical case for farm planning), the estimated CE is less than the EV of the risky strategy. Anderson and Dillon (1992) proposed an approximate classification of degrees of risk aversion based on relative risk aversion with respect to wealth $r_r(w)$ in the range 0.5 (hardly risk averse) to 4 (extremely risk averse). The decision rule for SERF is to rank the risky alternatives (within the decision maker's specified risk aversion coefficient) from the most preferred to the least preferred along the scale developed by Anderson and Dillon.

Simply put, the SERF analysis graphically shows which management action is superior for all levels of risk. Aaron's corn pricing example is shown in Figure 11.5. The CE is graphed for each of the three alternatives and for all possible risk aversion levels, ranging from neutral to extremely risk averse. Note that the cash marketing alternative has the highest CE for all levels of risk aversion. That is, no matter how risk averse or loving you are, you would prefer the cash market, since it is always the highest line. Therefore, in this case, risk does not make a difference. The lines never cross. Aaron knows that his risk aversion coefficient is less than 1, but it does not matter because the cash scenario is always best.

Sometimes one action is better than another at one level of risk aversion, but it is inferior at others. If, for example, the cash marketing line dipped under the hedging line in the graph, let's say at a risk aversion coefficient of 2, then you would choose the cash marketing alternative if you thought your aversion level was less

* We use the following equation to compute the CE for each level of risk aversion: $CE_r = LN(1 - \exp \text{utility})(-r)$, where r is the aversion level being evaluated and LN is the natural logarithm.

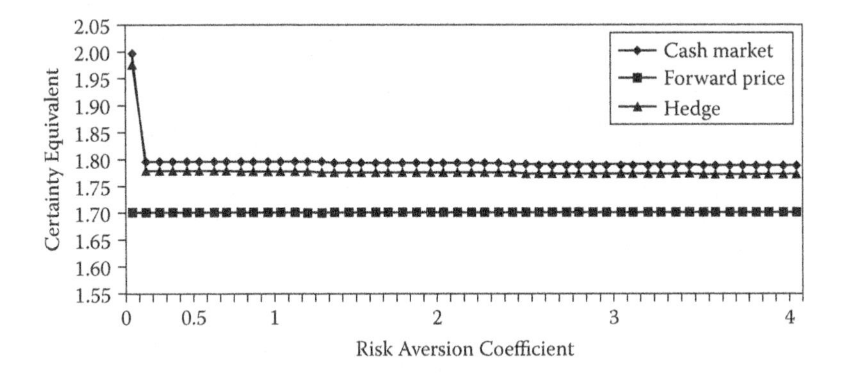

FIGURE 11.5 Stochastic efficiency with respect to a function (SERF) example for Aaron Sprague's corn marketing decision.

than 2 and hedging if you were more risk averse than 2. For Aaron, the cash market always dominates hedging and hedging always dominates forward pricing. He need not worry about his risk aversion because it has no impact. You can calculate your risk tolerance, or preference level, in Chapter 6. We discussed the fact that these estimates are difficult to make, but SERF gives you some extra confidence because you can see at what range of risk aversion your rankings hold, thus relieving you from having to pin down your exact risk aversion level. Many, if not most of the time, the ballpark around your estimated risk preference is all you need to know.

11.2.3 Ranking Risks Using a Payoff Matrix

Fortunately, many methods have been developed to rank different risk management actions. We will limit the criteria we consider for now to those that we can use with the payoff matrix approach. Let's start with the risk profile example for corn price developed in Chapter 10. As previously discussed, people generally want higher means and less dispersion (variance or standard deviation) for the outcomes of risky decisions. The risk profile can be used to compare risk management actions based on their statistical properties. For Aaron, the mean corn price was $1.97 and the standard deviation (SD) was 0.38. Suppose that Aaron had to choose between a management action with this distribution for selling on the cash market and another action with a different distribution. Further suppose that we had four hypothetical distributions with different mean–SD combinations for comparison. We make them hypothetical for now so we can cover these four distinct possibilities, as shown in the following chart:

	Mean	Standard Deviation
Cash sale	1.97	0.38
Alternatives		
Same mean, lower deviation	1.97	0.19
Same mean, higher deviation	1.97	0.76
Same deviation, lower mean	2.50	0.38
Same deviation, higher mean	1.50	0.38

Sometimes we can rank one distribution over another by using these risk profiles. We can do so from the numerical example above or with a visual inspection of the probability density functions (PDFs) and cumulative distribution functions (CDFs). For example, by looking at the numbers in the previous chart or a graph of the PDFs, we can see that the first alternative distribution is better than the cash sale because variation is reduced without sacrificing the mean. The second is never preferred because risk is increased without any gain in the mean. The third distribution is never preferred because the mean is reduced with no reduction in risk. Finally, the fourth distribution is preferred over the cash sale because the mean is increased with no change in risk. We can rank each of these distributions compared to the cash sale distribution; however, we cannot rank two distributions where the risk is better but the mean is worse or vice versa. In these cases, a decision maker would have to choose whether the gains or losses in risk outweigh the gains or losses in the mean. We would then go to tools that account for risk personality.

We can also use graphical displays for the same purpose. For example, we show the distribution of the cash sale in Figure 11.6. The mean is kept constant in the top panel and the variance is widened for comparison. If we increase the standard deviation (defined in Chapter 10) the curve flattens out, increasing the variability (low to high) of the outcome. If we decrease the standard deviation then the curve becomes thinner. A risk-averse person would choose the "skinniest" distribution if they all have the same mean. That is, if you could make a $100 average income regardless, you would prefer the distribution with the lowest range. Looking to the bottom panel of the figure, we show what happens when variance stays the same but the mean changes. A lower mean with the same variance is inferior, while a higher mean and the same risk is superior (shown as the curve shifting to

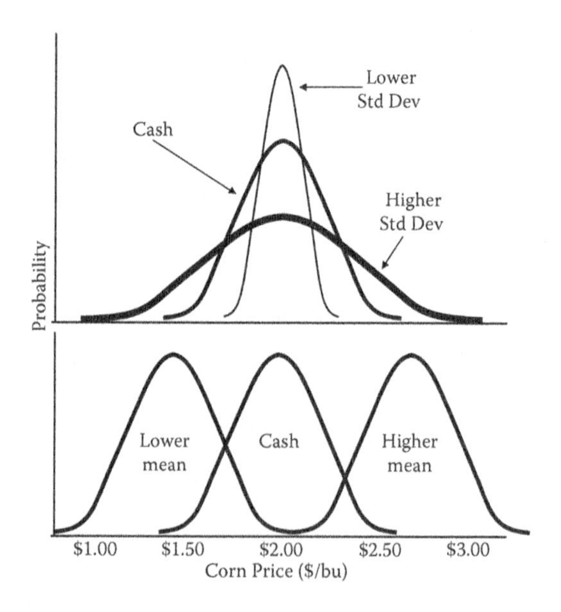

FIGURE 11.6 Comparing risk profiles.

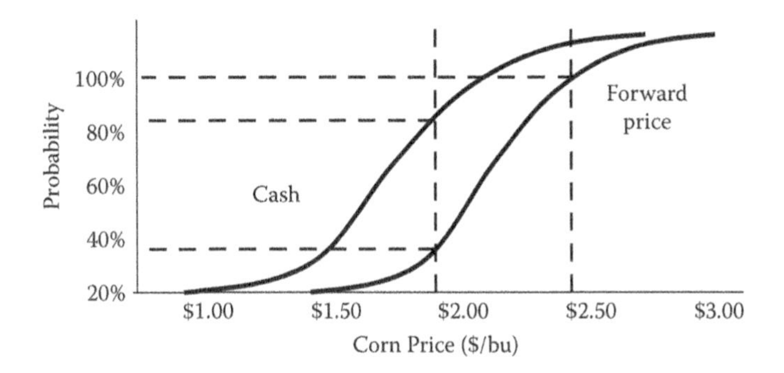

FIGURE 11.7 A comparison of CDFs for use in stochastic dominance.

the right). That is, if the risk stays the same, the highest average income possible would be preferred.

Risk Ranker provides you with the ability to visually compare the distributions from your payoff matrix, as shown in Figure 11.6. A complete example is shown in Part 3. Unfortunately, it is seldom the case that you can use just the PDFs to compare distributions. Sometimes you can simply move on to comparing the CDFs. A concept called *stochastic dominance* can help. Suppose two PDFs overlap; for example, one has a higher mean but a higher variance too. We can then turn to the CDFs for help. Take two distributions as shown in Figure 11.7. The CDF for the forward price management option starts at $1.50 per bushel and ends at $3.10 per bushel. The CDF shows that there is no chance of getting less than $1.50 or more than $3.10. Choose any value between these two values, for example $2. By drawing a line upward vertically from that point until hitting the CDF of the forward price, we can see there is about a 20% chance that the price will be below $2 and an 80% chance that it will be above $2. If we take the point where the same vertical line hits the alternative plan, cash sale, there is a 65% chance that the price will be below $2. Which would you choose?

As it turns out, when one curve is everywhere to the right of another, the curve on the right would be preferred no matter what your risk preference is when positive outcomes are good. That is, you could be risk averse, risk neutral, or even risk loving, and you would choose to forward price your crop. This is the case because at every value there is always a higher chance that the curve on the left will have a less favorable outcome compared to the curve on the right. This is called *first-degree stochastic dominance.* Stochastic dominance formally assumes a decision maker has utility for income (x) defined by a monotonically increasing function $u(x)$. Given two CDFs, $f(x)$ and $g(x)$, distribution $f(x)$ dominates distribution $g(x)$ by first-degree stochastic dominance when there is positive marginal utility of income for all x ($u'(x) > 0$), and for all x the cumulative probability under the f distribution is less than or equal to the cumulative probability under the g distribution (with strict inequality for some x).

Often the CDF curves cross and you can't rely on first-degree stochastic dominance. If you make the reasonable assumption that a person is risk averse, *second-degree stochastic dominance* will assure you that the curve on the right is better as long as they cross above the 50% level. You worry most about bad outcomes when you are risk averse; therefore, you apply first-degree stochastic dominance

Probability		Management Actions		
States	Probability	Cash Market	Forward Contract	Hedge
		Outcomes (Gross Revenue)		
Short U.S. Crop	0.65	**342500**	332700	336200
Normal U.S. Crop	0.35	287700	**305900**	295400
Expected Value		**323320**	**323320**	321920
Minimum Outcome		287700	***305900***	295400
Maximum Outcome		**342500**	332700	336200
Maximum Regret[a]		18200	**9800**	10500
Regret if Short[b]		0	9800	6300
Regret if Normal[b]		18200	0	10500
Hurwicz Index (α = 0.5)		315100	**319300**	315800

a = Maximum regret of short and normal crop for given management action
b = Difference between outcome and best outcome

FIGURE 11.8 Payoff matrix for selling corn on EWS Farms.

for everything that is below 50% and ignore everything above 50%. Formally, we can say that CDF $f(x)$ dominates CDF $g(x)$ by second-degree stochastic dominance under assumptions that a decision maker has: (1) positive marginal utility ($u'(x) > 0$); (2) diminishing marginal utility of income ($u''(x) > 0$); and (3) for all x, $f_2(x)$ is less than or equal to $g_2(x)$ with strict inequality for some x (where $f_2(x)$ and $g_2(x)$ are the second integral of f and g with respect to x).

Risk Ranker draws all of the PDF and CDFs for the actions that you are considering. You can look at the numerical table or at the graphs of PDFs or CDFs to compare your different options. If you are lucky, a simple visual inspection of the PDFs or using stochastic dominance with the CDFs will allow you to rank your best options.

Another, more reliable option, might be to compare the risk profiles numerically. While numbers may not seem as intuitive as graphics, the risk payoff matrix provides some very innovative ways to apply different preferences for looking at risk. Let's reexamine the simplified payoff matrix we introduced earlier for EWS Farms. The payoff matrix is reproduced in Figure 11.8 but with an added twist. At the bottom of the table we have reported the EV, minimum, maximum, and "regret" values of each management action. This information will be used to establish several more ranking rules.

We asked you earlier if you could choose which management action was best for you. Let's start by answering that question in the situation where there is no risk. The bolded outcomes in Figure 11.8 show what you would choose for different situations. For example, you would choose cash sales (bolded in the short crop row) if you knew the crop would be short and to forward price if you knew you would get a normal crop (also bolded). But what if you did not know whether there would be a short or normal crop? That's what risk is all about. Let's discuss some rules that can help Aaron and his family rank which management option best suits them when risk is present.

11.2.3.1 Maximize Expected Value

Expected value (EV) is the weighted sum of the outcomes using the probabilities of occurrence as the weights. The EV is what you would average if you stuck with a particular action for a long period of time. We can compute the EV for cash sales using the following probabilities: 0.65 chance of a short crop and 0.35 chance of a normal crop. When combined with the outcomes in Figure 11.8, the expected value for a cash sale is 0.65 x 342,500 + 0.35 x 287,700 = 323,320. The other expected values are calculated in a like manner. A risk-neutral person would base her choice on the EV. This person would not be concerned with variability in the outcomes but rather with the expected return over the long haul because it makes more money than any of the other strategies. A risk-averse person would not use this approach because it weights downside and upside risk equally. In this case, cash sales and forward contract have tied for the best outcome and are shown in bold.

This rule uses all the information available in the payoff matrix. It uses the probabilities and information about outcomes for all management actions considered. In that sense, this rule is better than some of the other rules we discuss below. It is appropriate for people who are risk neutral, but maintains a lot of ability for decision makers to experiment with risk. Decision makers can, for example, incorporate their expectations, experience, and biases into estimates of probability and even into how they estimate the outcomes given each state and management action. In this way they can experiment with how risk will impact them under different assumptions. For example, a risk-neutral decision maker could see whether changing the chance of a short crop from 65% to 60% or 70% would change the outcome. She might do so if she was unsure about her estimate. She could also experiment with the yield and price assumptions used to fill in the outcome for a short crop with cash sales to see if it makes a difference in which system ranks best.

11.2.3.2 Laplace Insufficient Reason

The Laplace insufficient reason rule, also called *equal likelihood*, is based on maximizing EV but it assumes that every probability state is equally likely. This is the same as the simple average. This rule is used when there is no reasonable basis for assuming that any one state is more likely than another. It is a reasonable criterion to use when no priors exist because it accounts for all the information that is known, albeit assuming everything has an equally likely chance of occurring. Expected value, in comparison, adds information about how likely each state of nature is compared to the others. This makes sense, *if* you have the information. If not, Laplace may be a reasonable rule to follow.

11.2.3.3 Maximax

Maximax is geared toward risk lovers; people who like the thrill of getting a high payoff with great risk. Under this criterion, a person looks through each management action and chooses the one with the highest possible payoff. That is, he or she is choosing the best of the best. To use a maximax ranking with a payoff matrix, first select the best possible outcome for each management action and then choose the action with the best of the best outcomes. For EWS Farms, the best outcome for the cash market sale is $342,500; for the forward contract it is $332,700; and for the hedge it is $336,200, as shown in the maximum outcome row of Figure 11.8. The best of all three of these outcomes is the cash market sale, which is bolded; therefore, the cash market has the highest potential return for EWS Farms.

Maximax and maximin (described next) are intuitively attractive, but one major limitation is that they do not take advantage of all of the information in the payoff matrix. Making decisions in a vacuum can lead to big mistakes.

11.2.3.4 Maximin

A more conservative person might focus on the worst-case scenario. A risk-averse person leans toward the best of the worst returns. This is called the *maximin strategy* because it is designed to maximize the minimum possible outcome. We are choosing the best of the worst that could happen. For EWS Farms, the worst outcome for the cash market sale is $287,700; for the forward contract it is $305,900; and for the hedge it is $295,400. The best of all three of these worst case outcomes is the forward contract; the maximum value is bolded in the minimum outcome row of Figure 11.8.

11.2.3.5 Most Likely

Sometimes an individual has information about what is most likely to happen in a given situation. In such a case, he or she usually would ignore historical probabilities or simply assume that the state with the highest probability will occur. For example, if you had information about a current weather pattern, like El Niño, you might not put much weight on historical weather data. You would decide that a wet year is most likely and make decisions accordingly. In another situation, you might subscribe to a market research newsletter and become convinced that there is going to be a short crop. In this case, you would pick the cash sales, which has the highest return for a short crop. That is, you would pick the highest return in the Short U.S. Crop row.

The most likely rule (also called the *modal index*) has the same limitations as maximax and maximin in that it does not use all available information.

11.2.3.6 Minimax Regret

In many situations, producers have to make decisions before they could possibly know the full extent of the outcome. Down the road, you may regret the choice you made and perhaps wish you had done something differently. Minimax regret, created by Leonard Savage, is a tool that can actually minimize your regrets. Some people call it the "could've, would've, should've strategy"—a strategy that assumes the decision maker will make a decision that does not result in the best outcome and then calculates the difference between the best outcome and the others. The difference is the regret. This decision process computes the maximum (max) regret you could have for each decision, then chooses the management action that has the minimum (mini) of the maximum regrets. It takes four steps:

Step 1: Look at the payoff for each state (short crop and long crop); find the highest payoff and bold it.

Step 2: Compute the regret for each row. The highest payoff for a short crop is the cash sale (bolded). If the crop was short and you had chosen cash sale, the regret is \$0. You made the best decision. If you had chosen to forward contract, however, you might have earned \$9,800 more if you had chosen cash sale (\$342,500 – \$332,700). That is, you could regret that you did not choose a cash sale to the tune of \$9,800. Your regret could be as high as \$6,300 if you had chosen to hedge. If you repeat this process for the normal crop row, you get regrets of \$18,200, 0, and \$10,500. These regrets are shown in Figure 11.8, under regret if "short" and regret if "normal."

Step 3: Identify the maximum regret for each management action. That is, compare the regrets between short and normal crops for each management action. This turns out to be \$18,200, \$9,800 and \$10,500 for the cash, forward price, and hedging management actions, respectively.

Step 4: The last step is to choose the minimum of these maximum regrets, which means that forward pricing is the preferred strategy, as bolded in Figure 11.8.

While this strategy is difficult to understand, it is the strategy that psychologists say most people use when making risk decisions. It links economics with emotions. Choosing forward pricing ensures the fewest regrets about what should have been chosen after the fact.

11.2.3.7 The Hurwicz Index

The Hurwicz index combines optimistic and pessimistic indexes, like maximax and minimax, respectively, in proportions determined by a coefficient α (optimistic) and $1 - \alpha$, (pessimistic). We assume $\alpha = 0.5$ in Navigator, which means we assign an equally likely chance of the worst outcome and best outcome happening for each management scenario. Then we choose the one that has the highest payoff. For example, in Figure 11.8, the Hurwicz index for the cash sale is 0.5 x 342,500 + 0.5 x 287,700 = 315,100. As shown by the bold text, forward contracting yields the maximum Hurwicz index.

This index uses more information than maximax, maximin, and most likely, but it still only uses two values in the payoff matrix, the high and the low. Therefore, a decision maker is still at risk of missing important information provided in the payoff matrix when using this criterion.

11.2.3.8 Minimum Standard Deviation/Coefficient of Variation

Variability in outcomes can be measured using the standard deviation (SD), which is defined as the square root of the variance of a set of potential outcomes (see Chapter 10). It is a measure of dispersion, or spread of the outcomes, and therefore indicates risk in the sense that the wider the dispersion is, the more risk there is. SD cannot be displayed in the example used here (since there are only two observations for each management action), but Risk Ranker does provide this information. As shown in Chapter 10, a person will realize an outcome within one SD from the mean two thirds of the time, and will realize a return within two SDs about 95% of the time. Standard deviation is not a perfect measure of risk since its spread includes both good and bad outcomes. Most people are not worried about the risk of getting more than the average income. Nevertheless, it is an indicator of risk since it measures how far the worst outcomes may fall below the mean. Using this measure, the decision rule is to minimize the SD, that is, to reduce the dispersion of potential outcomes to its lowest possible level. We could not compute the SD with only two outcomes in the EWS Farms example.

The coefficient of variation compares the EV to the SD and is defined as SD divided by EV. This risk measure combines variability and return in a very practical way. It still suffers from the same problem as minimizing the SD, since good and bad outcomes are both included in dispersion; however, it is very intuitive to compare the amount of dispersion to the EV. For example, if the SD for soybean prices was $2 and the EV was $10, the price would be expected to fall between $8 and $12, two thirds of the time. The coefficient of variation would be 0.2. A riskier prospect with an SD of $4 would have a coefficient of variation of 0.4. Or, said in reverse, a soybean price with a coefficient of variation equal to 0.4 would be between $6 and $14 two thirds of the time. This rule also cannot be applied to our example. If it could, we would choose the lowest coefficient of variation. You might look back at Figure 8.3 to see the coefficients of variation the Economic Research Service estimated for corn yield across the United States.

11.2.4 Ranking Risks without a Payoff Matrix

11.2.4.1 Value at Risk

Value at Risk (VaR) is a popular method for capturing the downside risk in financial decision making. It is an evaluation of what you stand to lose with an investment. VaR answers the questions "What is my worst-case scenario?" or "How much could I lose in a really bad month?" (Harper, 2008). This strategic tool considers only the undesirable parts of dispersion, as opposed to the SD, for example, and determines how much value there is in the undesirable part of the PDF tail.

For a confidence interval of 95%, VaR looks at the worst loss expected 5 times out of 100. Usually, the VaR is related to a specific time period since it is used so much in investing. For example, suppose you could make an investment with an expected return of $1 million where the SD was 4%. Then, suppose you wish to know the VaR for a 95% confidence level. You already know that you could make more or less than $1 million, which is represented by a PDF. Find the investment return that leaves 5% in the left tail. If this were a standard normal function, the 5% tail is the (mean) − (1.65 times the mean). To compute VaR, multiply $1.65 \times 0.04 \times \$1$ million = $66,000. In this case, there is a 5% chance of earning $66,000 or less on an investment in which you expect to earn $1 million. A complete example of this method is provided below in the discussion about the Risk Navigator Value at Risk tool.

11.2.4.2 Safety First and Stoplight Charts

Methods that rely on evaluating CDFs may be difficult for many people to understand. A *stoplight graph* is a more visually appealing depiction of probabilistic information. The stoplight procedure (Richardson et al., 2006) calculates the probability of a measure (e.g., mean gross margin) exceeding an upper cutoff value, being less than a lower cutoff value, or having a value between the upper and lower cutoff values. Like a stoplight, the three ranges are assigned colors of red (less than the lower cutoff value), yellow (between the upper and lower cutoff values), and green (exceeding the upper cutoff value). For example, you might determine that you need at least $250,000 in gross sales to keep out of trouble with your banker, and you would prefer to get over $300,000. The red zone for you would be below $250,000. The green zone would be over $300,000 and the yellow zone would fall in between. The underlying information for the stoplight comes from the CDF. For example, using the CDF on the right in Figure 11.7, you could create a lower cutoff of $2 per bushel and an upper cutoff of $2.50 per bushel. You would expect to get below the cutoff price 20% of the time, in the yellow (between $2 and $2.50) 60% of the time, and in the green (above your upper cutoff) 20% of the time. Part 3 of this chapter provides a graphical presentation of this situation.

11.2.5 Making Your Decision

Several methods have been presented to help you decide which management action to pursue, but you should pick the tool that matches your risk personality. Use minimax to avoid bad outcomes at all cost. Maximize EV if you are risk neutral. Use a SERF analysis if you want to maximize returns while considering your risk tolerance and risk preference. Most importantly, remember that every tool has its limitations. SERF, for example, won't protect you against worst outcomes. Minimax, minimax regret, and the Hurwicz criteria each use only a portion of the risk information. This can be, but is not necessarily, problematic. For example, suppose you had a choice between two actions, A and B. Would you prefer action A, where there is a 50-50 chance of getting 100 or 0, or B with a 100% chance of getting 1? Maximin would always choose B, since the maximum of the minimums is 1, even though B is clearly inferior for most people.

States	Action A	Action B
1	100	1
2	0	1

All of the rules have potential problems like the example of maximin above. The key is to understand the nuances of each rule and apply the one that best fits the situation and your risk personality. The Risk Navigator tools are easy to use and it is not costly to try and compare different methods.

It is also advisable to read books and periodicals about these risk rules, such as *Management Decision Theory: Managing Pragmatically* by Howard Flomberg (2008). There is even some very good information on the Internet that is a little easier to follow if you are so inclined. Finally, we have provided some innovative ways to expand beyond the boundaries of the SRM process in Chapters 13 through 18. Chapters 13, 14, 15, and 16 discuss methods to manage risk in the five risk categories, production, market and price, financial, institutional, and human. These chapters discuss techniques commonly used to address each type of risk and don't necessarily fit into the SRM process. Finally, we show you how to use more sophisticated tools like Excel and @Risk in Chapter 18, which allow you to build your own customized plans.

11.3 PART 3: RIGHT RISK NAVIGATOR MANAGEMENT TOOLS

Three tools have been designed to help you with ranking alternative risk management actions: Risk Ranker, Value at Risk (VaR), and the Risk Efficiency Tool (RET). Risk Ranker helps you determine how influential risk is in your decision by computing the CE index, allows you to compare risk profiles side by side, and ranks risk by seven different ranking rules. VaR is a stand-alone tool. SERF and stoplight graphs can be created using the RET.

11.3.1 Risk Ranker

The payoff matrix is the first item to appear in the Risk Ranker tool (Figure 11.9); it is designed to help you compare alternative management practices. Note that the second and third columns (probability and cash outcomes) are exactly the same as the information needed to create the risk profile we developed in the Risk Profiler tool in Chapter 10 (see Figure 10.12). Each probability scenario in the first column of the payoff matrix can be named. We have created five probability scenarios: low prices, weak prices, average prices, good prices, and high prices. The first management action is cash for cash sales, and four alternative actions are listed for comparison: hedge 1, hedge 2, forward price 1, and forward price 2. To complete the payoff matrix, the outcomes have to be entered for each probability–management combination. For example, the Sprague's said a low price of $1.26 would occur 10% of the time and that 20% of the time they would get a price of $1.67.

As shown in Figure 11.10, the first tab on the main menu in Risk Ranker is "Compare Profiles." The payoff matrix is reprinted in the upper left corner. The risk profile for each management action is determined from the probability column combined with the outcome column. This allows Risk Ranker to plot all PDFs and CDFs in one graph. Ranker also prints summary statistics for each profile. If you are

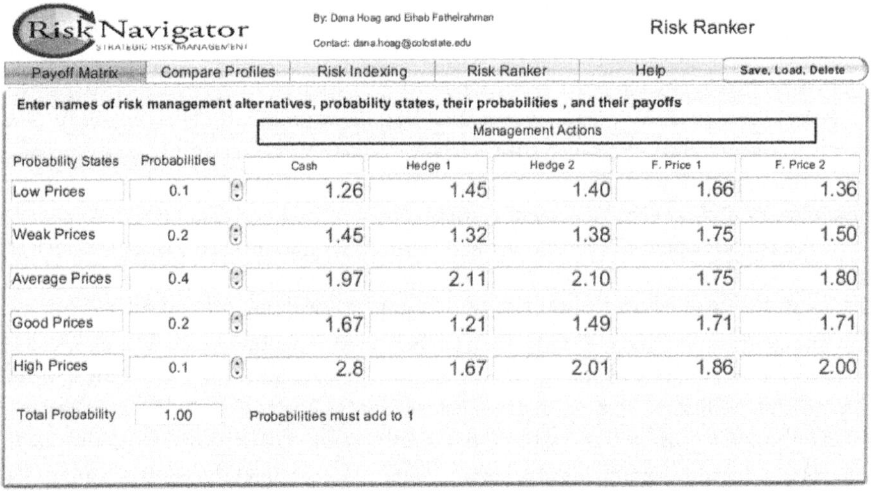

Risk Navigator
STRATEGIC RISK MANAGEMENT
By: Dana Hoag and Ehab Fathelrahman
Contact: dana.hoag@colostate.edu
Risk Ranker

Payoff Matrix | Compare Profiles | Risk Indexing | Risk Ranker | Help | Save, Load, Delete

Enter names of risk management alternatives, probability states, their probabilities , and their payoffs

		Management Actions				
Probability States	Probabilities	Cash	Hedge 1	Hedge 2	F. Price 1	F. Price 2
Low Prices	0.1	1.26	1.45	1.40	1.66	1.36
Weak Prices	0.2	1.45	1.32	1.38	1.75	1.50
Average Prices	0.4	1.97	2.11	2.10	1.75	1.80
Good Prices	0.2	1.67	1.21	1.49	1.71	1.71
High Prices	0.1	2.8	1.67	2.01	1.86	2.00
Total Probability	1.00	Probabilities must add to 1				

FIGURE 11.9 Risk Ranker: The payoff matrix, with EWS Farms example.

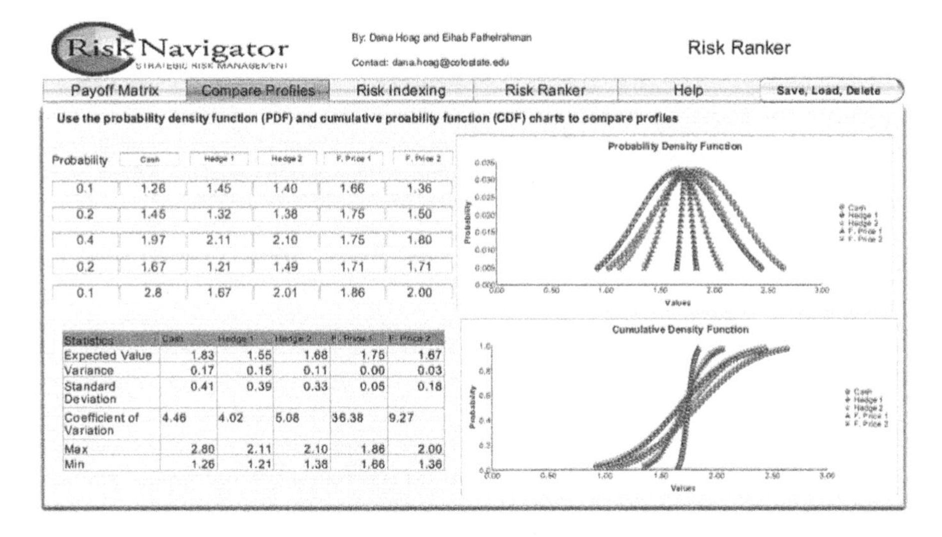

Probability	Cash	Hedge 1	Hedge 2	F. Price 1	F. Price 2
0.1	1.26	1.45	1.40	1.66	1.36
0.2	1.45	1.32	1.38	1.75	1.50
0.4	1.97	2.11	2.10	1.75	1.80
0.2	1.67	1.21	1.49	1.71	1.71
0.1	2.8	1.67	2.01	1.86	2.00

Statistics	Cash	Hedge 1	Hedge 2	F. Price 1	F. Price 2
Expected Value	1.83	1.55	1.68	1.75	1.67
Variance	0.17	0.15	0.11	0.00	0.03
Standard Deviation	0.41	0.39	0.33	0.05	0.18
Coefficient of Variation	4.46	4.02	5.08	36.38	9.27
Max	2.80	2.11	2.10	1.86	2.00
Min	1.26	1.21	1.38	1.66	1.36

FIGURE 11.10 Risk Ranker: Compare Profiles, with EWS Farms example.

lucky, you can determine the best management alternative from this analysis; however, there usually will be too much overlap to make reliable judgments. Can you tell which management action is best based on the PDFs or CDFs?

The next tab in the menu is "Risk Indexing." Risk indexing reveals how influential risk is for a particular problem. The index table, Figure 11.11, includes the CE index (CEI) and RP index (RPI) for risk aversion levels from almost risk neutral (0.5) to extremely risk averse (4.0). CEI and RPI are the CE and RP, respectively, divided by EV. We have entered the risk profile for the cash sale. The CEI ranges from 99 to 92%. Since Aaron is not very risk averse, he would look at the CEI for something

Risk Navigator
STRATEGIC RISK MANAGEMENT
By: Dana Hoag and Ethab Fathelrahman
Contact: dana.hoag@colostate.edu
Risk Ranker

Payoff Matrix | Compare Profiles | Risk Indexing | Risk Ranker | Help | Save, Load, Delete

Is your risk premium large enough to justify risk analysis?

Probability	Cash
0.1	1.26
0.2	1.45
0.4	1.97
0.2	1.67
0.1	2.8
Expected Value (EV)	1.82

Risk Indexing Results (Somewhat risk averse — Highly risk averse)

Risk Preference	0.5	1	2	3	4
Certainty Equivalent (CE)	1.80	1.77	1.74	1.71	1.68
Risk Premium (RP)	0.02	0.04	0.08	0.11	0.14
CE Index (% of EV)	99%	98%	96%	94%	92%
RP Index (% of EV)	1%	2%	4%	6%	8%

Enter your risk profile in the first two columns. Next, read down the column under your risk preference to find your certainty equivalent and risk premium. The bottom row shows the percentage risk premium you would pay to avoid this risk.

FIGURE 11.11 Risk Ranker: Risk Indexing, with EWS Farms example.

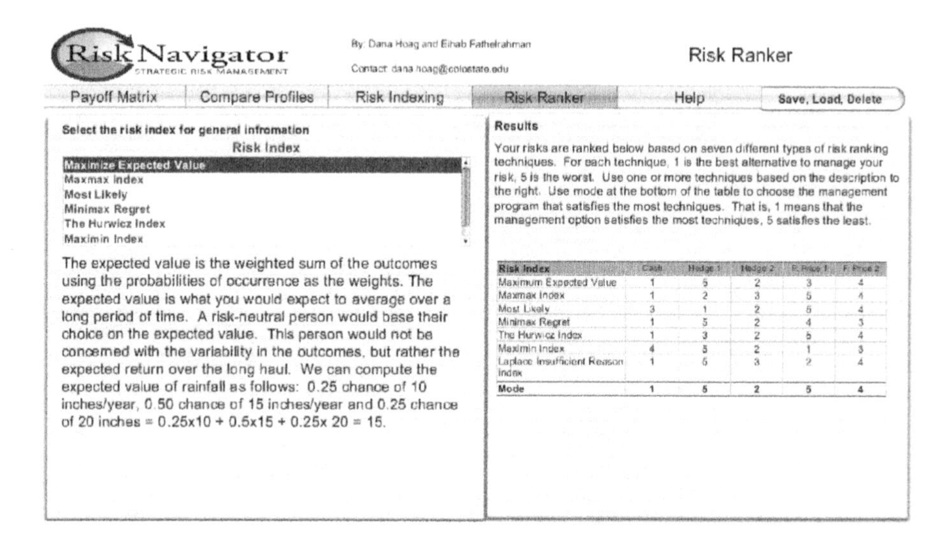

Risk Index	Cash	Hedge 1	Hedge 2	F. Price 1	F. Price 2
Maximum Expected Value	1	5	2	3	4
Maxmax Index	1	2	3	5	4
Most Likely	3	1	2	5	4
Minimax Regret	1	5	2	4	3
The Hurwicz Index	1	3	2	5	4
Maximin Index	4	5	2	1	3
Laplace Insufficient Reason Index	1	5	3	2	4
Mode	1	5	2	5	4

FIGURE 11.12 Risk Ranker: Risk Ranker, with EWS Farms example.

like a 0.5 or 1.0 risk aversion level, where the RP index is between 99% and 98%. At this level Aaron decides that it is not worth tackling a complicated risk management program when he is only willing to pay a maximum of 2% of his EV to avoid risks.

The next tool tab is the "Risk Ranker." This tool, shown in Figure 11.12, displays seven different risk ranking measures for the payoff matrix you entered: maximize EV, maximax, most likely, minimax regret, Hurwicz, maximin, and the Laplace insufficient reason index. This table makes it handy to compare rankings quickly and easily, but you need to match the decision rules to your risk personality in order to make any final judgments about which management action is best.

11.3.2 Value at Risk (VaR)

The VaR tool is simple to use and involves only one screen as shown in Figure 11.13. Enter your information in Step 1, including the size of your investment and ten observations for the time period you are considering. For example, enter yearly data on corn sales for ten years. View the variability of your investment in Step 2, and find your VaR in Step 3. We compute VaR for the 90%, 95%, and 99% confidence intervals.

11.3.3 Risk Efficiency Tool (RET)

The RET has two major risk management tools: a SERF analysis and a stoplight diagram (to compare management options). Start by entering data in the "My Data" section as shown in Figure 11.14. Assign a name to your management actions such as cash market, forward price, and hedge. Next, enter ten values for each action. In this example, we have entered ten corn prices. The prices in the cash marketing column represent those that might be expected if the crop was sold on the cash market, and

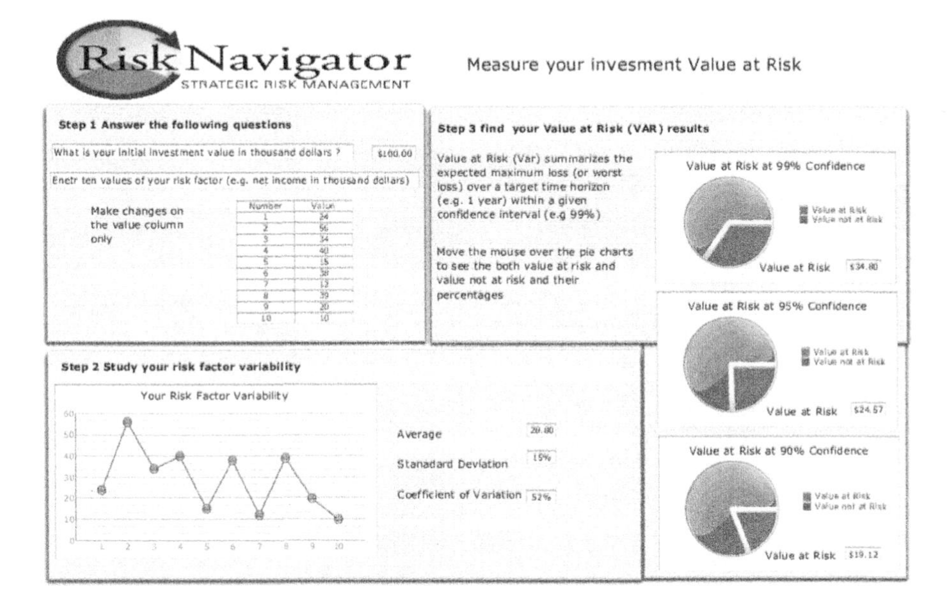

FIGURE 11.13 Value at Risk tool.

My Data **Risk Efficiency Tool (RET)**

Enter Variable Names	Variable 1 Cash Market	Variable 2 Forward Price	Variable 3 Hedge
Enter scenario names, variable names, and 10 data values only in the yellow cells. DO NOT enter data and or information in the blue and white cells.	Enter data for two or three variables in the columns below. Number of observations in each column must be equal. Home		Next
Data Observations	**Cash Market**	**Forward Price**	**Hedge**
1	2.55	2.21	2.25
2	2.47	2.24	2.18
3	1.91	2.05	2.01
4	1.48	1.70	1.85
5	1.56	16.00	1.75
6	1.84	1.70	1.85
7	2.12	1.84	2.11
8	2.33	2.55	1.95
9	1.70	1.92	1.83
10	1.77	1.74	1.65

FIGURE 11.14 Risk Efficiency Tool: My Data, with Examples for EWS Farms.

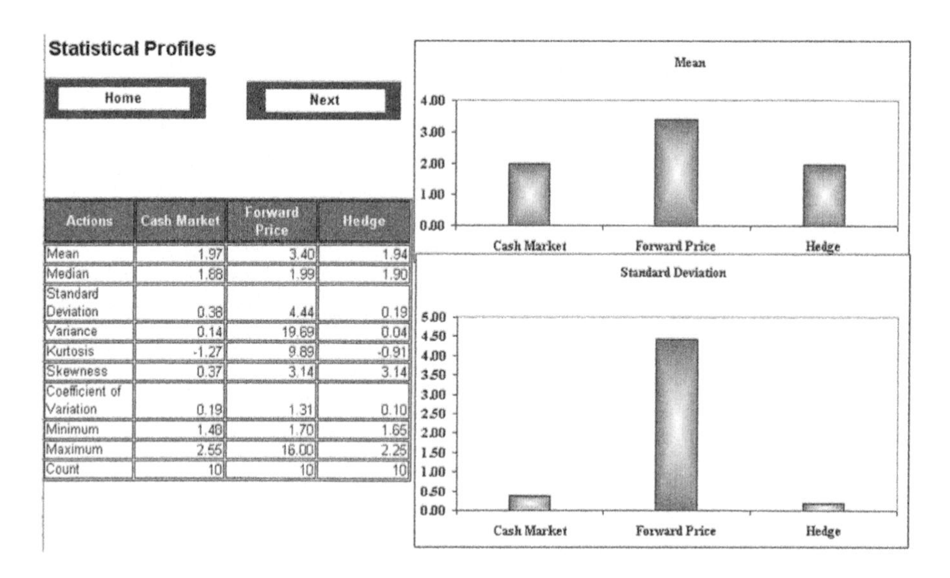
Statistical Profiles

Home Next

Actions	Cash Market	Forward Price	Hedge
Mean	1.97	3.40	1.94
Median	1.88	1.99	1.90
Standard Deviation	0.38	4.44	0.19
Variance	0.14	19.69	0.04
Kurtosis	-1.27	9.89	-0.91
Skewness	0.37	3.14	3.14
Coefficient of Variation	0.19	1.31	0.10
Minimum	1.48	1.70	1.65
Maximum	2.55	16.00	2.25
Count	10	10	10

FIGURE 11.15 Risk Efficiency Tool: Statistical Profiles, with Examples for EWS Farms.

so forth. You must enter ten numbers for each action considered. These will be used to create a PDF. Click next when you have entered all of your data. This takes you to "Statistical Profiles."

The statistical profiles for the EWS Farms case study are shown in Figure 11.15. The mean and SD are graphed and more statistical information is provided in the table. Note that forward pricing has the highest mean, but it also has the highest variance. The cash sales scenario has a slightly lower mean but much lower variance. Hedging has a much lower mean and only minimally lower variance. This is the same information that was shown in the PDFs in Risk Ranker.

When you click next on the statistical summary, you will be taken to the stoplight diagram, as shown in Figure 11.16. The stoplight is a very intuitive tool. Set a lower boundary that represents a minimum outcome and an upper boundary value as the positive threshold. Aaron entered $2.00 per bushel for his lower value and $2.50 per bushel for his upper boundary. The stoplight provides the probability that you will fall below the lower cutoff (colored red), above the upper boundary (colored green), and between the upper and lower boundaries (yellow). Note that the cash marketing and hedging strategies have a 60% chance of going below the lower boundary of $2.00. With forward pricing, the Spragues have a 20% chance of getting more than $2.50. There is no chance of getting more than $2.50 for the hedging scenario.

The last tool is a SERF analysis. The SERF will compare three management alternatives given continuous levels of risk aversion, from neutral to extremely risk averse. Click the blue button to run the SERF when arriving at this screen. As shown in Figure 11.17, a graph will appear with the ranks of the three alternatives for each risk aversion level. Find your risk aversion level and choose the alternative with the highest line for that aversion level. In this case, cash marketing is always the highest value. That means that you would choose it no matter how risk averse you are.

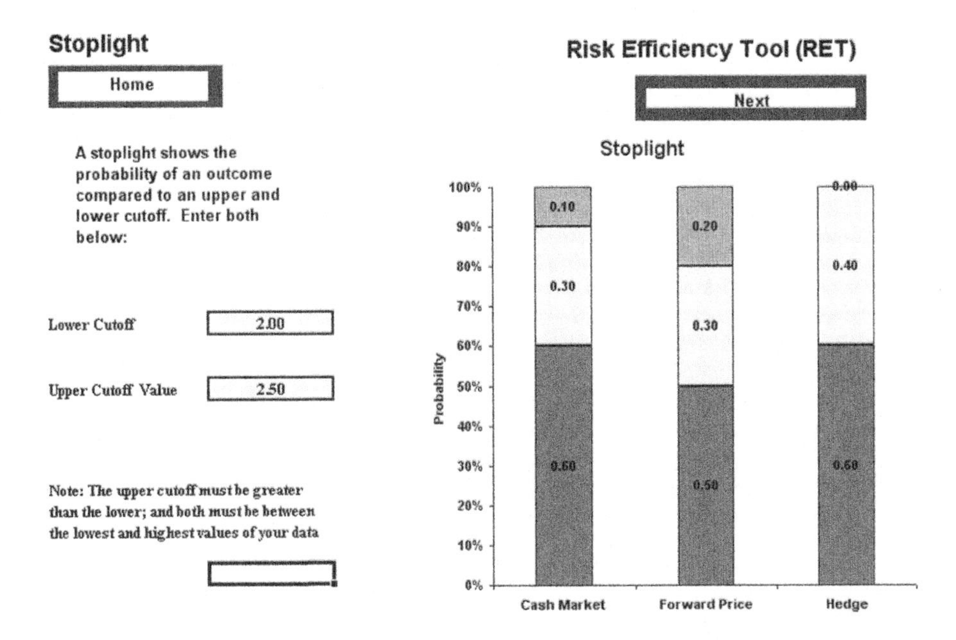

FIGURE 11.16 Risk Efficiency Tool: Stoplight Diagram, with Examples for EWS Farms.

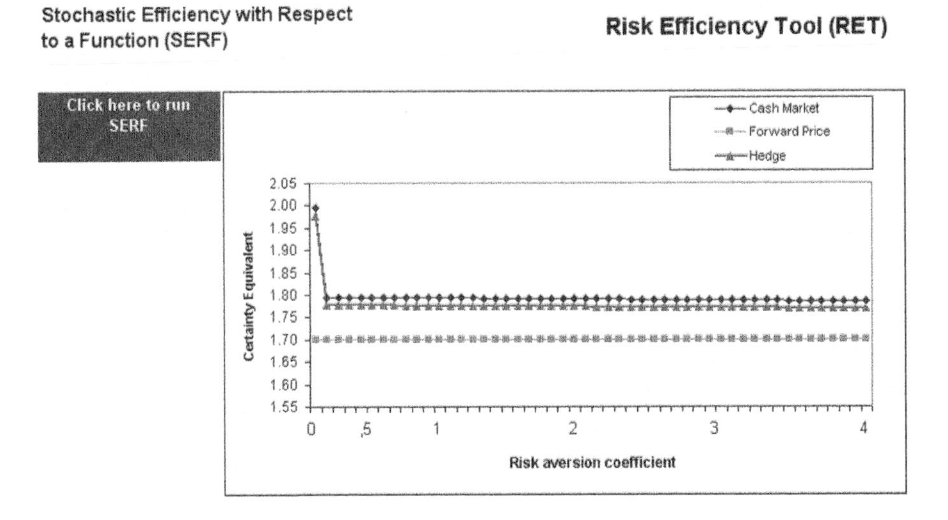

FIGURE 11.17 Risk Efficiency Tool: SERF Analysis, with Examples for EWS Farms.

Aaron would compare the CEs between 0.5 and 1. The advantage of this technique is that Aaron does not have to worry too much about whether the Risk Preference Calculator in Chapter 6 provided an accurate reading of his personal risk preference. Suppose for example that the CEs crossed at a risk aversion level of 2.0. All Aaron would have to worry about is that his index is below or above 2 and he can be confident in his decision.

REFERENCES

Anderson, J. R., and J. L. Dillon. 1992. *Risk analysis in dryland farming systems*. Farming Systems Management Series No. 2. Rome: Food and Agriculture Organization of the United Nations (FAO).

Flomberg, H. 2008. *Management decision theory: Managing pragmatically*. San Diego, CA: University Readers.

Harper, D. 2008. "Introduction to Value at Risk (VaR)." Investopedia Web site. Accessed November 5, 2008 from http://www.investopedia.com/articles/04/092904.asp.

Kastens, T. 2001. Risk and reward: How do farm returns stack up? Should farm managers invest in the stock market?" Kansas State University Extension. Accessed on January 9, 2009 from http://www.agmanager.info/farmmgt/finance/investment/LandvsStocks.PDF.

Musser, W., and G. Patrick. 2002. How much does risk really matter to farmers? In *A comprehensive assessment of the role of risk in U.S. agriculture*, edited by R. Just and R. Pope, 537–556. Boston: Kluwer Academic Publishers.

Richardson, J., K. Schumann, and P. Feldman. 2006. *Simetar (Simulation & Econometrics to Analyze Risk). Users Guide, 2006* (see http://www.simetar.com for purchase information and access to user's guide).

FURTHER READING

Arrow, K. J. 1965. *Aspects of the theory of risk-bearing*. Helsinki: Academic Bookstore.

Harwood, J., R. Heifner, K. Coble, J. Perry, and A. Somwaru. 1999. *Managing risk in farming: Concepts, research, and analysis*. U.S. Department of Agriculture Economic Research Service, Market and Trade Economics Division and Resource Economics Division, Rep. 774. Washington, DC: U.S. Department of Agriculture.

12 Steps 8 through 10
Manage Operational Risks

John P. Hewlett

CONTENTS

The best-laid schemes o' mice an' men often go awry.

—Robert Burns

The last three steps of the Strategic Risk Management (SRM) process are to: (8) Implement Plans, (9) Monitor and Adjust, and (10) Replan. The descriptions of these operational steps are combined in this last chapter about the SRM process. Chapters 13 through 18 describe risk outside of the SRM process in order to further advance your understanding and ability to manage risk.

12.1 PART 1: THE EWS FARMS

The operational stage of risk management is difficult and unfamiliar to many managers. Assigning tasks, monitoring work flow, making mid-course strategic corrections, and constantly reassessing both strategy and performance is an intense task. This level of the SRM process literally puts the plan to work.

12.1.1 Objective

Drawing from the work accomplished in Steps 1 and 2 (determining financial health and risk preferences) and coupled with the risk goals established in Step 3, *the objective of the operational stage is to describe the day-to-day activities a producer must perform to carry out the tactics analyzed and selected in Steps 4 through 7 of the SRM process.*

Strategic planning assembles an inventory of resources available and develops the plan itself. The tactical level looks at how the plan will be accomplished by examining various management alternatives and choosing one to pursue. The operational level makes the selected management alternatives happen and provides for the internal processes needed to sustain the effort through time.

12.1.2 Strategy

The eighth step of the SRM process is to implement the plans developed back in the tactical phase. The process of implementing a strategic plan may take many forms, depending upon the culture of the organization, the history of previous efforts, the number of individuals involved, the geographic scope, and the number and diversity of the production enterprises. Management style and degree of structure within the business also affect the process. The strategy selected also plays a large role in how implementation occurs. For example, a number of alternative strategies are available to develop or maintain a competitive advantage. Possibilities include creating entry barriers, competitive pricing, technological change and innovation, and adjusting firm architecture or personnel management. Each of these overarching strategies implies a very different method for implementation.

In its essential form, implementation is focused on three fundamental activities: resource acquisition, resource flow, and resource coordination. Resources, in this context, are the raw materials used to create the products or services sold by the firm. Traditionally, these resources are grouped into three broad categories: land, labor, and capital.

Implementation includes making sure high-quality resources are available when and where they are needed. Managers must have enough resources to provide services for the necessary period of time at the required intensity and skill level. The flow of resources into and out of the business must be carefully managed for successful implementation. Meeting resource demands by more than one enterprise, providing services within a limited window, or redirecting resources in response to changing conditions are all examples of the need for coordination. Strategic planning can assist in this process to some extent through development of detailed operational plans; however, day-to-day oversight is required to ensure success.

Good solid operational plans may well support accomplishment of the tactical objectives, but execution of such plans seldom goes exactly as envisioned. Delays in receipt of raw materials, failure to make progress as intended, unforeseen weather events, or market shifts can cause plans to change. The second step in the operational level, Step 9 (monitor and adjust), is intended to minimize these risks through monitoring resource performance and making adjustments as needed. Step 9 provides two types of control: informational and behavioral control. Informational control is focused on doing the right things. Behavioral control is concerned with doing things right or, put simply, making sure tasks are completed on time, by the right people, and in the most effective way possible. Listing these two types of control separately does not imply that they can be separated or that they might be assigned to different individuals. Rather, separating the two functions allows for discussion of each individually. As a matter of practicality, each type of control would be used when and as often as management deems it necessary, usually on the fly in the thick of day-to-day decision making.

As the production year unfolds, management will monitor and adjust as needed to keep the business on course and functioning smoothly. Such mid-course corrections, however, do not provide the strategic control necessary when the end goal changes. The Replan step (Step 9) considers fundamental shifts in the environment that require a course correction or an entirely new course of action.

The last step of the SRM, Replanning (Step 10), encompasses big changes in the landscape like retirement of key personnel, inclusion of new partners or children into the organization, starting or stopping an enterprise activity, the opportunity to purchase a neighbor's farm, or an estate transfer. Smaller changes are also considered in the replanning process, such as increases in debt capital due to unfavorable market conditions, higher than expected crop yields due to favorable weather, lower than expected feed costs, or faster than anticipated harvest due to higher labor efficiency.

12.1.3 EWS Farms Implementation

EWS Farms has a number of people involved in the operation. The two most current generations are now charged with operational management. Russell and his wife

Kimberlee are in charge of the decision making for the operation. They also have five children ranging in age from 28 to 19: daughters Desiree and Brianne, who are the oldest and youngest, and sons, Aaron, Russell, and Dustin.

The three families provide all of the operation's labor and management. Historically, however, the father has been in charge of both business management and production. When necessary, Kimberlee or the children contributed extra labor. Although this has worked well, the operation is in a time of change due to the addition of the two new families. The growth is seen as a very good opportunity to increase the efficiency of the operation by using the various talents and skills that son Aaron and son-in-law Aaron possess. Russell, on the other hand, brings extensive experience and is seen as a great resource in avoiding mistakes. The current goal of EWS Farms is to increase the efficiency of operations with the infusion of additional family members and their skill sets into the business.

As expected, the addition of the new families to the operation increased demands on the resources of the farm. Additional family members have increased the need for cash withdrawals for family living but profit margins have remained the same. The family fears that without increased resources to help generate more profit and cash flow the business will lack sufficient financial performance in the long run. In order to accommodate the recent changes, management has decided to plan and execute an expansion. With this in mind, the business will continue to monitor financial performance, as well as evaluate new opportunities for expansion.

Family dynamics can have considerable influence on business decisions. Monitoring and adjusting to the changes in family dynamics and adjusting financial resource use are critical during the expansion. It is not uncommon for one person to be in charge of making an important decision, another person to have the most influence on the decision, and yet another person to be expected to evaluate the resulting consequences of the decision. Proper alignment of these roles and responsibilities is critical to EWS Farms' success.

The Spragues have developed a detailed evaluation of EWS Farms' financial performance, which is presented in Chapter 5. The first three measurements are taken from the set of coordinated financial statements prepared for EWS Farms. They are net business income, annual net cash flow, and change in equity from beginning to end of year. The other two measurements are ratio calculations made using the information contained in the set of coordinated financial statements prepared for EWS Farms. These are the debt-to-asset ratio and the rate of return on business assets. The accompanying coordinated financial statements also reveal a rudimentary look at the overall debt structure of the farm.

Just as the use of human and financial resources must be monitored over time, other resources under the control of EWS Farms must be similarly managed. This includes managing soil fertility, tilth, weeds, pests, machinery, and water. If strategic plans, tactical objectives, or operational plans call for adjustments in resource quantity or timing, those changes must be implemented with constant monitoring and adjustments in usage rates over time. Where warranted, replanning (at least annually) provides feedback to the process, as well as a chance to step back and check the scope of the operation to ensure things are still on the right track.

12.2 PART 2: THE FUNDAMENTALS OF THE OPERATIONAL PHASE

The strategic planning approach is used by organizations around the globe to plan for and transition into the future. However, while planning is difficult, the real challenges come when attempting to execute those plans. Putting the rubber to the road, assigning tasks, monitoring work flow, making mid-course corrections, and constantly reassessing both strategy and performance is the difficult work of the operational level.

The operation phase is designed to assist the manager in putting SRM plans to work. Drawing from the work accomplished in Steps 1 and 2—determining financial health and risk preferences—coupled with the risk goals set in Step 3, operational steps describe the day-to-day activities needed to carry out the tactics analyzed and selected in Steps 4 through 7 of the SRM process.

12.2.1 A Model for Change

Under decision-driven change (Smith, 1999), once a strategic plan is completed it usually becomes the job of leadership to make the necessary business changes. This includes making tough decisions about policy and plan interpretation, communicating what the changes mean for all individuals involved, winning buy-in from the people, and keeping commitment levels high. A key assumption under this model is that all resources, skills, information, and other essential elements are already in hand.

In cases where the organization may not have the skills or resources it needs, other models have been developed. The behavior-driven change model suggests that leadership must still make decisions, win buy-in from the individuals, and keep commitment. The difference is that under this model a large number of individuals are required to learn specific new skills, behaviors, and methods of operating. They may even be asked to develop new working relationships. Personal responsibility is critical to the success of behavior-driven change, meaning that all individuals involved must be willing to work toward change. It is important to note that behavior-driven change usually fails when decision-driven techniques are applied to a situation that requires behavior-driven solutions.

Ten Principles for Successful Behavior-Driven Change

1. Keep performance, not change, the primary objective of behavior and skill change.
2. Focus on continually increasing the number of people taking responsibility for their own performance and change.
3. Ensure that each person always knows why his or her performance and change matter to the purpose and results of the whole organization.
4. Put people in a position to learn by doing and provide them with the information and support needed just in time to perform.
5. Embrace improvisation.

6. Use team performance to drive change.
7. Concentrate organization designs on the work people do, not the decision-making authority they have.
8. Create and focus energy and communicate using meaningful language.
9. Harmonize and integrate the change initiatives in your organization, including those that are decision driven as well as behavior driven.
10. Practice leadership based on the courage to live the change you wish to bring about. (Smith, 1999)

12.2.2 Strategy Implementation Models

Traditional models of strategy implementation describe the process as a linear, step-by-step progression from goals and objectives to implementation, to a comparison of performance, to desired standards, followed by feedback (Dess, Lumpkin, and Taylor, 2005). This model works well for firms engaged in straightforward, static, single-enterprise activities; however, where industry conditions are changing, where there are multiple enterprises, or perhaps where decision-driven and behavior-driven change is required simultaneously just to keep up with the competition, traditional models fail to keep pace.

Traditional models for strategy implementation (Figure 12.1) are not responsive or are too slow to respond to situations where management and individuals in the system must make decisions on the fly. Strategy, and even more fundamentally, goals and objectives, may change as progress is made in response to conditions, opportunities, challenges, and threats. Although the agricultural industry may not be as turbulent, or as dynamic, as the high-tech or other high-change industries, managers need models that allow better management of agricultural firms in the face of increasing risk and competition.

More recently developed strategy implementation models (Dess, Lumpkin, and Taylor, 2005) like that shown in Figure 12.2 provide for greater flexibility in the face of change. The processes of formulating strategy, implementation, and providing strategic control are less regimented and more adaptable. The process maintains the three essential building blocks of the strategic approach: formulate strategy, implement strategy, and strategic controls. The model, however, is reorganized to include two loops of control: informational and behavioral.

Strategic control may be thought of as providing control of the process of putting strategic plans into action. The success of any plan is much less likely with inadequate strategic controls in place. Under the SRM process, implementation

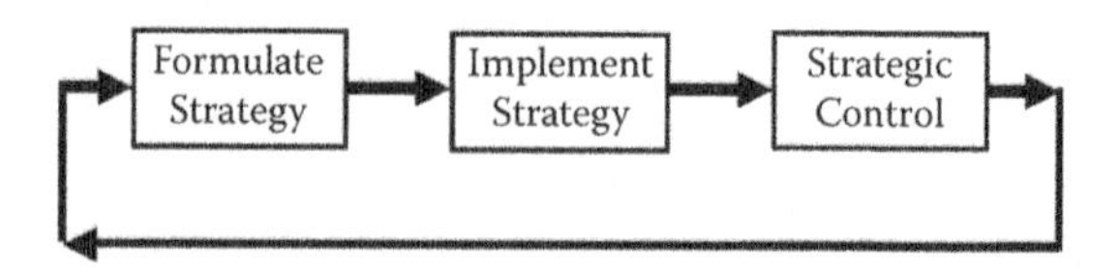

FIGURE 12.1 Traditional approach to strategy implementation.

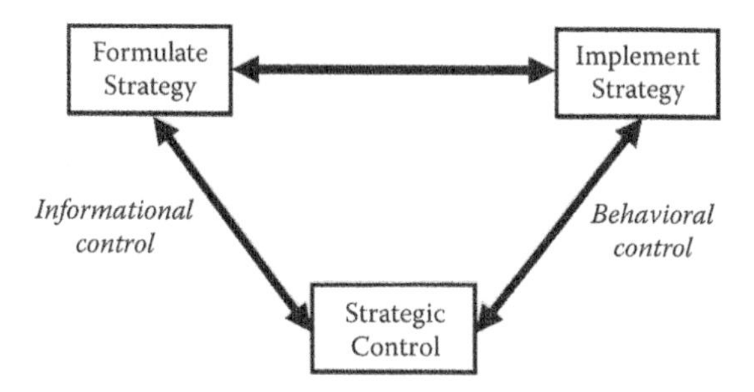

FIGURE 12.2 Contemporary approach to strategy implementation.

is concerned with maintaining a firm's competitive advantage using strategies that range from creating barriers to entry, competitive pricing or technological innovation to adjusting firm architecture, and personnel management (Boyes, 2004). Monitoring and adjusting along the way provides both informational and behavioral control. Strategic control culminates in the final replanning step of the SRM process.

Informational control is focused on doing the right things. As strategy gets implemented in the day-to-day activities of the firm, this control looks at whether these activities are the correct activities. Leadership considers whether changes in objectives or strategy should lead to a redirection of energy into new channels. This control allows for midstream rededication of effort and focus without waiting until the end of the quarter, the biennium, or some other arbitrarily established review point as used in traditional models.

Behavioral control falls under the monitor and adjust step of the SRM process and is concerned with doing things right. Behavioral control ensures tasks are completed on time, by the right people, and in the most effective way possible. This control also includes dimensions of the employee–employer relationship, communication mechanisms, and the incentive structures used in the organization. Effective behavioral control can be obtained by the balance of culture, rewards, and boundaries within the firm (Dess, Lumpkin, and Taylor, 2005). Together these three dimensions provide the environment, incentives, and effective limits for motivated and empowered employees to explore and implement solutions in a changing environment.

Key Features of Successful Control Systems

1. Focus on constantly changing information identified by top managers as strategically important.
2. Managers at all levels of the organization give the information received from the control systems frequent and regular attention.
3. Data and information generated by the control systems are interpreted and discussed in face-to-face meetings by the individuals in the organization.

4. The control systems serve as a catalyst for an ongoing discussion about underlying data, assumptions, and action plans. (Simons, 1995)

Both informational and behavioral controls are needed to keep the organization focused on the correct objectives (informational control) and working as effectively as possible to achieve those objectives (behavioral control). But consider the two parts separately: a great strategy has little value without implementation; whereas, a ready labor force will accomplish little without the guidance of a well-formulated strategy. While both controls are strategically necessary for success, they are not sufficient without adequate communication within the firm.

12.2.2.1 Communication

A culture of communication must be pervasive throughout the organization for this model to be effective. Constant monitoring, interpretation, and feedback are essential at all levels of the organization but especially for management.

In today's world of readily available and continuously connected forms of communication, such as e-mail, cell phones, voice mail, instant messaging, intranets, etc., a model requiring all individuals to be in constant communication with one another is not so far-fetched. Many firms in today's business world already use real-time diagnostic control systems to provide up-to-the-minute information on various critical measures of performance (Simons, 1995). Tools exist to organize these critical measures into reports that provide constant feedback to the operator or manager on critical performance, allowing for real-time decision making and on-the-fly, mid-course corrections.

The information and feedback Möbius strip offered in Figure 12.3 provides a visual representation of a system of communication with the desired features. Here the flow of information and feedback is continuous, seamless, and free-flowing. If you make a physical Möbius strip and inscribe the words of Figure 12.3 on it, you will find that the words communication and feedback are located on the same side of the strip as you move the strip between your fingers. This is an effective visual representation of just how linked the two concepts should be to provide the type of communication needed for the informational and behavioral controls to provide information, allow for interpretation, and supply feedback to all individuals in a seamless and continuous manner.

12.2.2.2 When Strategic Planning Does Not Work

Before launching into a more detailed look at the operational steps of the SRM process, it may be worth considering why strategic planning fails. Reid (1989) set out to

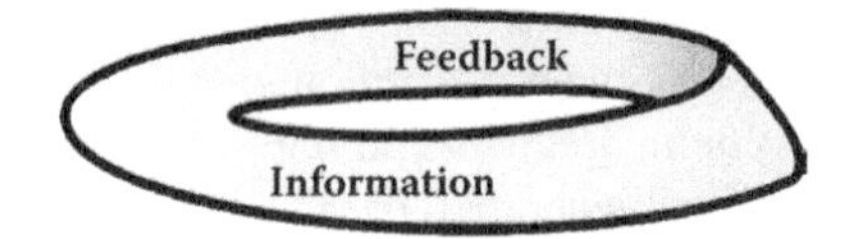

FIGURE 12.3 Information and feedback Möbius strip.

learn this by interviewing more than ninety chief executives and conducting thirty focus group interviews with senior executives. The firms represented by these individuals spanned the commercial and industrial spectrum. After analyzing the data collected, Reid offered the following reasons for strategic planning failures:

- Managers are not directly involved in the planning process.
- The planning process is not continuous and is unrelated (or perceived to be unrelated) to day-to-day operations.
- The strategic planning process does not stimulate strategic thinking.

Successful strategic planning is intended to and, in fact, must affect day-to-day activities of individual employees and management. As stated earlier, a great vision and set of strategic goals are worthless unless they are acted upon by a motivated and empowered workforce. However, as Reid relates, all too often strategic plans find their place on a shelf collecting dust rather than alive in the minds of the people doing the work. This may be due to the fact that people get caught up in the urgency of their work, rather than focusing on what is important (see Chapter 7). The SRM process is designed to help organizations avoid this problem through well-developed goal statements. The material presented in the following section will assist motivated and empowered individuals and teams to implement the process.

While time-consuming, a little diligence can translate portions of a strategic plan into action steps. Getting individuals involved in strategic thinking can be much more of a challenge. Many firms are governed by the age-old idea that "the tallest nail gets the hammer." In such an environment, it is no wonder little innovation occurs. Reid states, "What is clear is that companies do not engage in the type of fundamental thinking necessary to protect their futures … they continue to operate within existing arenas after the criteria for success have changed" (1989, 564). A culture of rewarding innovation and providing incentives for new, strategic ideas must be developed.

Organizations that are better at this tend to have a formalized mechanism to help overcome the don't-rock-the-boat resistance. The goal is to employ individuals who are empowered to take risks, who experiment, and who have the freedom to fail. Much literature has been written to assist managers in developing and sustaining this type of culture, and references are available at the Risk Navigator Web site.

Suggestions for Increasing the Success of a Strategic Plan

- Foster ownership of the plan for all individuals involved before implementation begins.
- The process should cause individuals to consider the strategic plan when faced with decisions. Not that the plan has all the answers, but in referring back to it, employees are able to recapture the essence of the organization's thinking and make decisions in alignment with that philosophy. (Goodstein, Nolan, and Pfeiffer, 1993)

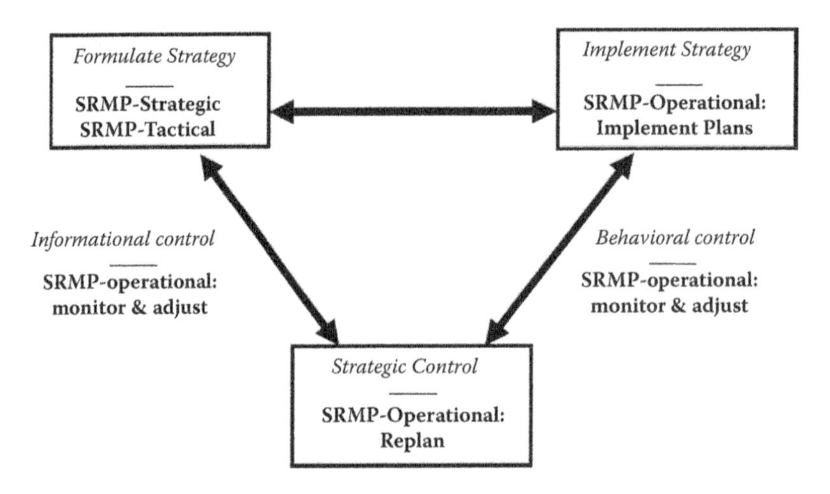

FIGURE 12.4 Operational steps in the SRM process.

12.2.3 Steps 8, 9, and 10 at the Operational Level

The operational level of the SRM process is concerned with putting the SRM plan into play. Where the strategic level of the process assembles an inventory of resources available and develops the plan itself, the tactical level looks at how the plan will be accomplished by examining various management alternatives and choosing one to pursue. The operational level makes the selected management alternatives happen and provides the internal processes needed to sustain the effort over time.

The operational level of the SRM process includes three steps: implement, monitor and adjust, and replan. Although, they may appear as a linear, step-by-step means of putting a strategic plan to work, given the preceding text, they would be better viewed as arranged in Figure 12.4.

The SRM strategic and tactical levels provide the strategy for the firm. The operational step "Implement Plans" provides for execution of the strategy, the operational step "Monitor & Adjust" provides both information and behavioral control, and the step "Replan" provides the strategic control function in the process. As presented earlier, this model provides for greater flexibility in the face of change and is less regimented and more adaptable.

Operational plans developed as part of the strategic planning process provide guidance on the general steps that should be taken; however, these plans do not bring about the desired actions unless they are acted upon. This acting, or doing, is the business of the three steps in the operational level of the SRM process.

12.2.4 Implementation

The process of putting a strategic plan into action may take many forms, depending upon the culture of the organization; history of previous efforts; size in terms of number of individuals involved, geographic scope, and number and diversity of enterprises; as well as management style and degree of structure within the business. In addition, the strategy selected plays a large role in how implementation occurs.

A number of alternative strategies are available to develop or maintain a competitive advantage. Possibilities include creating barriers to entry, competitive pricing, technological change and innovation, and adjusting firm architecture or personnel management (Boyes, 2004). Each of these overarching strategies implies a very different method of implementation.

In its essential form, implementation is focused on three fundamental activities:

1. Resource acquisition
2. Resource flow
3. Resource coordination

Resources are the raw materials needed to create the products or services offered for sale by the firm. In livestock operations, resources would include grass and other forages; stored feeds such as hay and grain; the livestock animals, buildings, and other improvements; the people involved in providing labor and management skills, etc. A similar list can be assembled for any business or alternative agricultural enterprise.

It may be helpful here to note that capital has three defining characteristics (Freeman, 1983).

Capital is durable, it provides services over a period of time, and is an input to the production of goods and services. The chief value of capital is derived from its use in the production of products or services. Capital is itself produced and is the output of some assembly or production process.

12.2.4.1 Resource Acquisition

First and foremost, implementation is about making sure the correct quantity of resources is available where and when they are needed. These resources must be the correct quality and must be in a condition to provide service for the necessary period of time at the level required. Resources may be obtained in a number of ways, including purchasing with equity capital, purchasing with credit, renting or leasing, via share arrangements, custom hiring, subcontracting, or trading for like resources or services.

Capital resources provide services over a period of time. As a result, accurately calculating the benefits and costs of capital resources requires consideration of the time dimension. This is especially true where the level of costs and returns varies over time. Capital budgeting is the term generally used to refer to these types of calculations. Although an in-depth presentation of capital budgeting is beyond the scope of this text, tools that put the computations within reach of the average user do exist. An inexpensive financial calculator is all that is needed. In addition, all current spreadsheet software has built-in functions for computing these costs and returns, and many ready-made spreadsheets are available from various universities and other sources on the Web. Capital budgeting is necessary to accurately compare alternative resource acquisition strategies.

Resource acquisition decisions are some of the most important decisions management can make, given the long service life of some capital resources. Purchasing a tractor that has high annual repair costs not only increases farm repair bills but also increases the risk of incurring delays or failure to perform certain operations.

All costs and returns should be carefully considered before making decisions about capital resource acquisition.

12.2.4.2 Resource Flow

The flow of resources into and out of the business must also be carefully managed for successful implementation. Not only is it critical that resources be available in the necessary quantity, but they also must be available when and where they are needed. Where resource levels fluctuate over the production year, these levels must be managed throughout the year. Resources that may fluctuate include stored feedstuffs, farm diesel fuel, rangeland forages, ranch labor, and so forth. The success of many farm and ranch enterprises hinges on the ability of management to constrain the use of resources to sustainable levels, while simultaneously generating optimum output. The Resource Flow Plan, a Risk Navigator tool, was developed to assist managers with the day-to-day and month-to-month planning and tracking of resource inflows and outflows.

Resources Used in Production and Their Attributes

LAND

Types: cropland, rangeland, irrigated land, field corners, etc.
Topography: slope and elevation
Weather: rainfall, frost zone, sunny days, wind exposure, etc.
Soils: soil types, textures, components, depth, etc.
Plants: native and nonnative plant populations
Biologic: earthworms, insects, bacteria, fungi, virus, and other populations

LABOR

Types: day workers, seasonal labor, part-time help, year-round employees, etc.
Relationship (family labor): aunts, cousins, spouse, children, grandparents, etc.
Functions: management, marketing, accounting, hiring, analysis, etc.
Qualities: skills and proficiencies, likes and dislikes, risk preferences, experience, willingness, health, enthusiasm, roles and responsibilities, etc.

CAPITAL

Equipment: tractors, implements, trucks, trailers, feeders, head gates, etc.
Buildings: shop, shed, calving barn, milking parlor, line shack, etc.
Improvements: fences, water tanks, sprinkler system, roads, etc.
Capacity: acres per hour, years of useful life, size, head per day, etc.
Livestock: breeding stock, commercial herd, roping horse, show animal, etc.
Debt versus equity sources: repayment capacity, risk bearing, returns, costs, etc.

12.2.4.3 Resource Coordination

Coordination of acquisition and flow of resources throughout the production year is an essential facet of implementation. Resource demands by more than one enterprise, services provided by subcontractors within a limited window, or redirecting a resource in response to changing conditions are all examples of the need for coordination. Strategic planning can assist in this process to some extent through development of detailed operational plans, but day-to-day oversight is required to ensure success.

Communication is a key dimension in the coordination function. Effective managers will develop plans for appropriate communication to be employed as operational plans unfold. Details in a communication plan include: who needs to know, what they should know, and when they should know it. Communication may take many forms, increasing the complexity of this important function. Modes of communication used in a typical agricultural operation may include: verbal, written, radio, hand signals, flags, shouting, waving, head nods, etc. In addition, a number of languages may be used, depending upon the ethnicity and backgrounds of the employees.

It goes without saying that not all individuals involved in an organization will communicate in the same way. Furthermore, not all individuals are equally skilled in this area. Personalities, age, gender, relationships, the size of the operation, and other factors will all influence how well individuals communicate. However, everyone can learn to become more effective in both giving and receiving communication.

Resource coordination also involves breaking plans into specific action steps, scheduling the order of those steps, and developing contingency plans. Some tools for scheduling action steps include: end-point scheduling, optimist/pessimist scheduling, calendars, Gantt charts, and project evaluation and review technique (PERT charts).

A Risk Navigator tool developed for time management was introduced in Chapter 7. Time management is critical for effective resource coordination. Individuals should attempt to shift emphasis away from activities that are not-important-but-urgent or not-urgent-and-not-important to activities that are important-but-not-urgent. The matrix for time management provides the user with a method for thinking about how time is used and which activities might be deemphasized in order to provide more time for other important work.

Time as a Resource

Fact: To control your operation and your life, you must control your time.
Fact: We have all the time there is.
Fact: You cannot save time—you can only spend it or invest it.
Fact: Putting in more hours is not the answer.

The Action Planning Worksheet Risk Navigator tool was also introduced in Chapter 7 and may be helpful in translating tactical objectives into the specific action steps of an operational plan. If operational objectives are developed using this worksheet, the tool becomes a helpful device for assisting with resource coordination.

Although resource acquisition and flow are extremely important to the success of any agricultural operation, in some ways these are the easy aspects of the implementation step. Coordinating resources as plans unfold takes more talent and is, perhaps, more difficult to balance. Effective and timely communication is an important dimension to this component, as are adequate planning and consideration of possible contingencies.

12.2.5 Monitor and Adjust

Although good solid operational plans supporting accomplishment of the tactical objectives may be developed, execution of such plans seldom go exactly as envisioned. Delays in receipt of raw materials, failure to make progress as intended, unforeseen weather events, or changes in markets can all cause plans to change. These events generate risks demanding a response from the manager. Step 9 in the operational level is intended to minimize these risks through monitoring resource performance and making mid-course adjustments as needed.

Keys to Implementation

Always check the assumptions.
Adjust activities as needed.
Always think down-board. (Goodstein, Nolan, and Pfeiffer, 1993)

Referring to the diagram presented in Figure 12.4, we see that the monitor and adjust step provides two types of control—informational control and behavioral control. Remember that informational control is focused on doing the right things. Behavioral control is concerned with doing things right or that tasks are completed on time, by the right people, and in the most effective way possible.

12.2.5.1 Informational Control

Informational control describes the responsibility of management to constantly check that the day-to-day activities of implementation are following the operational plans. This leads to accomplishment of the tactical objectives on the way toward achievement of the strategic goals, while honoring the spirit of the mission statement and core values of the people in the organization. Informational control describes both the action of checking to see that the right things are being accomplished and adjusting activities as needed to keep on course.

Posting the mission statement and strategic goals of the organization in a prominent place, as suggested in Chapter 7, can be an effective aid to keeping these critical ideas uppermost in the minds of management. Monthly or quarterly reviews might also be scheduled throughout the year to ensure that these ideals are repeatedly used as a reference point when informational control is exercised.

The action planning worksheet presented in Chapter 7 is a practical tool for implementation. It may also be used to monitor and adjust as plans are executed. This tool contains a line for one or more tactical objectives targeted by the series of action steps outlined on the balance sheet form. In this way, the tool helps to keep the tactical objectives in mind, as well as the operational plans intended to reach them. As such it can serve as an effective means to monitor business activities from one season to another. As plans unfold, making required adjustments in the action steps listed on the form provides a means for mid-course correction, as well as a record for later reference.

An adequate system of record keeping on all important dimensions of business performance is essential to allowing management to practice information control. With accurate, timely records, management will focus its attention on critical success indicators. Critical success indicators (CSIs) are measures developed within the strategic planning process to provide feedback on the degree to which the strategic plan is implemented (Goodstein, Nolan, and Pfeiffer, 1993). Another term often used to describe such measures is *performance benchmarks.*

A few other Risk Navigator SRM tools already presented and discussed in Chapter 7 are also helpful here. The Critical Success Indicator Worksheet provides a tool to plan and monitor enterprises within the business. These tools may provide managers with the critical management information necessary to adjust to changing conditions.

12.2.5.2 Behavioral Control

Behavioral control is focused on making sure plans are accomplished according to the philosophy of the organization. It is concerned with how things are done. Dess, Lumpkin, and Taylor (2005) suggest that behavioral control can be broken into three separate dimensions: culture, rewards, and boundaries. Under this concept, each of these dimensions must be in balance and be consistently applied for effective behavioral control. Much has been written for corporate managers on each of these dimensions, and a brief description of each is provided below as an introduction to the concepts.

Culture, in this sense, means the values and ideals that help shape how business is conducted. It includes a number of different boundaries, some written and many unwritten, that govern the day-to-day interactions of people in the business. The manner of dress, how promotions are awarded, the process for handling grievances, methods of communication, etc., are all part of the business culture. Many aspects of culture can be controlled directly by management; other dimensions must be modeled through the actions of managers themselves.

Rewards are likely the most powerful tool available to managers for influencing employee behavior. Although the system of remunerating employees can be a source of discouragement if not transparent and consistently applied, it can also be a prime motivator if correctly implemented. Additionally, the system of rewards and incentives must be in alignment with the culture, values, and goals of the business. Bonuses, performance-based incentives, promotion ladders, vacation days, benefits programs, etc., are all rewards.

Boundaries are the explicit rules governing the behavior of all employees in the business. These constraints are usually outlined in the company policy manual; however, many may also be unwritten and are only discovered by the employee making a mistake. Unfortunately, this is often the case for smaller agricultural organizations. To be effective, boundaries must be set with the business culture and system of rewards in mind. All three serve to govern and guide employee actions to achieve the objectives of the business.

A well-defined and clearly presented system for behavioral control communicated to employees would go a long way toward motivating the type of actions desired by management. Entire textbooks have been written on these subjects and practical tools for implementing the controls are available. To implement this control, management must monitor employee interactions with others inside the organization, as well as with suppliers and customers outside the organization. Reward systems must be administered fairly and consistently and boundaries enforced for effective behavioral control. Again, as operational plans are implemented, management may need to adjust one of the three dimensions of behavioral control to keep the business running smoothly.

Performance management is a broad concept that encompasses all communication between a manager and an employee intended to foster a high level of work performance. It must include exchanges about what to do, how to do it, how well it was done, and how to improve on the performance next time (Rosenberg et al., 2002). For more background and practical tools for managing agricultural labor, see "Ag Help Wanted: Guidelines for Managing Agricultural Labor" (www.RightRisk.org), which covers the roles and responsibilities of an agricultural employer, organizational planning, staffing the farm business, supervising agricultural work, managing employee performance, communication, and problem solving.

Another Risk Navigator tool, Resource Flow Plans (Figure 12.5), provides an excellent method for budgeting resource inflows and outflows. In addition, if used diligently, it provides the information necessary to monitor resource balances and feedback on how resources are being used and to what extent. Timely use of this plan can provide an opportunity for management to adjust resource use to follow previously set guidelines before too much damage occurs.

Management may want to consider developing CSI-based measures using the Critical Success Indicator Worksheet for performance of employees. Again, where adequate record keeping is available, critical success indicators can provide the means for clearly defined behavioral control, which is linked to the success of the business as a whole.

12.2.6 Replan

As the production year unfolds, management will monitor and adjust as needed to keep the business on course and functioning smoothly. However, such mid-course corrections do not provide adequate strategic control when the port-of-call changes. Replanning provides for the level of control necessary to consider fundamental shifts that require setting an entirely new course or perhaps drafting an entirely new map.

The SRM Replan step is a reduced version of the goal-setting process described in Chapter 7 and should be completed annually. Its essential elements include an evaluation of resource performance and reconsideration of strategic goals in light of past performance.

Resource Flow Plan For: ____________

Description	January	February	March	April	May	June	July	August	September	October	November	December	Total
BEGINNING Resource BALANCE:													
Resource PRODUCTION:													
Source #1													
Source #2													
Source #3													
Source #4													
Source #5													
Source #6													
Source #7													
Source #8													
Source #9													
Source #10													
Resource from other sources:													
Resource from other sources:													
TOTAL Resource AVAILABLE:													
Resource USE:													
Use #1													
Use #2													
Use #3													
Use #4													
Use #5													
Use #6													
Use #7													
Use #8													
Use #9													
Use #10													
TOTAL Resources REQUIRED:													
ENDING Resource BALANCE:													

FIGURE 12.5 Resource Flow Plan.

12.2.6.1 Resource Performance

Just as monitoring resource quantity, quality, and timing of inflow and outflow is important to the monitor and adjust step of the process, so it is with replanning. Assessing changes in the resource base from one year to the next can provide much-needed trend information, which is critical to making decisions about capital resource replacement, changes in the land base, or even changes in labor resources.

As mentioned before, adequate, accurate, and timely records are necessary for optimum resource performance. Annual reports may be compiled from this record-keeping system, including estimates for critical success indicators. Such reporting is routinely completed for the financial resources of a business because it is necessary for filing tax reports with the Internal Revenue Service. However, reports should also be compiled for the other resources within the business, providing for a more holistic evaluation of the entire unit.

The Critical Success Indicator Worksheet provides a basis for evaluating enterprise performance from one year to the next. In addition, it provides sections to record CSIs for land, labor, and capital resources over time. This instrument can be helpful in tracking business performance and can provide the basis for making adjustments in resource management as needed. Tools and techniques presented in Chapter 5 for determining financial health are also appropriate here. Performance of the financial resource is obviously critical to the long-term health of the entire organization. As such, management may wish to give performance of this capital resource special attention at the close of each operating year.

12.2.6.2 Reconsider Strategic Goals

Although strategic goals are intended to serve as long-term destinations for an organization, sometimes the landscape changes to the point where reaching a particular goal is no longer possible. Major events, such as the death of a key player, a heart attack, major injury, bankruptcy, or resource failure, might put strategic goals out of reach.

Marrying someone with different interests can alter the plans of a child intending to come back to manage the operation. Loss of interest in a particular enterprise due to changes in markets or increases in competition may change goals. And changes in the players, whether they are family members, new employees, spouses of the employees, or perhaps new service providers to the business should trigger a reevaluation of the strategic goals. As outlined in Chapter 7, the reevaluation should include all individuals who may be able to contribute to or be impacted by the goals.

When replanning, consider changes in risk exposure, reasons why target CSIs remain unmet, and whether tactical objectives or operational plans should be readjusted. In short, replanning should cause management to take a step back and consider where things stand, how much progress has been made toward long-term goals, and whether or not the long-term goals, tactical objectives, and operational plans are still relevant as written. This process should take place, at least in part, every year, with perhaps a more in-depth effort every three years. If the replanning step of the process is not given enough serious attention, management may discover one day that it has efficiently and effectively climbed the ladder as quickly as possible, only to discover the ladder of supposed success was leaning against the wrong wall.

12.2.7 Summary

The operational level of the SRM process provides a foundation for achieving the goals and objectives established at the strategic level. There are three basic steps at this level: implement, monitor and adjust, and replan. Each step is necessary to accomplish all the functions of the operational level.

In essence, the operational level is concerned with achieving the strategic goals and tactical objectives on a day-to-day basis. The operational level puts the plans of the organization to work on the ground, using the resources available, through the activities of the people in the business. It provides for monitoring and adjustment as the plans unfold, with periodic replanning to ensure the goals and objectives are still relevant and resource performance remains at the necessary levels. These steps of the process also take into account the risk exposure of the organization, methods of communication used within the business, assignment of tasks to individuals, and other dimensions of resource management that are key to success over time.

12.3 PART 3: APPLICATIONS AND WORKING LESSONS

While Part 2 introduced several alternative tools for the operational level of the SRM process, Part 3 covers the specific Risk Navigator tools provided to help accomplish this stage of risk management that were not covered in previous chapters.

12.3.1 Resource Flow Plan

The Resource Flow Plan (Figure 12.5) is one Risk Navigator tool developed to assist managers with the day-to-day and month-to-month planning and tracking of resource inflows and outflows. This concept is familiar to anyone who has used cash flow budgeting; however, many have not applied the idea to other areas of resource management. The upper section of the worksheet is used to record inflows of resources expected over each month from various sources. Outflows are recorded for all uses in the lower section of the worksheet. An ending balance is calculated as the difference of all inflows and outflows, and any carryover balance is brought forward to the beginning of the next month. This worksheet may be used in budgeting resource needs for the coming month or year, as well as to document actual resources used throughout a production season.

12.3.2 Resource Coordination Tools

12.3.2.1 End-Point (Backward) Scheduling (Figure 12.6)

A good way to schedule is to start with the time the objective is to be accomplished and work back to the present. The use of end-point scheduling helps us be realistic as it requires we start with the time we wish to finish rather than the time we can start. It is the finish time that is most important.

12.3.2.2 Optimist/Pessimist Scheduling (Figure 12.7)

With this tool, we first determine the optimistic parameter. How much time would it take to accomplish this objective if everything went well? Then we determine the

Project Plan for________________

Project Steps	Date
Project completion	9/xx/XX
Step 5	7/xx/XX
Task 5.2	7/xx/XX
Task 5.1	6/xx/XX
Step 4	5/xx/XX
Step 2–3	4/xx/XX
Step 1	3/xx/XX
Task 1.3	3/xx/XX
Task 1.2	2/15/XX
Task 1.1	1/10/XX
Today	x/xx/XX

FIGURE 12.6 End-point schedule.

Example Optimistic/Pessimistic Schedule

Project Steps	Week 1 – Week 7	Est. Completion Early Finish	Est. Completion Late Finish
Step 1.0			
sub-step 1.1	EF-2/LF-8	2	8
sub-step 1.2	EF-12/LF-17	12	17
sub-step 1.3	EF-5/LF-9	5	9
		19	34
Step 2.0			
sub-step 2.1	EF-13/LF-20	13	20
sub-step 2.2	EF-40/LF-45	40	45
sub-step 2.2.1	EF-14/LF-28	14	28
sub-step 2.2.2	EF-20/LF-22	20	22
sub-step 2.3	EF-6/LF-8	6	8
		93	123
Step 3.0			
sub-step 3.1	EF-45/LF-55	45	55
sub-step 3.2	EF-17/LF-25	17	25
		62	80
TOTAL Days:		174	237

FIGURE 12.7 Optimist/pessimist scheduling.

pessimistic boundary by asking ourselves how much time it would take to accomplish this objective if everything went poorly. Finally, since we know, not everything that could go wrong will go wrong, and that all that could go right will not go right, we pick what we think will be the most realistic time frame. This forces us to set an upper and lower boundary of our time estimates. We then focus our attention on the area within the boundary.

12.3.2.3 Calendars

A desk-top calendar that depicts the entire week in some detail allows us to write in when various action steps must be accomplished during the course of the year. A pocket calendar is not as detailed but does allow for easy portability. A wall calendar that depicts month by month allows us to see at a glance when action steps need to be accomplished. Day planners are more complex but allow us to review several projects and plan at the same time.

12.3.2.4 Gantt Charts (Figure 12.8)

A Gantt chart depicts the beginning and ending dates for an event, the sequencing of events, and the extent to which an event is completed. As action steps are accomplished, they are noted on the chart. Also barriers are identified. The timing of action steps is most important in Gantt charts. The technique involves graphically laying out the sequence of steps or events that must be accomplished in order to achieve the objective. The time to complete each event is clearly specified and a critical path is determined. The critical path is the longest sequence between the series of events. It is called the *critical path* because any delay in this path will delay the entire project.

12.3.2.5 PERT Chart (Figure 12.9)

The project evaluation and review technique, or PERT, is a method of presenting the plan for accomplishing a particular project by outlining the steps required, the

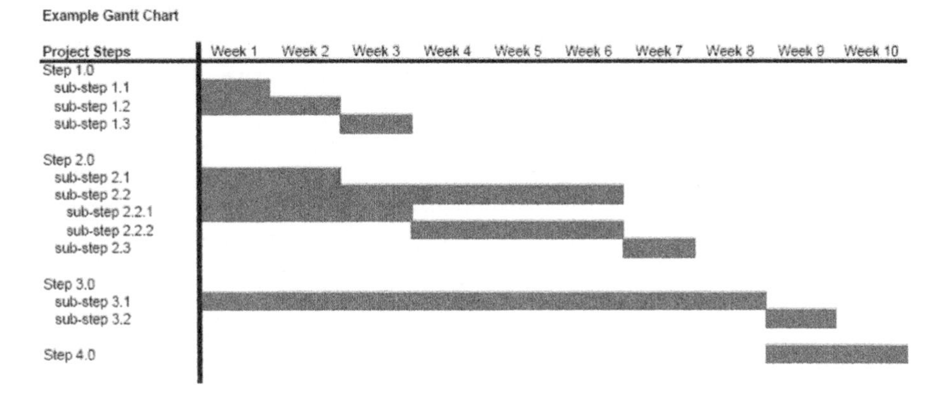

FIGURE 12.8 Gantt charts.

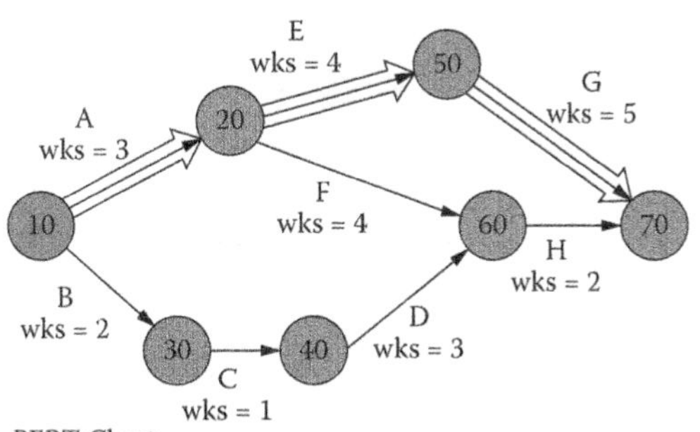

FIGURE 12.9 PERT charts.

time estimated for each task, and the minimum time needed to complete the entire project. PERT was developed to simplify the planning and scheduling of large and complex projects and is more of an event-oriented technique rather than a start/completion date approach.

12.3.2.6 Time Management Worksheet

The Time Management Worksheet Risk Navigator tool is a good way to assess how you and members of your management team spend the majority of your time. During a management team meeting, ask each individual to fill in the quadrants of the Time Management Worksheet with a listing of tasks or jobs he or she performs. The focus should be on the important tasks or jobs completed in the past month, quarter, year, or several years of a major project.

Spend some time discussing where each person placed the majority of their accomplishments on the Time Management Worksheet. Activities focused on accomplishing strategic goals and tactical objectives would usually be found in Quadrant II of the worksheet. These activities directly contribute to making the mission statement of the organization become a reality. Individuals should, where possible, shift their emphasis away from activities that are not-important-but-urgent or not-urgent-and-not-important. Instead they should focus on activities that are important-but-not-urgent to better focus the time spent in their professional and personal lives.

12.3.2.7 Action Planning Worksheet

The Action Planning Worksheet Risk Navigator tool provides assistance in planning the action steps required in greater detail for each tactical objective. This worksheet provides blanks for specific action steps, dates for completion of each step, the person or persons responsible, and the tracking or measuring system that will be used to determine if the actions have occurred as planned.

The Action Planning Worksheet might describe the operational plans in more detail than simply using the Strategic Planning Worksheet presented earlier. When completed, the Action Planning Worksheet can be a very helpful tool in assigning specific task responsibilities by individual and by set of activities across individuals for a week, a month, or a season of operation.

12.3.2.8 Critical Success Indicator Worksheet

The Critical Success Indicator Worksheet Risk Navigator tool provides a tool to plan and monitor enterprises within the business. The tool looks at enterprise income/inflow and expense/outflow for a production year and provides columns for describing changes in these important dimensions over time. Target CSIs may be entered in the target column for income or cost areas for which management has concern or is looking to more closely manage.

Lower portions of the form offer sections to plan the use of land, labor, and capital resources over time and challenge the user to consider the strengths, weaknesses, opportunities, and threats (SWOT) of the plan. CSIs may also be set for these areas of business performance, providing management feedback on other areas of performance outside the cash flow aspects covered in the upper portions of the worksheet.

While the Critical Success Indicator Worksheet may be an effective tool for enterprise-by-enterprise business activity planning, it can also serve as an ideal instrument for monitoring business performance and can provide the basis for making adjustments as needed.

12.3.2.9 Risk Management Worksheet

The Risk Management Worksheet (Figure 12.10) is another Risk Navigator tool for contingency planning and monitoring over time. Each source of risk is considered on an enterprise-by-enterprise basis, challenging management to consider the potential risk exposure each enterprise represents and various methods and plans for mitigating those risks. A risk checkbox column offers a place to indicate the sources of greatest risk or potential advantage for each enterprise.

The lower portion of the worksheet offers a place to consider the SWOTs posed by each enterprise to the whole operation. Comparing worksheets over time offers information on how risk management and contingency plans may be evolving. This could provide managers with the information necessary to adjust to changing conditions.

Risk Management Worksheet

Refer to the USDA document "Introduction to Risk Management, How to Manage Risks: Production, Marketing, Legal, Financial, Legal, Human Resources" at: http://agecon.uwyo.edu/agecon/Programs/RiskMgt/GeneralTopics/GENERALDEFAULT.htm for more information about these categories of risk.

Source of Risk:	Enterprise #1 ______	Risk Check box	Enterprise #2 ______	Risk Check box	Enterprise #3 ______	Risk Check box
Market/Price Risk:						
Current Potential Risks from This Source:						
Methods & Plans for Addressing These Risks:						
Production Risks:						
Current Potential Risks from This Source:						
Methods & Plans for Addressing These Risks:						
Human Resource Risks:						
Current Potential Risks from This Source:						
Methods & Plans for Addressing These Risks:						

Legal Risks:						
Current Potential Risks from This Source:						
Methods & Plans for Addressing These Risks:						
Financial Resource Risks:						
Current Potential Risks from This Source:						
Methods & Plans for Addressing These Risks:						

SWOT Analysis Strengths			
Weaknesses			
Opportunities			
Threats			

Use the risk check box to indicate the source of greatest risk or potential advantage for each enterprise. Use the SWOT section to *think* about potential holes or opportunities these risk present for the overall operation.

FIGURE 12.10 Risk Management Worksheet.

REFERENCES

Boyes, W. 2004. *The new managerial economics*. New York: Houghton Mifflin Company.

Dess, G. G., G. T. Lumpkin, and M. L. Taylor. 2005. *Strategic management: Creating competitive advantages*, 3rd ed. New York: McGraw-Hill Irwin.

Freeman, A. M. 1983. *Intermediate microeconomic analysis*. New York: Harper & Row.

Goodstein, L., T. Nolan, and J. W. Pfeiffer. 1993. *Applied strategic planning: How to develop a plan that really works*. New York: McGraw-Hill, Inc.

Reid, D. M. 1989. Operationalizing strategic planning. *Strategic Management Journal* 10: 553–567.

Rosenberg, H. R., R. Carkner, J. P. Hewlett, L. Owen, T. Teegerstrom, J. E. Tranel, and R. R. Weigel. 2002. *Agricultural help wanted: Guidelines for managing agricultural labor*. Western Farm Management Committee. Accessed June 2, 2009 from http://AgHelpWanted.org

Simons, R. 1995. Control in an age of empowerment. *Harvard Business Review* 73: 80–88.

Smith, D. K. 1999. *Make success measurable! A mindbook-workbook for setting goals and taking action*. New York: John Wiley & Sons.

FURTHER READING

Gable, C. 1998. *Strategic action planning NOW!: A guide for setting and meeting your goals*. New York: St. Lucie Press.

Section III

Advanced and Customized Risk Management Programming

The Strategic Risk Management (SRM©) steps were designed to make it easy to complete a risk management plan from A to Z. While this helps people organize their thoughts, it also constrains analysis to those things that fit nicely into the ten-step format. Risk is dynamic and the tools to manage it need to be too. In this section, we turn things around and discuss traditional methods decision makers use to manage the types of risk identified by the Risk Management Agency—marketing and price risk, financial risk, production risk, human risk, and institutional risk. These chapters refer to the SRM process and Risk Navigator tools when appropriate, and introduce a few new Risk Navigator SRM Internet tools; however, the focus is on discussing methods authors have used to address different types of risk.

At some point, the organization that benefits the SRM becomes a liability. As your experience grows, you will increasingly want more flexibility than is allowed. You might want to compare more options than our format allows, utilize your own accounting software, or develop customized risk management simulations. Therefore, we conclude this section with two chapters that are meant to take your skills to the next level. In Chapter 17 we discuss how to use the Ag Survivor scenario for EWS farms in order to expand your understanding of risk. This is one of many scenarios available on our Web site that lets you practice making risk management decisions in a simulated and safe environment. We conclude in Chapter 18 with a brief discussion about how to use advanced features in Microsoft Excel™ and special risk management add-ins, like @RISK™ and Semitar© to build your own customized management plans.

13 Marketing and Price Risk

James Prichett

CONTENTS

Commodity marketing is a set of tactical management alternatives that focus on price risk management. The function of commodity marketing is to reach the broader business goals set in the Strategic Risk Management (SRM) process. EWS Farms specializes in grain production. Grains fit the definition of a commodity because it's difficult to differentiate grain from one farm to the next. Unfortunately, commodity producers can't set grain prices; if they tried, competitors would offer a lower price. Instead, prices arise from a process of balancing the year's harvest and carryover stocks against buyers' grain demands. EWS Farms' managers believe risk management improvements can be made in their grain marketing.

Price and marketing risk management must time crop sales so that prices received, from a forward contract for example, would be quite different from the price received at harvest. Of course, the goal is to receive a higher overall price. If managers knew when the market high price would occur, the entire crop would be sold that day! In truth, no one knows when this will happen, but understanding the nature of how prices vary will help avoid making pricing mistakes.

The purpose of this chapter is not to beat the market; rather, it is to introduce concepts that will help you avoid marketing mistakes and generate reasonable returns for the existing economic conditions. I provide examples from EWS Farms as much as possible to help make the concepts more understandable.

13.1 EWS FARMS: THE PAYOFF FOR IMPROVING MARKETING SKILLS

The EWS Farms marketing strategy is straightforward: sell grain at harvest. Over 15 years, the farm has received an average harvest price of $2.19 per bushel for their corn grain and $2.97 per bushel for their wheat. Figures 13.1 and 13.2 show a histogram of historical corn and wheat harvest prices in northeastern Colorado. Between 1970 and 2000, cash corn prices ranged from a low of $1.14 per bushel to a high of $3.35 per bushel (Figure 13.1) Wheat prices also have a wide range, from $1.20 to $4.78 per bushel (Figure 13.2). It's clear from the figures that local cash grain prices can vary a great deal over time. Price risk is a real concern.

13.1.1 Price Impacts on Gross Revenue and Net Income

Managing prices can be a very important tool for increasing overall farm profits. Let's see how an improvement in average crop prices can alter the bottom line for EWS Farms. In the second column of Table 3.1, you'll see that EWS Farms' gross revenues are $315,615, assuming a 15-year average for the price and yield. In the next column, prices have been increased by 5% and gross revenues recalculated to be $331,396. Not surprisingly, the price increase leads to 5% higher gross revenues as is written in the last column. Perhaps more surprising is the increase in net income found in the next row. With average prices and yields, the farms' net income is $41,702. After the 5% increase, net income grows by 37.8%.

Why does net income increase dramatically more than the 5% increase in gross revenue? The answer is that while revenue goes up, costs stay the same. That is, once

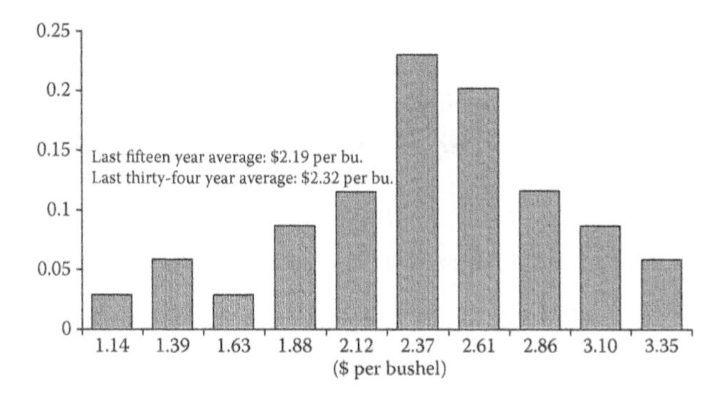

FIGURE 13.1 Harvest cash corn price histogram for EWS farms (1970 to 2005).

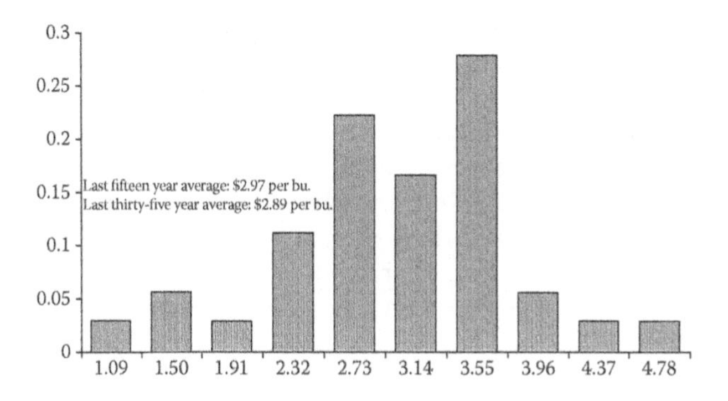

FIGURE 13.2 Harvest wheat price histogram for EWS farms (1970 to 2005).

TABLE 13.1
Financial Measures with Average and Improved Prices

Financial Measure	With Average Prices	With Price Improvement	Percent Difference
Gross revenue ($)	$315,615	$331,396	5.0%
Net income ($)	$41,702	$57,483	37.8%
Asset turnover ratio	20.3%	21.3%	1.02%
Rate of return to assets	2.5%	3.6%	1.02%

a producer has a crop, net income, the part he keeps, can be dramatically improved by investing some time to increase prices. Increasing net income improves key efficiency ratios such as the asset turnover ratio and the rate of return on assets (last two rows of Table 13.1). Clearly, avoiding pricing mistakes can have a big impact on a farm's bottom line.

13.1.2 The Catch: Market Risk and Its Impact on Revenues

EWS Farms faces significant variability in crop prices from year to year, so it won't be easy to stabilize revenues and avoid low prices. We will discuss risk management techniques soon, but first let's discuss natural variability. Income variability can be reduced naturally by the price–yield interaction in the marketplace. This "natural hedge" varies from one location to the next. The strength of this natural reduction influences the need to manage price risk and the tools chosen to manage it; so it is important that we discuss it first.

13.1.2.1 The Natural Hedge

EWS Farms yield and price variability is illustrated in Figure 13.3, which charts yields (as bars) and cash prices (as a line) for irrigated corn. Notice that most of the time, but not always, high yields are associated with low cash prices; a good example is in 1994. Likewise, when the cash price of corn is high, yields tend to be low, as was true in 1995. Because prices and yields are negatively correlated, cash revenue tends to be less volatile than either of these. That is, since revenue is equal to price times yield, raising one variable while the other goes down keeps revenue constant. This "smoothing" of revenue is called a *natural hedge*.

The strength of the natural hedge varies with geography. In a place like the Corn Belt, farm-level yields are highly correlated with national yields, so these yields

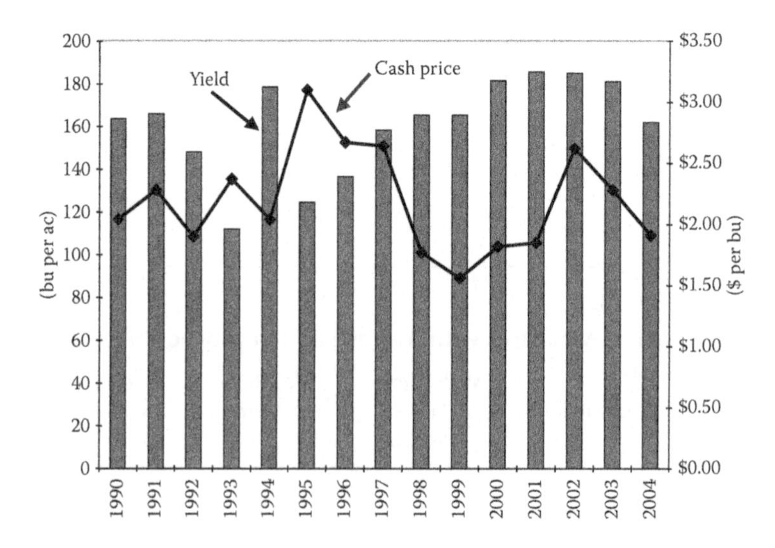

FIGURE 13.3 EWS Farms' corn prices and irrigated yields.

have a strong negative correlation with national price. A strong negative correlation means the natural hedge is quite strong, and Corn Belt farmers face less revenue variability than other locations. EWS Farms is outside the Corn Belt, where yields have little influence on national prices, so it faces more revenue variability. The correlation between EWS Farms' yields and the local price for irrigated corn is −0.47, dryland corn is −0.43, and dryland wheat is −0.28, so it faces more revenue risk than producers in the Corn Belt where price and yield correlations are often lower that −0.7.

The strength of the natural hedge also varies by crop. While the domestic corn crop largely determines U.S. prices, U.S. wheat prices are closely tied to international supply and demand. Because U.S. production is a small share of the worldwide crop, the natural hedge is weak, making wheat income more variable than corn income (Harwood et al., 1999).

The strength of the natural hedge will influence how well risk management tools stabilize revenues. All things being equal, the weaker the natural hedge, the more likely risk management tools will reduce revenue volatility. For EWS Farms, price risk management may benefit wheat more relative to corn, but the choice of a specific crop's risk management tool will also depend upon the tool's cost.

In sum, this section focuses on the payoff from improved marketing skills and the natural protection markets can give. Profitability improves with better marketing skills, but variability in revenues will still exist. Revenue variability isn't as dramatic as price variability, however, because of the natural hedge. The stronger the natural hedge, the less effective risk management techniques will be at reducing volatility. The remainder of the chapter focuses on price risk and price risk management tools. But, as has been discussed in Chapter 9, eliminating risk is not without cost. A risk–return trade-off exists: as revenue (or price) volatility is reduced, so too is the expected return. Farm managers should consider if the reduction of revenue (price) volatility is sufficiently better than the natural hedge. For this reason, price risk management strategies will always be compared against a benchmark—cash sale at harvest.

13.2 PRICE RISK MANAGEMENT

Price risk can be divided into three categories: the risk derived from worldwide supply and demand conditions (futures price risk), the risk that local factors might create cash price volatility (basis risk), or the risk that, over time, prices might not compensate a producer for storing a commodity (spread price risk). The first two categories are linked by a simple pricing equation:

$$\text{Local Cash Price} = \text{Futures Price} + \text{Local Basis}$$

That is, the cash price received by farmers is the sum of the futures price and the local basis.

Futures prices fluctuate according to changes in worldwide supply and demand conditions, while local basis varies with transportation costs, the availability of storage, and local supply and demand conditions. So, let's consider marketing alternatives that reduce the volatility of the local cash price by fixing the futures price, the local basis, or both.

Spreads are a third category of price risk. The commodity's spread is the difference between two futures contract prices quoted on the same day but with different expiration dates. Several tools can be used to manage spread risk, including the storage hedge, but these have been omitted from this chapter in lieu of the references found at the end of the text.

13.2.1 Futures Prices and Contracts

Futures prices are closely linked to the local cash price for grain. Cash market grain buyers determine their offer price using a publicly reported futures price, and then they make adjustments (called the basis) according to their local conditions such as transportation and storage costs. The futures contract price is quoted from a commodity exchange, an organized market where futures contracts are traded. A futures contract is a commitment to make or to take delivery of a specific grade and quantity of a commodity at a designated future time and place. The commodity exchange (i.e., the Chicago Board of Trade) establishes the contract and its specifications.

13.2.1.1 Who Uses Futures Markets?

A variety of professionals trade futures contracts. Speculators buy and sell contracts because they seek profits from trades. These traders seek to "buy low and sell high," and carefully watch for evidence of world supply and demand changes. Speculators are willing to take price risk in order to profit, and their trades mean they are a ready buyer for any seller in the market. Speculators' bid prices reflect their market information, meaning they are a good indicator of underlying supply and demand conditions on any given day. In this way, futures prices are closely tied to the actual, worldwide cash prices for commodity grains.

Hedgers are a second group of commodity futures traders, and this group includes farmers, grain elevator managers, livestock feeders, and processors. The goal of a hedger is to shift the risk of adverse price change to the market. For instance, EWS Farms may seek to hedge part of its soon-to-be-planted corn crop. For the farm, prices that decline throughout the growing season are of particular concern. In order to hedge against this risk, they will create a futures market position that will be profitable if prices fall. That is, the farm will establish a "short" market position so that they can sell high and buy low if prices fall. In this way, EWS Farms will profit in the futures market while prices are falling in the cash corn market. (An example is found in the next section.) Likewise, feedlot managers are wary of increasing prices that drive up their cost of production. These managers adopt "long" market positions (so they can buy low and sell high), where profits in the futures market from a price increase offset lost profits because the price of cash corn, and, as a result feed costs, have increased.

To sum up, the commodity exchange, its futures contracts, and the futures prices that result from trades are all part of an important economic process. The prices themselves reflect information about worldwide supply and demand conditions. Traders observe new market information, evaluate its importance, and incorporate their subjective opinions into bid prices. As a result, futures contract prices represent a consensus opinion about how important economic shocks are. And as is discussed in the next section, futures contracts are useful for shifting risk from producers to speculators because of the economic tie between the futures contract and the grain commodity on which it is written.

13.3 HEDGING PRICE RISK WITH FUTURES CONTRACTS

Hedging is a risk management practice used to transfer cash market price risk from a commodity grain producer to the futures market. The idea behind hedging is that the grower establishes a futures market position that profits when cash market prices are declining. Hedging in the futures market is successful because futures contract prices are closely tied to local cash prices. Hedging effectiveness depends on how tightly correlated futures prices are with cash prices during the hedging period.

For example, expected profits for EWS Farms will decrease if corn prices fall between planting and harvest time, but holding a crop for future sale may also provide big dividends if prices go up. To avoid losses, the farm manager might establish a desirable price for the growing crop by taking the following steps that utilize a hedge:

Step 1: In the spring, the farmer sells a new crop futures contract (hedges) with an expiration date after harvest. The futures market position (sell or short) initiated in the spring is exactly the same action the farmer will perform with the cash crop at harvest, many months from now.

Step 2: At harvest (or when the farmer is ready to sell the cash crop), the hedge is lifted when the cash grain is sold to a buyer and a futures contract is purchased to offset the futures commitment. The net price received for the grain is the cash price paid for grain at the elevator plus or minus the gain or loss from the futures contract.

Examples of hedging when prices go up versus when they go down are shown for EWS Farms in the next section.

13.4 EWS FARMS T-ACCOUNT FOR ESTABLISHING PRICE

Figure 13.4 is a T-account, which is a useful way to illustrate what occurs in the cash and futures markets. The columns of the T-account are divided into three pricing components, local cash market (cash), the futures market (futures), and the basis. The rows of the T-account are points in time. The T-account begins in the row marked May 20, the time that EWS Farms would like to establish a price for the growing crop. As is written in the far right column of this row, the EWS managers expect the harvest basis (in October) to be $0.20 under the December corn futures prices. The EWS Farms' basis expectation is derived from historical records of the local basis, but the basis at harvest time might be different, which is why it is written as an expectation. On May 20, December corn futures prices are trading at $2.70 a bushel, and this is written in the futures column. Given this futures price, the expected cash

Begin Here

	Cash	Futures	Basis
May 20	Expect cash value to be $2.50/bu	Sell 5,000 bu of Dec Corn futures @ $2.70	Expect basis to be 20 cents under futures
Oct 20	Sell corn at elevator for $2.00/bu	Buy back Dec futures @ $2.20/bu	Basis is 20 cents under futures
Summary	Sold cash @ $2.00	Gain in futures +$.50	

Net Price Received = $2.50/bu (before any futures commission or hedging costs)	They reached their goal of pricing 5,000 bu of corn at $2.50/bu. The decline in the cash price was exactly offset by a decline in futures

FIGURE 13.4 EWS Farms T-account for establishing price.

price is $2.50 per bushel ($2.70 futures price plus –$0.20 expected basis), which is noted in the cash column.

In order to initiate the hedge, the manager of EWS Farms sells a December futures contract at $2.70 per bushel as is written in the futures column. If prices fall between May 20 and the day on which the hedge is lifted, the futures position will be profitable. These futures profits will offset any losses in the cash market. Note that the sell action in the futures market on May 20 is the exact action the manager will take in the cash market at harvest.

At harvest (October 20), the manager sells corn in the local cash market for $2 per bushel, $0.50 below what was expected on May 20. December corn futures contracts are trading at $2.20 per bushel on that day, and the manager buys a December futures contract to offset his futures market position. The difference between the cash market and the futures market is the basis, which is $0.20 under as anticipated.

The last row of the T-account summarizes the hedging action by market and will be used to calculate the net price received. Corn was sold in the cash market for $2 per bushel. The futures market position garnered $0.50 profit per bushel because the selling price ($2.70) is $0.50 greater than the buying price ($2.20). The net price received is the local cash grain price plus the gain in the futures market, $2 + $0.50 = $2.50 per bushel. The net price received is exactly what was anticipated on May 20. The manager was able to meet the pricing goal because the decline in the cash price was exactly offset by profits generated from a declining futures price.

13.4.1 Hedging and a Price Increase

The manager is satisfied with the short hedge because it protected the farm against downward price movements in the cash market, but what if prices had increased? Figure 13.5 examines hedge performance when prices increase from May 20 to October 20.

	Cash	Futures	Basis
May 20	Expect cash value to be $2.50/bu	Sell 5,000 bu of Dec Corn futures @ $2.70	Expect basis to be 20 cents under futures
Oct 20	Sell corn at elevator for $3.20/bu	Buy back Dec futures @ $3.40/bu	Basis is 20 cents under futures
Summary	Sold cash @ $3.20	Loss in futures −$.70	

Net Price Received = $2.50/bu (before any futures commission or hedging costs)	They reached their goal of pricing 5,000 bu of corn at $2.50/bu. The increase in the cash price was exactly offset by an increase in futures

FIGURE 13.5 Pre-harvest pricing and increasing prices.

The May 20 row of the T-account is the same as the previous example: December futures contracts are selling at $2.70 per bushel, the expected basis is $0.20 under the December futures contract, and the expected cash price at harvest is $2.50 per bushel. The manager initiates the hedge by selling a futures contract (5,000 bushels) at the prevailing price.

Between May 20 and October 20 prices increase, perhaps because of drought in the Corn Belt. The potential supply reduction means that the December futures price has increased from $2.70 per bushel (the May 20 price) to $3.40 per bushel. The basis is $0.20 under the futures contract, so the local cash price is $3.20 per bushel.

Even though cash market prices increased, the futures market position lost money. A contract sold at $2.70 per bushel and another was bought at $3.40 per bushel to offset the position. The loss is $0.70 per bushel as is written in the last row of the T-account. The net price received is the sum of the cash market price and the loss in the futures market, or $2.50 per bushel.

The $2.50 per bushel is familiar; it's the expected cash value for May 20. Just as in the case of a price decrease, the established price of $2.50 is realized when prices increase. In this sense, the expected price of $2.50 per bushel is locked in. The manager of EWS Farms has transferred price volatility to the futures market by establishing a price of $2.50 per bushel, but at the same time the transfer has eliminated the opportunity to profit from upward price movements. The establishment of a pre-harvest price, but loss of upside potential, is an example of the risk–return trade-off.

13.4.2 Hedging and Basis Risk

The manager has established the price on the futures market, but the local basis has not been fixed or held constant. In fact, basis is not known until the grain is marketed, and it may change from what was expected. A changing basis can make

What's Different?

	Cash	Futures	Basis
May 20	Expect cash value to be $2.50/bu	Sell 5,000 bu of Dec Corn futures @ $2.70	Expect basis to be 20 cents under futures
Oct 20	Sell corn at elevator for $2.08/bu	Buy back Dec futures @ $2.20/bu	**Basis is 12 cents under futures**
Summary	Sold cash @ $2.08	Gain in futures +$.50	

Net Price Received = $2.58/bu

(before any futures commission or hedging costs)

They exceed their goal of pricing 5,000 bu of corn at $2.50/bu, and net $2.58 WHY? Because on Oct. 20, the cash price was 12 under rather than the expected 20 under

FIGURE 13.6 Pre-harvest pricing and basis risk.

a hedge more or less effective, and for this reason, a hedger is a basis speculator. Figure 13.6 is an example of a basis speculation.

The T-account in Figure 13.6 begins with an expected cash price of $2.50 per bushel and a basis of $0.20 under the December futures price. The same action is taken on May 20; a December corn futures contract is sold at $2.70 per bushel. Between May 20 and October 20, the futures price has fallen to $2.20 per bushel; however, during the same period the basis strengthened from an expected $0.20 under to $0.12 under the December futures contract. The result is an October 20 cash price of $2.08 per bushel.

The summary of Figure 13.6 illustrates the impact of a changing basis. The cash price received for grain is $2.08 per bushel, and the profit from the futures position is $0.50 per bushel. The net price received is $2.58 cents a bushel—$0.08 higher that was expected.

In this case, a stronger basis (from $0.20 under to $0.12 under) worked to the advantage of EWS Farms, as it improved the net price received. The important point is that a short hedge will establish a price for a growing crop, but the basis (and hence the local cash price) will still fluctuate. In this sense, a farm producer with this type of hedge is a basis speculator. Basis is generally much less volatile than futures prices. And as you'll see in Section 13.6, "Additional Marketing Tools: Cash Contracts," local marketing alternatives exist to fix the futures price of a commodity, the basis, or both.

To recap, a stronger basis makes the short hedge more effective, but a weaker (wider) basis makes the hedge less effective. For long hedges (i.e., a feedlot whose long hedge protects against higher prices for feed ration grains) a weaker basis makes the hedge more effective, while a stronger basis makes the hedge less effective.

13.5 OPTIONS ON FUTURES CONTRACTS

Hedges can be initiated with options on futures contracts, too. In contrast to the hedge in the previous example, options allow the hedger to capture gains from increasing prices, and option buyers need not post margin like they have to in order to initiate a hedge. Options traders do pay an up-front premium, however, to hold the option. In this way, options hedges are similar to crop insurance. The option holder pays a premium up front and then may or may not choose to exercise price insurance based on market conditions. Futures hedgers do not pay a premium.

The world of commodity futures options can be very complex, but we will focus on two simple hedge options that protect against falling prices: the short hedge with put options and the fence strategy (out-of-the-money put options combined with a long call).

13.5.1 Option Characteristics

To begin, let's briefly focus on option characteristics and jargon. First, an *option* is the right, but not the obligation, to buy or sell a futures contract at a predetermined price anytime within a specified period. Options are derivative instruments because the option is written on an underlying asset—in this case, a futures contract.

In general, two types of options are available: a *call option* is the right to buy the underlying futures contract at a predetermined price prior to expiration, while a put option is the right to sell an underlying futures price prior to expiration. *Put options* are most frequently chosen by hedging producers to protect against falling prices, and we'll examine a short hedge using put options a little later.

The predetermined price at which an underlying futures contract may be bought or sold is called the *strike* or *exercise price*. The strike prices for options are set in predetermined multiples (e.g., $0.10 per bushel for corn) by the commodity exchange. When an option is first listed, a strike price is set near the price at which the underlying futures price is currently trading, and then a predetermined number of strike prices are set above and below this strike price. Additional strike prices are added as futures prices trade up or down from their initial level.

Figure 13.7 is an example of an option for a corn futures contract. The underlying asset is the December corn futures contract; the strike price is $2.30 per bushel, and the option type is a call. In this case, the buyer (or holder) of this option has the right, but not the obligation, to sell a December corn futures contract at $2.30 per bushel, even if the futures contract is currently trading at a different price.

The lower portion of Figure 13.7 lists call and put option premiums for the same December corn futures contract. Currently, the December corn futures price is trading at $2.16¾ per bushel. Both put and call options have been established at $0.10 strike price intervals ranging from $1.90 per bushel to $2.40 per bushel. The premiums for call options are listed in the second column, while the premiums for put options are listed in the last column.

Notice the premiums for call options tend to decrease with increasing strike prices; the premium for a $1.90 call option is greater than that of a $2.20 call option. Why is that? A call option is profitable if the underlying futures price is greater than

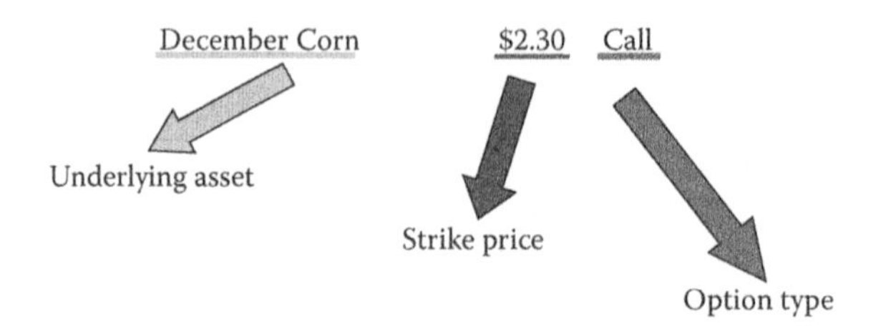

Strike Price (cents per bu)	Call Premium (cents per bu)	Put Premium (cents per bu)
190	20¾	8
200	13¾	11
210	8¾	14
220	5½	19
230	3¼	21
240	2	31½

FIGURE 13.7 Option and option premium example for corn futures.

the strike price. For the $1.90 call option, the holder of the option has the right, but not the obligation, to purchase a futures contract at $1.90 per bushel. If exercised today, the option holder can buy the futures contract at $1.90 per bushel, and then sell the contract at the existing market price of $2.16¾ per bushel. Exercising the option is profitable because the holder can buy low (at $1.90 per bushel) and sell high (at $2.16¾ per bushel). Options that are profitable to exercise are said to be in-the-money. An option is out-of-the-money if it is not profitable to exercise, and an option is at-the-money if its strike price is approximately the same as the underlying futures price. Put options are in-the-money if the underlying futures contract price is trading at a level less than the strike price (sell high, buy low).

The holder of an option may do any of the following prior to expiration:

Exercise the option
Offset the position (buy a put option to offset writing [selling] a put)
Allow the option to expire

The commodity exchange makes sure that for every option holder there is an option writer (seller). The option writer receives a premium from the option buyer and is obligated to take the opposite position if the option is exercised. As a result, option writers must post margin. Option buyers do not have to post margin because they can always allow the option to expire if market conditions are not favorable. In contrast, option writers do post margin because they must take an unfavorable futures market position when the option is exercised.

Option premiums are determined in an open outcry auction. Consequently, option premiums change daily with new market information. Because traders adjust their premium bids in a competitive market setting, economists are able to forecast a probability density function for the price of futures contracts at expiration (or more traditionally stock prices) from option premiums.

When an option is in-the-money, it is said to have intrinsic value. The greater the option's intrinsic value, the greater its premium. So, intrinsic value is one part of the option premium. But even options that are out-of-the-money may have value to traders. These traders have some inkling that the option may be in-the-money before the expiration date and are willing to a pay a premium based on this likelihood. This value is called the option's time value and is also a component of the option premium. Therefore, the premium of an option is:

$$\text{Option Premium} = \text{Intrinsic Value} + \text{Time Value}$$

Options that are in-the-money have time value; after all, prior to expiration they may go even deeper in-the-money.

Option premiums are great indicators of market volatility; as markets become more volatile, options are more likely to become in-the-money prior to expiration, so they have a greater time value. The greater the time value, the greater the option premium will be. So, if an option hedge is used as price insurance, it will become more expensive as market volatility increases (the premium will go up). Likewise, the greater the protection level desired by the trader, the greater the cost. Increased price protection involves choosing options that have strike prices close to the underlying futures contract price. This means these options have the best chance of being in-the-money at expiration. These options have greater time (and perhaps intrinsic) value, and their premiums are larger as a result.

13.5.2 A Short Hedge with Options: EWS Farms Wheat

A specific hedging example will illustrate these points. In this section, we focus on a short hedge to protect against falling wheat prices. Option hedges are compared against a benchmark sale at harvest and a futures hedge in this example.

In March, EWS Farms examines the following alternatives for pre-harvest pricing on wheat: a cash sale at harvest, a short hedge with futures contracts, a short hedge with a \$3.20 per bushel put option, a short hedge with a \$3.40 per bushel put option, and a fence strategy (buy a \$3.20 put option and write a \$3.90 call option). In order to compare the alternatives, a payoff matrix is created in a spreadsheet (Figure 13.8).

Price information is written at the top of the figure. The current futures price for a July wheat contract is trading at \$3.46 per bushel, and the expected harvest basis is \$0.60 under the futures price. To the right, three options are listed along with their

RightRisk Marketing Payoff Matrix
E.W. Sprague Wheat Enterprise
Taxes are ignored

Today

Current July Futures	$	3.46
Expected Basis	$	(0.60)

Option Type		Strike Price		Premium
Put Prem. @	$	3.40	$	0.24
Put Prem. @	$	3.20	$	0.15
Call Prem @	$	3.90	$	0.15

If July Futures Price is ($/bu)		Expected Cash Price Rec'vd		Net Price With Futures Hedge		Net Price with $3.40 Put Option		Net Price with $3.20 Put Option		Net Price Buy $3.20 Put and Write $3.90 Call Fence
$	3.00	$	2.40	$	2.84	$	2.54	$	2.43	$ 2.58
$	3.10	$	2.50	$	2.84	$	2.54	$	2.43	$ 2.58
$	3.20	$	2.60	$	2.84	$	2.54	$	2.43	$ 2.58
$	3.30	$	2.70	$	2.84	$	2.54	$	2.53	$ 2.68
$	3.40	$	2.80	$	2.84	$	2.54	$	2.63	$ 2.78
$	3.50	$	2.90	$	2.84	$	2.64	$	2.73	$ 2.88
$	3.60	$	3.00	$	2.84	$	2.74	$	2.83	$ 2.98
$	3.70	$	3.10	$	2.84	$	2.84	$	2.93	$ 3.08
$	3.80	$	3.20	$	2.84	$	2.94	$	3.03	$ 3.18
$	3.90	$	3.30	$	2.84	$	3.04	$	3.13	$ 3.28
$	4.00	$	3.40	$	2.84	$	3.14	$	3.23	$ 3.28
$	4.10	$	3.50	$	2.84	$	3.24	$	3.33	$ 3.28

Initial Margin	$ 865.00	Future Hedge Costs	$	0.02
Hedge Months	6	Options Hedge Costs	$	0.02
Interest Rate	10%	Fence Hedge Costs	$	0.02
Brokerage Fee	$50.00			

FIGURE 13.8 Wheat alternatives payoff matrix and assumptions.

respective premiums: a July wheat futures put option with a $3.40 strike price, a $3.20 put option on July wheat, and a call option with a $3.90 strike price.

Assumptions for hedging costs are found at the bottom of the figure. In this case, hedging costs are comprised of brokerage fees and an interest charge on margin deposits that might be used elsewhere in the operation. The required margin deposit is $865 for a single contract, and this margin is held for 6 months with an assumed 10% interest rate. The brokerage fees are $50 per transaction, or $100 for a sell and then buy round turn. To the right, hedging costs have been rounded to $0.02 per bushel.

The payoff matrix is located in the middle of the figure and is organized with the marketing alternatives in columns from left to right. The potential July futures prices are listed in the rows. The potential prices range from $3 to $4.10 per bushel. Note that in this payoff matrix, all the potential futures prices are listed as if they have an equal probability of occurrence; that is, it is just as likely that a $3-per-bushel futures price will result in July as a $4-per-bushel futures price.

The second column is the marketing alternative, the benchmark of a cash sale at harvest. Notice that when the futures price is $3 per bushel, the cash price is $2.40 per bushel because the local cash price is assumed to be $0.60 under the July futures price ($3 + (–) $0.60 = $2.40). The cash price received increases with each row as the futures price increases in $0.10 increments.

The third column of the payoff matrix is a short futures hedge. A futures contract is sold at $3.46 per bushel (the current futures price), and an offsetting transaction is made in July. In the first row, the net price received is $2.84 per bushel, and is calculated as the cash price received plus the futures gain/loss minus hedging costs:

Net Price = Cash Price + (Sell Futures Price – Buy Futures Price) – Hedging Costs

or

$2.40 + ($3.46 – $3) – $0.02 = $2.84.

As the futures price increases (moving down to the next row), the net price received is locked in at $2.84 because the gain in the cash market is exactly offset by a loss in the futures market.

The next two strategies are the result of a short hedge with put options at different strike prices. The steps for this hedge are as follows:

Step 1: When the hedge is initiated, the farmer buys a put option on a new crop futures contract with an expiration date after harvest. The put option gives the producer the right to perform exactly the same action in the futures market that he or she will perform with the cash crop at harvest.

Step 2: At harvest (or when the farmer is ready to sell the cash crop), the hedge is lifted when:

a. the cash grain is sold to a buyer;
b. the option is exercised; (If so, the producer sells a futures contract at the strike price and immediately buys a futures contract of the same expiration month to offset the position. The transition is profitable because the put option strike price was greater than the prevailing futures contract price.)
c. *or* the option is allowed to expire or be sold if any time value remains. (In this case exercising the option would not be profitable because the prevailing futures price is greater than the put option strike price.)

Step 3: The net price received for the grain is the cash price at the elevator plus the gain from the options position, less the option premium, less the cost of the hedge.

A specific example will illustrate the put options hedge. Consider the case in which the put option has a $3.40 per bushel strike price, and the July futures contract price is $3 per bushel. The price received for cash grain is $2.40 per bushel. The option is in-the-money (has intrinsic value) because the trader can exercise the option, sell a futures contract at the strike price of $3.40 per bushel, and complete the round turn by buying a futures contract at $3 per bushel. (This is the first row, fourth column of Figure 13.8.) Exercising the option and completing the round turn creates profits of $0.40 per bushel. The premium for the put option is $0.24 per bushel and hedging costs are $0.02 per bushel. Therefore, the overall proceeds from this marketing alternative are:

$$\$2.40 + (\$3.40 - \$3) - \$0.24 - \$0.02 = \$2.54 \text{ per bushel}$$

As long as the option is in-the-money (strike price is greater than the futures price), the net price received is locked in at $2.54 per bushel. In this case, the put option hedge has established a price floor more readily calculated as:

$$\text{Price Floor} = \text{Strike Price} + \text{Basis} - \text{Premium} - \text{Hedging Costs}$$

If the put option is out-of-the-money, the net price received is above the price floor and the option is allowed to expire. (We will assume that no time value exists for these calculations.) The net price received is calculated in this case as:

$$\text{Net Price} = \text{Cash Price} - \text{Premium} - \text{Hedging Costs}$$

or

$$\$2.64 = \$2.90 - \$0.24 - \$0.02 \text{ when the futures price is } \$3.50 \text{ per bushel}$$

It is revealing to chart the alternative marketing payoffs as shown in Figure 13.9. In this chart, the net price received is measured on the vertical axis, and the July futures price is measured on the horizontal axis. Proceeds from a sale in the cash market are charted as the diagonal line that begins at $2.40 per bushel when the futures price is $3 per bushel and ends with a net price received of $3.40 per bushel when the futures price is $4.10 per bushel. The futures hedge is charted as the horizontal price line at $2.84 per bushel. The $3.40 put option alternative begins as a horizontal line at $2.54 per bushel and then begins to increase once the futures price reaches $3.40 per bushel.

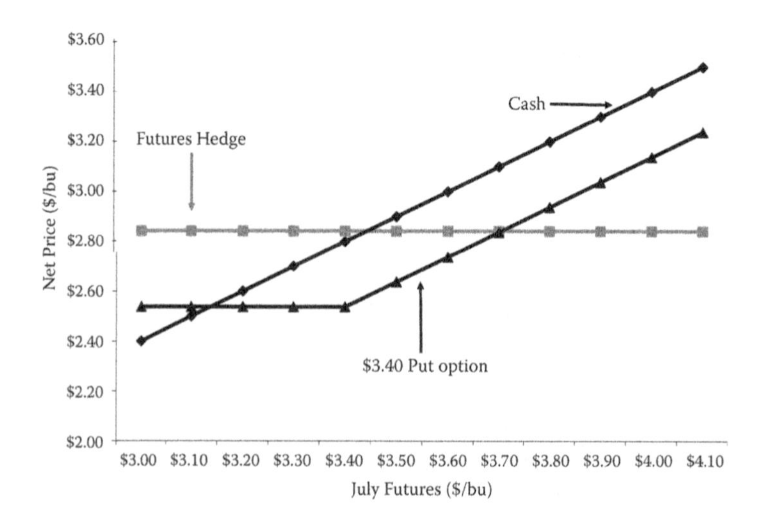

FIGURE 13.9 Three marketing alternatives for EWS Farms.

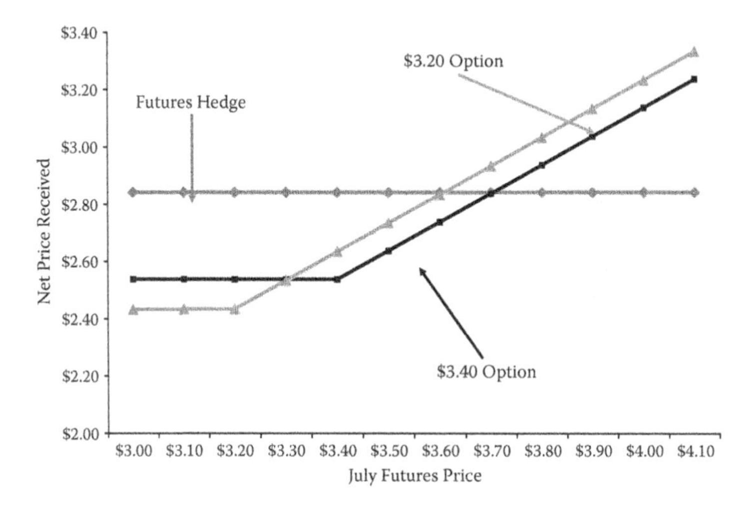

FIGURE 13.10 Option hedges and cash marketing for EWS Farms.

The put option strategy's advantage is that it establishes a price floor but still allows the producer to take advantage of increasing prices, outdistancing the futures hedge when futures prices are greater than $3.70 per bushel. The cost, though, is the option premium of $0.24 per bushel, which is why the option price floor is below the guaranteed price of the futures hedge. The futures hedge provides the greatest net price received until the futures price is $3.44 per bushel, which is the point at which the benchmark strategy dominates the others. This strategy, a sale in the cash market, has both the lowest and highest net price received and is the strategy with the most variability in net price received.

Let's return to Figure 13.8 and the $3.20 put option hedge (Column 5). The payoffs found in this column are calculated in the same way that they were for the $3.40 put option hedge, but in this case the premium is $0.15 per bushel and the strike price is $0.20 lower. The net result is a lower price floor established at $2.43 per bushel but with greater upside potential relative to the $3.40 put option hedge. The lower floor and greater upside are best seen in Figure 13.10 in which both option hedges are charted.

13.5.3 Fence Strategy

A final option is considered in the last column of Figure 13.8, the fence strategy. A fence strategy occurs when the trader establishes a hedge by buying a put option but then defrays a portion of the put option by writing a call. When the trader writes a call, he now has to post margin and is responsible for taking a long position in the futures market at the strike price if the option is exercised. The trader also receives the call premium, regardless of whether the option is exercised or not.

If EWS Farms initiates the fence strategy, a price floor will be set at $2.58 per bushel and a price ceiling is established at $3.29 per bushel. The price floor is in effect when the put option is in-the-money, and the floor is calculated as:

$$\text{Price Floor} = \text{Put Strike Price} + \text{Basis} - \text{Put Premium} + \text{Call Premium} - \text{Hedging Costs}$$

or

$$\$2.58 = \$3.20 + (-)\ \$0.60 - \$0.15 + \$0.15 - \$0.02$$

The price ceiling is set at $3.28 per bushel and results when the $3.90 call option is in-the-money. When the call option is in-the-money, the option is exercised, and the trader must take a long position in the market to offset the option buyer's futures transaction. The price ceiling for the fence strategy can be calculated as:

$$\text{Price Ceiling} = \text{Call Strike Price} + \text{Basis} - \text{Put Premium} + \text{Call Premium} - \text{Hedging Costs}$$

or

$$\$3.28 = \$3.90 + (-)\ \$0.60 + \$0.15 - \$0.15 - \$0.02$$

Between the floor and ceiling of the fence strategy, neither the $3.20 put option nor the $3.90 call option is in-the-money. In this area between the floor and ceiling, the net price received is simply:

$$\text{Net Price Received} = \text{Cash Price} - \text{Put Premium} + \text{Call Premium} - \text{Hedging Costs}$$

or when the futures price is $3.50 per bushel

$$\$2.88 = \$2.90 - \$0.15 + \$0.15 - \$0.02$$

The fence strategy is charted in Figure 13.11, along with the $3.20 put options hedge. The price floor for the fence strategy is $0.15 higher than the price floor of the $3.20 put hedge, which is exactly equal to the premium received for the call. The fence strategy remains $0.15 above the put hedge until the $3.90 call is in-the-money. From this futures price forward, the fence strategy plateaus, but the $3.20 put options hedge continues its upside potential.

Five alternative marketing strategies are considered in the payoff matrix of Figure 13.8, and all of the strategies have advantages and disadvantages. Which strategy should EWS Farms choose?

13.5.4 Choosing among Marketing Strategies

No single marketing strategy dominates all others, and the optimal strategy depends upon the EWS Farm decision maker's preference for risk and the ability of the farm to withstand risk. If the farm is in a good equity position and can withstand year-to-year cash flow fluctuations, a cash sale at harvest might be preferred as it nets the greatest potential return. If the farm is highly leveraged, pre-harvest pricing

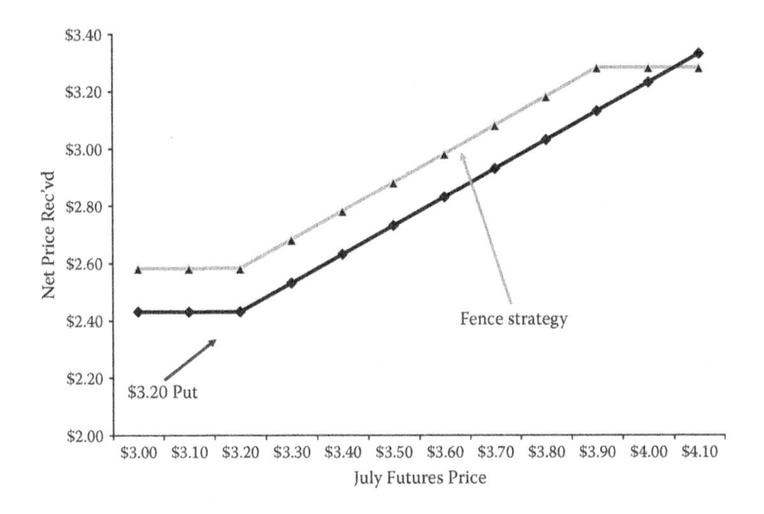

FIGURE 13.11 Fence strategy for EWS Farms.

might be pursued more aggressively with a futures hedge. A put options hedge may provide price insurance with upside potential, and the price floor might be set to cover the crop's cost of production. This type of insurance is more costly than a futures hedge. A fence strategy might be useful when a price floor is desired, but the premium required for a put options hedge is too large to justify the expense. In this case, writing a call may offset the premium cost, but the trader will need to post margin.

Cash flow might prove important when choosing among the marketing alternatives. Is the farm willing to post margin and make margin calls? Perhaps more importantly, is the lender to EWS Farms willing to answer margin calls? The lender's preferences for price risk management may influence EWS Farm's decision. Finally, EWS may choose to avoid expensive put option hedges if they do not have the funds to pay option premiums up front.

Market information influences the choice of pricing tactics. As an example, if futures market prices are trending upward, futures hedges tend to be expensive relative to put option hedges because of margin calls. In a sideways or choppy market, an options program may be more expensive because the upfront premium costs outweigh margin calls.

Basis risk is assumed away in the strategies depicted in Figure 13.8. Futures price risk is the only risk category considered. Basis risk may also be important, and the next section describes risk management tools used for managing basis risk, as well as futures price risk.

13.6 ADDITIONAL MARKETING TOOLS: CASH MARKET CONTRACTS

Cash market contracts are written agreements between the grower and a local buyer, such as the grain elevator, processor, or feedlot. Cash contract terms, in contrast to

TABLE 13.2
Cash Contracts and Futures Contract Differences

Futures Contracts	Local Cash Contracts
Standardized terms	Negotiated terms
Value fluctuates	Value fixed (or formula)
Organized by third party	Locally offered
Liquidity	Less liquidity
More heavily regulated	Obligation more rigid
Transferable obligation—flexibility	

futures contracts, may be negotiable and can vary from one contract to another. The contract terms include setting a price or price formula (for example, $0.20 above the December futures contract price quoted on November 1). The delivery quantity, location, and time are also stipulated within the contract. The differences between cash market contracts and futures contracts are described in Table 13.2.

Much like futures contracts, local cash price contracts can be used to reduce the price volatility of a growing crop. Contracts, because of the flexible terms mentioned above, can fix all or a portion of the components that determine local cash prices. These contracts can be used to fix only the futures part of the pricing equation, fix only the basis portion of the pricing equation, or fix the local cash price.
Cash contract types include:

- Traditional cash contracts
- Forward contracts
- Basis contracts
- Delayed pricing contracts
- Hedge-to-arrive or futures-only contracts
- Options-based contracts
- Minimum price contracts
- Minimax contracts

The focus of cash contracts is often the price, but other risks exist. For example, these contracts have a specified quantity that must be delivered. If the grower has low yields, then he may need to purchase additional grain to fulfill the contract requirement or cash settlement is made. Production risk is an important consideration in local cash contracts, while it is not an issue with futures contracts.

Another risk is counter-party risk. In this case, the grain buyer cannot fulfill his obligation to pay the producer. Managing counter-party risk means becoming familiar with the buyer's business and its health, being skeptical of "good" deals, asking questions, understanding the terms and conditions of the contract, understanding who holds the title to the commodity prior to delivery, and recognizing how disputes will be handled (e.g., by arbitration or mediation).

13.6.1 Forward Contracts

Forward contracts negotiate the specific sale of a predetermined quantity and quality of grain for future delivery. Forward contracts fix all three components of the pricing equation (futures price, basis, and cash price), so they are effective at locking in prices and require little management. Disadvantages for this contract include missed opportunities if prices should increase prior to delivery, yield risk for a growing crop, and counter-party risk.

13.6.2 Basis Contracts

Basis contracts include terms for the sale of a predetermined quantity of grain for future delivery. Rather than fixing all pricing components in advance, only the basis is fixed and the producer sets futures prices at a later date. This manages basis risk and allows for flexibility in setting the futures price, an advantage in an upward trending market. A disadvantage is that futures prices can be volatile, and these prices need be monitored so the price can be set at a preferred time.

13.6.3 Delayed Pricing Contracts

Rather than a price risk management tool, delayed pricing contracts are more of a logistical tool to bring grain into the marketing channel. In this case, the grain is delivered and the producer can lock in the cash price at a later date, or perhaps within a particular calendar window. This contract is an alternative to on-farm storage but does not provide any price risk management.

13.6.4 Hedge-to-Arrive (Futures-Only) Contracts

Hedge-to-arrive contracts allow the producer to establish the futures price at contract signing, and then the basis is locked in at a later date or within a specified window of time. This contract manages futures price risk, which is often the most volatile component of the pricing equation. Producers still face basis price risk, however, and must actively manage the contract to lock in the basis during the pricing window. Much like basis contracts, effective use of hedge-to-arrive contracts requires knowledge of seasonal basis and futures price variation.

13.6.5 Option-Based Contracts

Some cash contracts mimic options hedges. Two are considered here: minimum price contracts and minimax contracts.

Minimum price contracts: Minimum price contracts function much the way a put options hedge does. A minimum price is guaranteed for the crop, and the producer is allowed to lock in a higher price should markets trend upward. Minimum price contracts are not subject to basis risk, as both the futures price and basis are locked in to ensure the minimum price. These contracts generally have a fee associated with them.

Minimax contracts: Minimax contracts are very similar to a fence option strategy: they establish a minimum price with limited upside potential at a lower cost. Relative to the fence strategy, minimax contracts lock in both the futures price and the basis, so they have less price risk. Producers are generally charged a fee for participating in minimax contracts, but the fee is less than a minimum price contract.

Cash contracts are an alternative price risk management tool to trading in the futures markets. These contracts generally have more terms that can be negotiated (like the quantity) but may introduce yield and counter-party risk. Futures- and options-based cash contracts allow the producer to fix a portion of the pricing equation but also allow other terms to fluctuate during a pricing window. Flexibility to lock in the futures and/or basis means that the contract needs to be actively managed and a producer must be familiar with seasonal futures price and basis patterns to make the best use of these contracts.

The previous sections have outlined tools used to manage price risk but have stopped short of addressing a tactical marketing plan with price and date objectives. The next section presents a systematic way to perform tactical market planning.

13.7 RISK NAVIGATOR SRM TOOL: MARKETING PLAN

Marketing Plan is a tool available on the Risk Navigator SRM Web site. A marketing plan is a road map outlining how much grain will be sold, when it will be sold, and how it will be sold (e.g., via a forward contract). More specifically, the marketing plan is a written outline of systematic pricing tactics with quantity and price triggers. Writing down a marketing plan before the crop is planted allows us to check our logic and then play "What if." What if the farm produces only half of what we expect? What if prices are higher than we expect? When the marketing plan is reviewed, it indicates both past success (what works) and past failure (what doesn't work).

A marketing plan is an outline of price and quantity objectives used to generate a reasonable return for the business given the existing market conditions. Let's break this definition down; examining a few words will lend meaning. The marketing plan has price and date objectives, which are really triggers for a marketing action. When a price or date trigger is met, the producer sells the quantity listed in that objective. The objectives act as a road map for when the crop will be sold and how much will be sold. Additionally, a marketing plan should generate a reasonable return—reasonable in the sense that the return is consistent with the current pricing environment (existing market conditions), that it will generate something close to the market average, and that it will cover the costs of the farming operation.

Figure 13.12 represents a very simple marketing plan for corn. (A more complex marketing plan may be built using the Marketing Plan Risk Navigator SRM tool.) The arrow represents a timeline from January to harvest. The marketing plan has several date triggers: March 15, May 1, and harvest. It also has several price triggers associated with the dates: $2.30 per bushel and $2.25 per bushel. The quantity objectives are to sell 20% of expected production on March 15 (perhaps using forward contracts), sell an additional 20% on May 1, and sell the remainder at harvest.

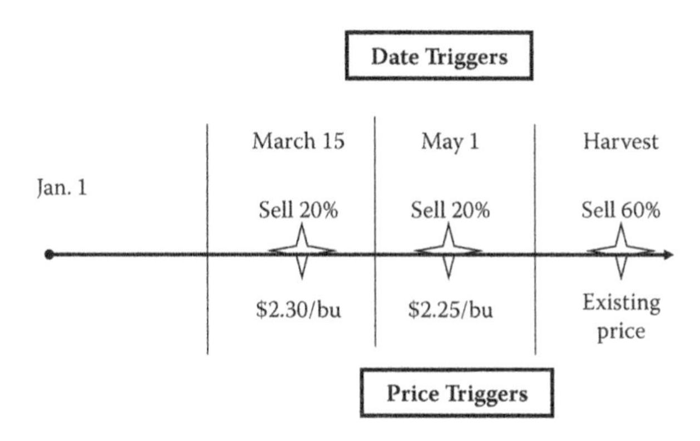

FIGURE 13.12 Simple marketing plan diagram.

13.7.1 Marketing Plan: The Advantages for EWS Farms

A written marketing plan introduces discipline into the commodity marketing process. In the example, the producer has decided to sell 20% of the expected crop by March 15 under a price trigger of $2.30 per bushel. What happens if the forward contract price is greater than $2.30 per bushel prior to March 15? The producer pulls the trigger and sells 20% of the crop. What happens if March 15 arrives and prices are below $2.30 per bushel? The producer still pulls the trigger and sells the crop. In this manner, the producer guarantees that he or she is taking what the market is willing to give, even if prices are not at their best. (Note: Prices might be at their high on March 15 relative to the rest of the year, and selling the crop means that the producer ensures at least a portion of the crop receives this high. Spreading sales throughout the year is a risk diversification strategy.)

A marketing plan also allows a manager to check her logic. Formally writing the plan allows the producer to examine price, date, and quantity objectives and to ask the following questions: Are the date objectives set at times when prices are at their historical best? Are the price triggers reasonable given history and current market information? For example, is this the time of year when basis is narrow and stable or wide and uncertain? Are the quantity objectives consistent with what I can produce? Will the quantity objectives meet my cash flow needs?

Writing down a marketing plan detaches you from the decision-making process. Detachment is necessary because we often sell or choose not to sell our crop based on emotion. Have you ever thought, "I won't sell because I deserve more for the crop. Look at the work I put into it" or "Prices are really rising now; I'll wait until tomorrow to get a higher price," but the higher price never materializes? The marketing plan detaches you from the decision by introducing decision rules.

Another advantage of a marketing plan is that it gives the proper perspective by placing the marketing decision in the context of costs and the current economic climate. As discussed previously, the marketing plan enforces discipline in decision making and allows you to check the logic behind your decisions. Finally, you can ask the important "what if" questions. For example, what if prices aren't as high as I expected? What if I only produce 50% of expected production?

13.7.2 An Example Marketing Plan for EWS Farms

Let's review a marketing plan using an EWS Farms example in Risk Navigator's Marketing Plan tool. We cannot present graphics for this tool as we have in previous chapters due to the size of the information. You will have to follow the example on our Web site (http://www.risknavigatorsrm.com) by clicking on tools, then on market plan.

Six components exist in this marketing plan, and each component has its own worksheet. The components are titled: The Relationship between the Strategic Risk Management Plan and the Marketing Plan; Production History and Expectations; Expected Prices; Production Costs; Price, Date and Quantity Targets; and Review and Evaluation. We'll examine these components in turn.

13.7.2.1 Putting It in Context

The first step is relating the marketing plan to the SRM process. Remember, the SRM process establishes the overall strategy, and the commodity marketing plan is a tactical part of this strategy. To link tactics with the SRM process, EWS Farms answers questions posed on the first sheet of the marketing plan. For example: How will the marketing plan fulfill the business plan's objectives? How do the marketing plan's objectives relate to the business plan's strategy? Is it intended to serve the financial goals of the farm?

The marketing plan needs to be consistent with the business plan's long-term objectives and short-term goals. What are other important considerations that deserve attention when writing the marketing plan? Is cash flow a concern? Are financial resources needed for other business enterprises? How will the marketing plan impact the farm's tax liability? Note these and other considerations as appropriate.

13.7.2.2 Production History

Knowledge of the farm's production history is very useful when setting the quantity objectives for the marketing plan. In the marketing plan example, click on the Production History tab and note that EWS Farms' harvested acres and yields are recorded in the shaded cells. What is the average number of acres harvested over time? And what is the average yield over time? (Note: The total production and average production are calculated for you.) Is the total production predictable from year to year? How variable are yields? For EWS Farms, irrigated corn has averaged 172 bushels per acre with a high of 195 bushels and a low of 118 bushels.

The production information on the right is useful when setting the quantity targets. How much grain can the farm realistically expect to produce? Given the variability of yields, how much should the farm be willing to contract? The planting information and the likelihood of a loss will be useful as quantity objectives are set.

13.7.2.3 Prices

Price expectations can play an important role when setting price and date triggers. A first step is to assess the historical situation: What is the average price at different points during the year? (You may wish to think about futures prices and basis separately.) Does it make sense to set a date objective when prices are historically at their lowest point? How variable are prices during the year? Does it make sense to plan on selling the entire crop when prices are the most variable?

When examining futures prices, it's often the case that pricing opportunities occur in the early spring (in or around the U.S. Department of Agriculture's planting intentions report), just after corn planting begins in the Corn Belt and just prior to the pollination period. At the same time, low prices are often seen during harvest with recovery occurring throughout the remainder of the winter. Every cropping year is different, however, so it is best to assess the current situation and adjust the marketing plan to fit conditions.

University extension specialists and the U.S. Department of Agriculture (USDA) are good sources of current price and outlook. These specialists generally assess the situation but seldom make recommendations. Another source of information and pricing recommendations are market advisors who often provide a service to help producers "beat the market." But can someone really beat the market? AgMAS® has tracked the performance of market evaluators, and their periodic reports are located at http://www.farmdoc.uiuc.edu/agmas/.

So what is the difference between following the advice of a marketing service and using the marketing plan? One difference is in perspective. A market advisor believes that he or she has information other traders don't have. If the advisor acts on the information, then he or she can "beat the market" and consistently garner profits above the market average. A marketing plan has a different focus; it seeks to average returns and avoid mistakes by using a variety of methods to spread out grain sales. The market advisor actively manages grain sales using price and quantity information to forecast short-term results. The marketing plan is passive—you only worry about current prices as they relate to price triggers, and you only make sales when the price or date triggers are met.

Return to the Marketing Plan spreadsheet and notice the cells in which high, average, and low prices can be entered. In addition, the target price or the loan rate for the crop can be entered next to the Farm Bill heading. The prices are an assessment of potential prices in the future and will be used to set price targets and calculate revenues.

13.7.2.4 Costs

It is important to set price and quantity objectives in the context of our production costs, so click the worksheet titled Costs. What costs are important?

Let's look at the corn example about halfway down the worksheet. In the short term, we need to cover at least our variable costs, so use a per-acre estimate of corn variable costs under the heading Revenue Required Per Acre next to the Variable Cost label. In the longer term, we need to pay for land rent (or mortgage interest) and make payments for (or in anticipation of) new machinery. Therefore, it's good to write down the per-acre sum of variable costs, land payments, and machinery. Finally, family living expenses are often covered by the farming operation, so add the per-acre contribution of family living expenses to the variable cost, machinery, and rent payment sum and write the total next to the appropriate label.

Once the totals have been entered, the price needed per bushel to meet each of the costs is placed in the column titled Needed Price/bu @ the Expected Yield. To meet corn variable costs in the example, a price per bushel of $1.41 is needed, and the needed corn price increases to $2.01 per bushel and $2.13 per bushel as costs are added. Are these prices attainable given the price expectations you entered? The price expectations are found at the bottom of the corn table, and the medium price for EWS Farms is $2.25 per bushel. After the same cost information has been entered for wheat, the farm's expenses are summed at the top of the sheet assuming the EWS Farms' acreage distribution.

13.7.2.5 Price, Quantity, and Date Targets

Production, price, and cost information provide a good context so that actual price, date, and quantity objectives can be set on the worksheet titled Price, Date and Quantity Targets. The first step is to identify the date triggers in the far left-hand column. What dates seem to fit the historical pattern of prices? Are there periods when prices tend to remain high, and then decline? Setting a date trigger during the "high" period is advantageous. How does the trigger date relate to the expected production total? (For example, selling the entire crop before it is planted might be considered risky.)

Next consider the method of contracting/sale. How does the method of sale influence the firm's risk position? Is there a delivery requirement, or is the pricing based on a futures hedge? Does the method of sale fix all pricing components (e.g., a forward contract) or does it allow a portion of the pricing formula to fluctuate (e.g., a basis contract)? The method of sale influences the riskiness of the revenue generated by pre-harvest and post-harvest sales.

Move to the far right column and write down the pricing objectives in the appropriate columns using your price expectations (in the upper right-hand portion of the worksheet). Are the price triggers acceptable (will they meet costs)? Given the expected price, are the trigger prices attainable?

Finally, set the quantity objectives. How much should be sold each period? Are you comfortable contracting that amount? The cumulative quantities are summed under the column "% priced." How much of the crop is left after each sale is made? Are you pricing the largest of portion of the crop at a time that is best for the farm business?

13.7.2.6 Monitor and Review

Notice that all of the objectives on the Price, Date, and Quantity Objectives sheet have been transferred to the pink cells on the Evaluate and Review worksheet. Use this sheet to write down when you actually make a sale, and then compare your performance over time. What sales are you pleased with? Why? Are the circumstances that made that sale successful likely to continue into the following year? Did you learn any new information from the sale? Why were mistakes made?

Be sure to stay disciplined and follow the price and quantity objectives. If conditions change, however, take note of this and adapt the marketing plan to fit conditions. For instance, poor planting weather may make you revise the quantity objectives if you fail to plant the expected acreage. The best marketing plan is one that is reviewed often so that lessons learned from one year are carried over to the next.

13.8 A LITTLE PHILOSOPHY

Finally, a last bit of marketing philosophy. Even when you make the best plans, unexpected events happen; bad outcomes occur. A marketing plan will help you avoid some common marketing pitfalls, and for that reason it is useful. Don't abandon the marketing plan just because you missed the market high! Remember, the marketing plan is designed to be passive, to help you obtain a market average. The plan may prevent you from hitting the market high, but it also prevents you from hitting the market low. Finally, be sure to design a plan that fits your operation. A marketing plan that works well for one operation may not be suitable for another.

REFERENCE

Harwood, J., R. Heifner, J. Perry, A. Somwaru, and K. Coble. 1999. Farmers sharpen tools to confront business risk. *Agricultural Outlook*. U.S. Department of Agriculture, Economic Research Service, March.

FURTHER READING

Carter, C. 2003. *Futures and options markets: An introduction*. Upper Saddle River, NJ: Prentice Hall.

Hamilton, N. D. 1995. *A farmer's legal guide to production contracts*. Philadelphia, PA: Farm Journal, Inc.

Purcell, W., and S. Koontz. 2000. *Agricultural futures and options*, 2nd ed. Upper Saddle River, NJ: Prentice Hall.

Scnepf, R., R. Heifner, and R. Dismukes. 1999. Insurance and hedging: Two ingredients for a risk management recipe. *Agricultural Outlook*. U.S. Department of Agriculture, Economic Research Service, April.

14 Production Risk

Jay Parsons

CONTENTS

14.1 INTRODUCTION

Production risks emanate from the producer's inability to accurately forecast input productivity and output yields. Economists spend a significant amount of time calculating optimal input and output levels in various agricultural production settings. These calculations are usually based upon an estimated production function (or response curve) and expected values for input and output prices. Experiments are conducted to determine the shape of a production function. For example, soil and crop scientists test the response of crops to various fertilizer levels. Animal scientists and nutritionists test the response of livestock to various feeding regimes. All of this is important work and, when combined with input and output prices, can provide valuable economic information to producers. However, to be of optimum use, it all needs to be translated into the producer's production setting, which is typically filled with risks and uncertainty. Risk and uncertainty can quickly turn a well-intentioned and otherwise accurate scientific result into meaningless, or sometimes even misleading, information. It is important that producers look at all of the information and consider their risks when making production decisions.

Sources of production risk include weather, pests, diseases, weeds, genetic variations, wildlife, quality of inputs, machinery efficiency, and the timing of operations. Anything that can pose a risk to the quantity and/or quality of production from an agricultural operation is considered a production risk.

Producers can use a number of different control strategies to address production risks. Some of these controls address the risk directly from the input side by attacking the source. For example, a producer can mitigate the risk of low rainfall or drier than usual weather by using irrigation systems. Other control strategies are a little more indirect or passive in their approach. For example, crop insurance will do nothing to contribute to higher yields, but it is an excellent tool for diminishing the risk of low production. Some of these different risk management alternatives have already been discussed in Step 5 of Chapter 9, "Identify Risk Management Alternatives."

In the sections that follow, we will thoroughly discuss production risks from both an input and output viewpoint. I will follow that with a discussion of specific risk management strategies that can be used to address agricultural production risks. Several useful examples are included in the discussion, including a Risk Navigator stochastic crop budget tool. We'll begin with a discussion of the difference between input and output production risk.

14.2 INPUT VERSUS OUTPUT RISK

The distinction between input and output production risk is pretty straightforward. Output production risk comes from uncertainty related to output yields. Input production risk comes from uncertainty surrounding inputs. This sounds simple, but there is a gray area in the middle that can be confusing. In this case, that gray area involves the production process and the level of detail at which the decision maker is working.

The cause of output yield variability generally can be traced back to changes in specific input factors. For example, variability in sunflower yields for a dryland farmer can usually be traced back to specific inputs like soil moisture and growing temperatures. Technically, all production risk can be classified as input risk if the decision maker wants to dig down to the level of the detailed input factors that cause the variance in output. It is often more practical, however, to focus on the various output possibilities that result from the production process and simply think in terms of output production risk for analyses purposes.

Although there are many circumstances that can influence crop production, weather is usually cited as the major contributing factor. In fact, weather combines with a number of other factors such as plant genetics, timing, fertilizers, and pesticides to determine the final yield and quality of output. If we take the quality and quantity of the other inputs as a certainty, then it is easy enough to focus on the weather and its effect on output. This is usually done in the context of output production risk because weather is not an input that producers can generally exercise control over in the decision making process. In other words, weather is not a control or decision variable; it is a risk factor that ultimately affects the quantity and quality of output. Dealing with it in the context of an output production risk seems to make sense in most cases.

On the other hand, in some production settings the components of weather, like temperature and rainfall, can be thought of as controlled inputs. Confined livestock operations and greenhouses certainly seek to closely control them in their intensely managed environments. In this context, the variance in the growing environment or climate might make more sense if it is treated as an input risk. For example, the risk of a failure or suboptimal performance of the climate control systems would adversely affect output.

The ability to control or to take action to control variances in the input factor is one of the key components involved in deciding whether or not to treat the situation in the context of an input risk or an output risk. In the two examples given above concerning climate, it is pretty clear that in some cases growing climate can be treated as an input risk and in others there is very little control over it, so it makes more sense to treat it in the context of output risk. One must also bear in mind that even though the ability to control an input risk exists, it may not be economically feasible to do it to such a degree that variance is eliminated.

Consider the dairy farmer who is relying on quality hay to feed his livestock and obtain high production numbers. The farmer has the ability to test the hay for forage quality and can carry those testing procedures to an extreme level so as to practically eliminate the risk of using poor quality hay. In the manufacturing sector, this concept is called *statistical quality control* and can be implemented on both the input and output side in the production process. It makes sense for the dairy farmer to think of the quality of the hay as an input production risk; however, forage quality tests cost money and the producer may choose to accept the production risk as an unknown factor in the process. Output production risk is then a convenient way for the producer to simplify the production risk analysis by focusing on variances in output yield and quality without worrying about the specific bundle of input factors that produce the results. Deciding whether to treat the production risk as an input risk or an output risk ultimately resides with the decision maker, who has the ability to take the analysis to a level deep enough to isolate the input risk causing the variability in output and their ability to control that input risk.

Production cost is a parameter to analyze when trying to understand production risk. The cost per unit of production is an excellent economic measure, and variability in this measure can be used as a measure for production risk in a more general sense without getting too carried away with defining the source as an input or output risk. In most cases, however, it ignores issues pertaining to output quality.

14.3 INPUT RISK

A production risk can be defined as an input risk if the variability in the production process can be attributed to one of four causes: (1) uncertainty in the quality of one or more inputs, (2) uncertainty in the quantity of one or more inputs, (3) uncertainty in the timing of one or more inputs, or (4) uncertainty in the price of one or more inputs.

14.3.1 Quality of Input Risk

When making an input decision in production agriculture, the decision maker often knows the price of the input, the quantity needed, and when it will be used. Once these parameters are known, the input risk disappears if the decision maker also knows the quality of the input. But in production agriculture, there are several examples where uncertainty in the quality of one or more inputs exists. Take genetics for example. Even though a livestock producer may know an animal's genetic history, there still is some uncertainty about what exactly came out of the mating process and how that animal will perform in the production setting.

Labor is another input where uncertainty about quality can create a risk to the producer. This situation crosses over into the human risk category but, from a production standpoint, the quality of available labor is an important factor to consider when making production decisions. When hiring new labor, an uncertainty exists, creating a risk that producers must accept.

As Robison and Barry show (1987), risk associated with quality of inputs generally leads decision makers to use larger quantities of inputs than would be used if the quality were known with certainty. This is a result derived from diminishing marginal returns and it depends upon the additivity assumption of input risk. Consider the production function graphed in Figure 14.1 with input L. Diminishing marginal returns means that for each addition unit of input the increase in output is less than it was for the previous unit of input added. Diminishing marginal returns is the reason for the curvature, or more specifically, the concavity of the production function graph.

Suppose there is uncertainty about the quality of input L in Figure 14.1. Let's say L is units of labor hired to work. Uncertainty about the quality of input L involves variability in the services provided by L units of labor. Let ε represent a normal random variable with a mean of zero and a standard deviation of σ_ε. When L units of labor are hired, L + ε units are actually received in terms of the service provided. Then, the output from hiring L units of labor is actually $f(L + \varepsilon)$, which may be

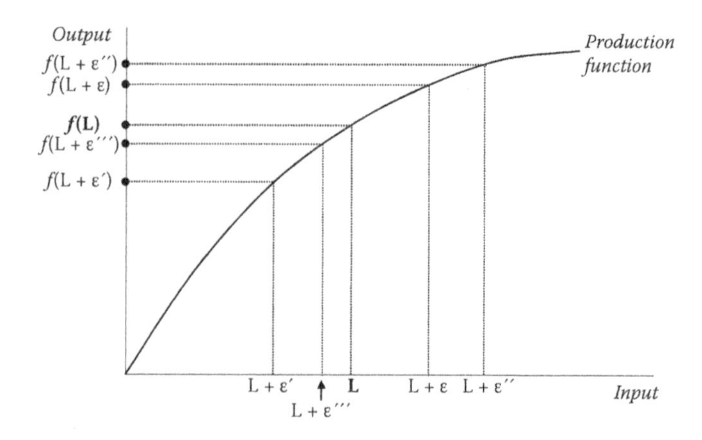

FIGURE 14.1 Range of outputs with input level L and random risk ε.

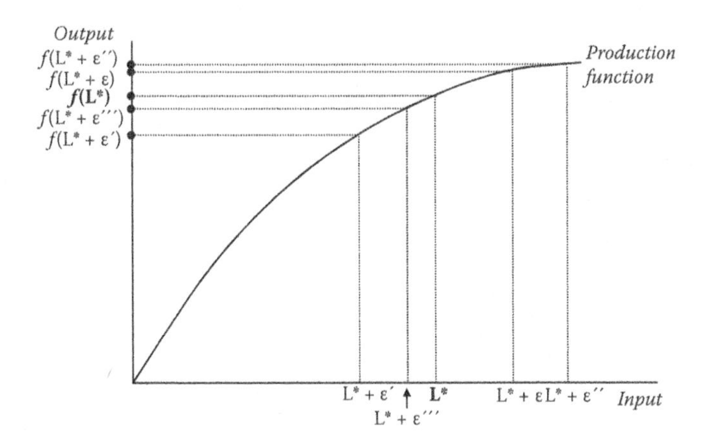

FIGURE 14.2 Range of outputs with input level $L^* > L$ and random risk ε.

greater than or less than $f(L)$, depending upon whether ε is greater than or less than zero. Figure 14.1 shows several possible production outcomes using L units of input and random quality draws of ε, ε', ε'', and ε'''. The seemingly random quantities of production output symbolized by $f(L + \varepsilon)$, $f(L + \varepsilon')$, $f(L + \varepsilon'')$, and $f(L + \varepsilon''')$ capture the essence of quality of input risk.

Leaving the random risk unchanged with draws of ε, ε', ε'', and ε''', Figure 14.2 shows the effect of increasing the level of input from L to L*. Notice how much more tightly bunched the possible output levels are in Figure 14.2 in comparison to the possible output levels in Figure 14.1. Increasing the level of input from L to L* had the effect of decreasing the variance in output. Therefore, input L is called a *risk-decreasing input.* Because of this, decision makers will sometimes use more of an input when uncertainty is present. We'll discuss this more in the next section.

14.3.2 Quantity of Input Risk

Figures 14.1 and 14.2 can also be used to explain quantity of input risk. A producer usually knows the quantity of each input he uses in the production process; however, there are inputs, such as rainfall, over which the producer has very little control. Also, there are technology risks with input applications that could result in variable quantities being applied. If the random variable ε is used to represent this uncertainty in the quantity of input supplied, then Figures 14.1 and 14.2 provide pictures depicting the variable output that can result from quantity of input risk associated with intended levels of L and L*, respectively. Note that the difference between quality of input risk and quantity of input risk is more contextual than theoretical. Figures 14.1 and 14.2 show that in theory, both concepts involve uncertainty about the *effective* quantity of input provided.

Application of nitrogen fertilizer is an example where quantity of input risk can exist. Field trials with nitrogen fertilizer applicators (Hanna et al., 2002) have shown considerable variance in the rates at which anhydrous ammonia flows out of the individual knives on the applicator. As technological improvements take place, these variances are diminishing, but the producer is then faced with the decision of when and how to upgrade equipment. In the meantime, a producer wishing to apply a fixed rate of nitrogen fertilizer to a field could see considerable fluctuations, resulting in variable rates being applied throughout the field and contributing to crop yield variability.

Water is an important input when growing crops and forage. For nonirrigated land, rainfall comes in variable quantities and has a profound effect on production. For irrigated land, the producer usually controls the total quantity. However, when water supplies are tight, producers sometimes find themselves on the wrong end of the water rights list and short of the quantity they need to grow their crops.

In addition, the quality and reliability of an irrigation system is important to insure proper water delivery to the crop. An unreliable system can leave a growing crop without water at a critical juncture in the growing season and pose a tremendous risk to yields. Like fertilizer, the uniformity of water delivery is important and is dependent upon the technology used. Studies from the 1980s and early 1990s (Dinar, Letey, and Knapp, 1985; Feinerman, Shani, and Bresler, 1989; Dinar and Letey, 1991; Knapp, 1992) show that increasing irrigation uniformity results in more uniform yields, less overall water use, and increased profits. It is no wonder that tremendous advances have been made in the last 20 years to improve the quality and reliability of irrigation systems, especially those of the center pivot variety common in the western United States.

In general, some inputs can be thought of as risk increasing and some as risk decreasing (Chambers and Quiggen, 2000). The most commonly accepted definition of a risk-decreasing input is an input where the variance in output declines when you add more of the input. An empirical example would be the use of pesticides. Pesticides are used to control losses when a disease or insect outbreak occurs. In the absence of an outbreak, they generally do not increase yields. The distribution of outputs is diminished on the downside while remaining unchanged on the upside. This, in effect, dampens the dispersion and decreases output variance. Thus, pesticides are generally classified as a risk-reducing input.

On the other hand, fertilizers can have both a positive and negative effect on yields. If the weather cooperates, yields can increase significantly; however, if the weather is not favorable, using fertilizers can decrease yields. The result is a wider dispersion of possible output yields brought about by using fertilizer. Because of this, fertilizer is commonly classified as a risk-increasing input.

Classifying an input as risk reducing or risk increasing solely on the basis of its effect on output variance is a dangerous proposition. It relies on the notion that variance is the chosen measure for describing risk and that less is better. More importantly, it ignores the changes in mean and median output levels brought about by the use of the input. If using more input drives up the output levels, while also increasing variance, it may be a trade-off the decision maker is willing to make and one he views as a risk-neutral or risk-reducing situation. It all depends upon the trade-off and that is why the coefficient of variation (the ratio of the mean to the standard deviation) is often viewed as a better measure of risk. A risk-reducing input would then be defined as one that decreases the coefficient of variation with increased use of the input.

14.3.3 Input Timing

Input timing is a critical component to consider when it comes to input risk. Just like a recipe needs to be followed to get a cake to rise perfectly in the oven or the meringue on top of that pie to turn out perfectly, a producer is keenly aware of the importance input timing can have on the resulting output. When it comes to rainfall for dryland crops and forage, a well-timed drink from Mother Nature is just as important as the amount received. Similarly, reliable irrigation systems and water supplies are critical to the success of an irrigated crop producer in delivering well-timed water inputs to a growing crop. Timing may not be everything, but it is a critically important component when it comes to input risk and the production process.

14.3.4 Price of Inputs

One of the most often cited sources of input risk is variation in input prices. While technically this is a market risk on the input side, it is sometimes discussed in terms of being a production risk. When a producer makes a decision regarding production, he may not know with absolute certainty all of the inputs and associated costs, and this exposes him to the risk of unexpectedly high input costs. For example, a producer with a growing crop in the field may encounter unexpectedly high harvest costs due to a spike in fuel prices, or a livestock producer may find himself with unexpectedly high trucking costs for the same reason. Either one of the examples could result in a significant increase in the cost of production despite having no impact on yields.

14.3.5 Stochastic Budgeting Example

A helpful technique for analyzing risky production decisions is called *stochastic budgeting*. A stochastic budget is typically built off of a conventional budget put together under assumed certainty then tweaked to account for variability in one or more of the values (such as yield, price of fertilizer, or price of output) in the budget. The conventional budget used to construct the stochastic budget can be a partial budget for a given enterprise or a whole farm budget for the entire operation. It can be a short-term budget or a long-term budget covering multiple periods. It can be a cash flow budget or a profit budget. Or it can be a budget that focuses on returns on investment. The type of budget depends upon what the decision maker is trying to decide and how she wants the information organized for analysis.

The idea behind stochastic budgeting is to take a budget drawn up in deterministic terms and extend it by adding stochastic components. For example, consider the conventional partial budget for dryland wheat production given in Figure 14.3. (Note that the budget does not include government payments.) Direct costs related to pre-harvest and harvest operations are given with certainty in the budgeting process. At first, the budget is like any other. Stochastic budgeting can be introduced into the picture on the input side by specifying a probability distribution for one or more of the budget variables deemed to be a risk factor. This action turns a known value, such as crop price, into an uncertain value. The outcome from a stochastic budget is a distribution of possibilities for any key output variable that can be expressed

	Unit	Price or Cost/Unit	Quantity	Value or Cost per Acre
Gross Receipts from Production				
Hard Red Winter Wheat	BU	4.60	30	138.00
Crop Insurance Indemnity				
Total Receipts				**138.00**
Direct Costs				
Operating Preharvest				
Seed	Dols	6.39	1.00	6.39
Fertilizer	Dols	31.76	1.00	31.76
Herbicide	Dols	6.37	1.00	6.37
Custom Application	Dols	4.00	1.00	4.00
Crop Insurance	Dols			7.85
Fuel	Dols			10.84
Repair & Maintenance	Dols			8.24
Labor	Dols			2.23
Interest Expense	Dols	7%	38.84	2.72
Total Preharvest				**80.40**
Operating Harvest				
Fuel	Dols			4.83
Repair & Maintenance	Dols			7.88
Labor	Dols			1.48
Hauling	BU	0.14	30	4.20
Total Harvest				**18.39**
Total Operating Costs				**98.79**
Total Gross Margin				**39.21**
Property and Ownership Costs				
Machinery Ownership Costs	Dols			42.78
General Farm Overhead	Dols			10.00
Real Estate Taxes	Dols			1.80
Total Property and Ownership Costs				**54.58**
Net Receipts Before Factor Payments				**–15.37**
Factor Payments				
Land @ 4.00%	Dols			32.00
Return to Management and Risk				**–47.37**

FIGURE 14.3 Sample enterprise budget for dryland winter wheat (with fallow rotation).

graphically, such as in a probability density function (PDF) or Contributing Factor Diagram (CDF), or in a statistical summary form.

Consider a situation where operating costs are deemed to be somewhat unknown. The deterministic budget in Figure 14.3 has the cost at $98.79 per acre, which is what the decision maker deems as the most likely occurrence. However, the decision maker feels that costs could be as low as $90 an acre or as high as $113 an acre, depending upon fuel costs, machinery repairs, etc. We can introduce this uncertainty into the budget by specifying the total operating costs as a distribution with $90 as the low, $113 as the high, and $98.79 as the most likely occurrence. Tools like the Risk Profiler introduced in Chapter 10 can be used to generate a distribution of outcomes for the total operating costs. Figure 14.4 displays distribution output from the Risk Profiler for this situation.

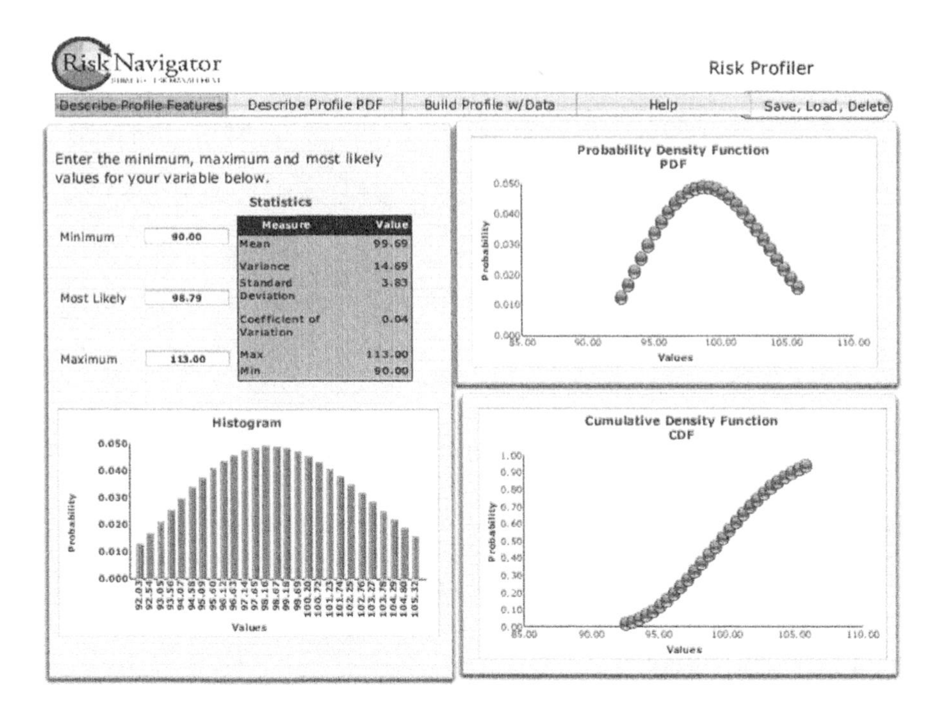

FIGURE 14.4 Sample dryland winter wheat total operating cost distribution.

The distribution of values associated with total operating costs depicted in Figure 14.4 can be plugged one at a time into the budget in Figure 14.3 to produce a distribution for key output variables such as total gross margin. For example, the low value of total operating costs equal to $90 would result in a gross margin of $48. The high value of total operating costs equal to $113 would result in a gross margin equal to $25, and so on and so forth. The stochastic variable for total operating costs produces stochastic output for the total gross margin.

Risk Navigator's Stochastic Crop Budget Generator tool can be used to gain a better understanding of these concepts. The Stochastic Crop Budget Generator is explained in greater detail in the last part of this chapter. For now, a graph of the key output total gross margin is given in Figure 14.5 in the form of a cumulative distribution function. Looking at the graph, you can see that because of the distribution associated with the operating costs, there is only a 20% probability that total gross margin for this crop will fall below $34.50. This could be an important component for producers to consider as they make production decisions.

The stochastic elements introduced into the dryland wheat budget and the results depicted in Figures 14.4 and 14.5 are very simplistic. Obviously, there are a number of areas in the budget where uncertainty exists and stochastic elements can be introduced. More complex budgets with greater detail provide even more opportunities; however, do not get too carried away with introducing multiple stochastic elements without accounting for correlated dependencies among variables. That is, if you vary price and yield, for example, you should account for the fact that a high price should

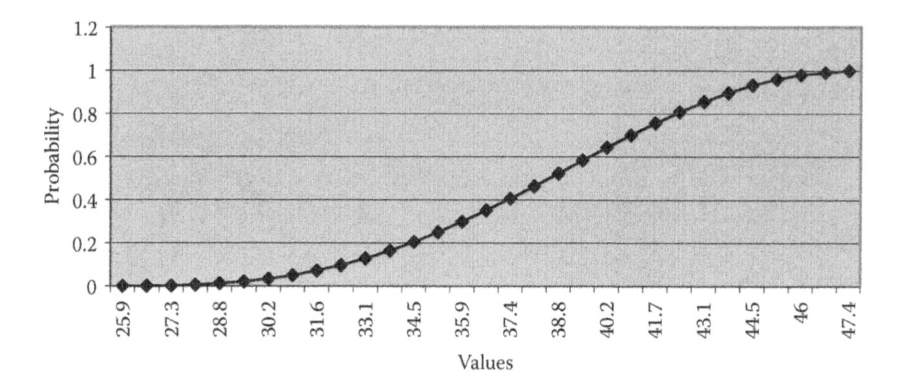

FIGURE 14.5 Cumulative distribution function for gross margin with uncertain operating costs.

not be randomly selected with a high yield, since this combination is not likely to be found in practice. We'll return to the stochastic budgeting example in the next section with a look at variances in the output side of production.

14.4 OUTPUT RISK

As mentioned earlier in the discussion of input versus output risk, production risk is quite often treated in the context of output risk as a matter of convenience and practicality. Several instances of input variation can be summarized neatly for analysis purposes into one variable measurement on the output side: yield. In the real world, yield variations are usually very difficult to attribute to a finite and defined set of specific input variations, let alone attributing them all to one input variation as we did in our earlier discussion. From a decision-making standpoint, this is often an unnecessary step anyway because the decision maker often has very little influence over some of the most influential input factors. For example, even though the managers of EWS Farms can select what wheat variety they plant in which fields, when they plant, and how much fertilizer to apply, they cannot influence the weather, which is likely the biggest influence on the yields they will experience. The influence of weather and other output production risks can materialize in a couple of different ways: variations in the quality of output produced and variation in the quantity of output produced.

14.4.1 Quality of Output Risk

Quality of output risk is sometimes characterized as a market risk because variations in quality will directly influence the price received for the product. Our discussion is limited in context to the sources of production risk that can lead to variable quality in output. This comes with an understanding of the fact that strategies designed to limit the impact of those sources are the best way to directly address this type of production risk. Market risk management strategies can play an important role in mitigating the final resulting impact.

In the introduction to this chapter, we listed several sources of production risk. All of these sources are at times capable of impacting production quality. For example, pests and diseases can lower the quality of grain and make it difficult for a grain producer to market their crop without accepting large price discounts in the form of dockage. A livestock producer can suffer from the negative impacts of poor genetics and produce some animals that perform very poorly on a carcass quality–based marketing grid. Harvesting a crop too early or too late can result in poor quality, which can have a negative impact on price received if marketed "as is" or drive up costs if steps are taken to condition the output to meet the demands of the market place.

In rare cases, quality of output risk can take the form of upside risk, and this partially offsets the negative impact of risk on the quantity produced. For example, wheat producers, like EWS Farms, are familiar with the impacts that low rainfall can have on yields. While low soil moisture stresses the crop and lowers the quantity of wheat produced, that same stress can drive up the wheat's protein content. Depending upon market conditions, the price premium paid for high-protein wheat can sometimes be a significant boost to the bottom line and help mitigate some of the negative impact of having less to sell. One of the strategies EWS Farms and other wheat producers can use to help address the risk of low soil moisture is to make sure they are selling into a market that provides good and reliable premiums for high protein content.

In a later section, we'll return to our stochastic budgeting example and address output risk using the Stochastic Crop Budget Generator to affect both quantity and price. As discussed above, output price variability can be used to address the impact of quality of output risk. However, as we cautioned at the end of the last section on input risk, do not get too carried away with introducing multiple stochastic elements into the budget without appropriately accounting for correlated dependencies among variables. This includes the situation depicted in the preceding paragraph where there was an inverse correlation between a producer's yield quantity and the quality of product produced. That inverse relationship resulted in a lower production quantity being partially offset by increasing quality and the price associated with higher protein content. Even in today's increasingly differentiated marketplace, this situation is fairly uncommon and not something the Stochastic Crop Budget Generator tool is designed to handle. The Stochastic Crop Budget Generator tool depicts a realistic situation for most commodity crop producers where their output price and quantity of production are assumed to be independent of one another. That is, a single producer's output isn't enough to influence the market price one way or another. So, while you can simulate quality of output risk with the Stochastic Crop Budget Generator by making output price stochastic, you cannot correlate that output price distribution with yield.

14.4.2 Quantity of Output Risk

Production risk is usually described as any risk that causes variability and/or uncertainty in the quantity of the output produced. We've chosen in our discussion to dig a little deeper into the causes and look at alternative ways to characterize the variability that leads to production risk. We've also looked at variability in the quality of output, which is often characterized as a market risk because of its association with variable output prices. Despite this broader and, perhaps, more in-depth look at production risk, we don't want to mislead the reader into thinking that variability in the quantity of output is not the focus of production risk. The convenient and simplifying nature of focusing an analysis of production risk on the variable quantities produced far outweighs the downside of any details that approach may be covering up.

For crop producers, especially dryland crop producers, the biggest production risk is the weather. Usually, yields for dryland wheat farmers on the eastern plains of Colorado can be correlated directly with the amount of moisture and the timing of when it is received. Farmers have no control over when it rains. So even though they know that the amount and timing of rainfall are the key variable inputs that will determine their yields, they rightfully begin their production risk management plans by focusing on the yields themselves and the impact variability in those yields could have on their operation. We'll discuss this in more detail later in the chapter, but crop insurance is the most common tool used to address this risk. With U.S. farm policy providing significant subsidies toward the premiums charged, most farmers find this to be a convenient and economically sound way to protect against the downside risk associated with yield fluctuations on a year-to-year basis. However, several bad years within any ten-year period can have a negative effect on the amount of coverage that can be purchased and limit its effectiveness.

14.4.3 Stochastic Budgeting Example—Variable Output

Stochastic budgeting plays an important role in decision making under risk. This becomes even more apparent when production risk is introduced in the form of output variance. The partial budget depicted in Figure 14.3 projects a yield value of 30 bushels per acre under the assumption that this is most likely to occur. But, suppose the past ten years' worth of production data for this operation is as depicted in Figure 14.6. There are a number of ways to incorporate this information into a yield distribution and to make the budget in Figure 14.3 stochastic instead of assuming the yield will be 30 bushels per acre per year.

Year	Yield
1	24
2	35
3	32
4	28
5	14
6	18
7	29
8	33
9	38
10	30

FIGURE 14.6 Ten years' worth of sample yield data.

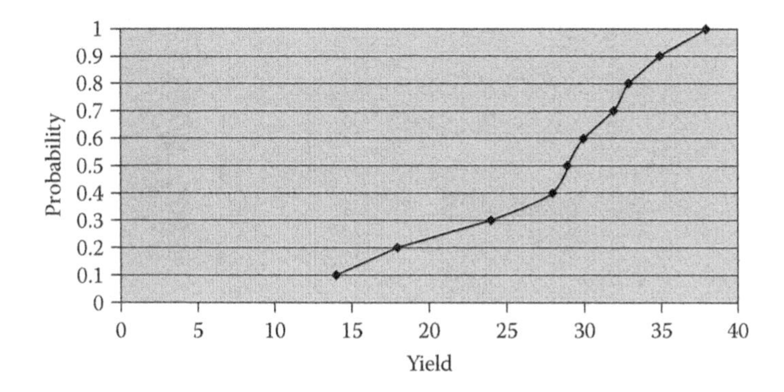

FIGURE 14.7 Discrete data cumulative distribution function for crop yield.

One way of incorporating this yield uncertainty is to take the historical yields and treat them as equally likely to happen in a discrete distribution. To get a grasp on how this would play out, we can sort the yield data in Figure 14.6 from least to greatest and graph them in a CDF as in Figure 14.7. So, for example, Figure 14.7 indicates that there is a 30% probability that yield will be 24 bushels to the acre or less and a 70% probability it will be higher. Plugging each of these yield values into the crop budget, we can create a CDF for the resulting ten total gross margin values (Figure 14.8). There is about a 25% probability that total gross margin will be negative and a 75% probability that it will be positive, according to Figure 14.8. (Going strictly from the discrete data we would make this a 20/80 split in probability.) Also, suppose our goal is to achieve a gross margin of $55 or higher in order to at least break even if property and ownership costs are figured in. Dropping a vertical line at $55 on the horizontal axis and estimating the point of intersection with the CDF gives a value of about 0.72 on the y-axis. This means the analysis shows a 72% probability of total gross margin being at or below the $55 per acre needed to cover both operating and fixed costs. This type of analysis could continue for some time looking at other key output variables for the crop budget, like return to management and risk, etc., in light of the variable yield data.

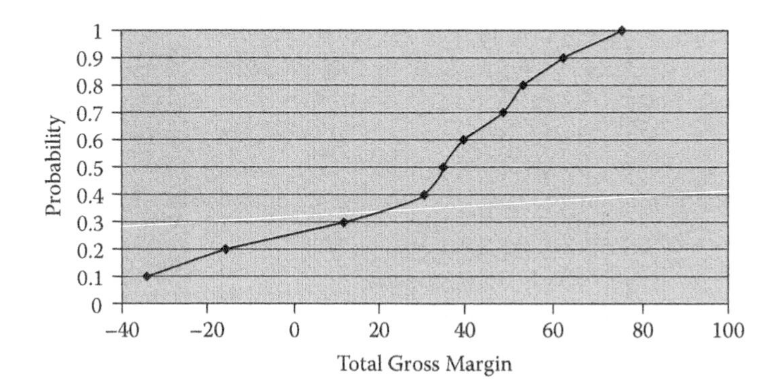

FIGURE 14.8 Discrete data cumulative distribution function for total gross margin.

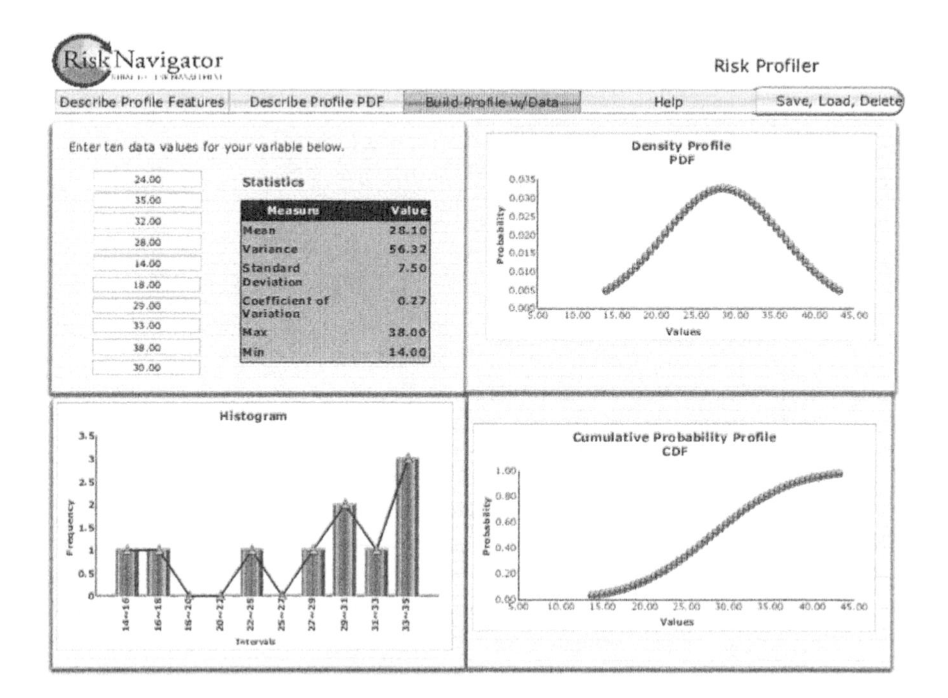

FIGURE 14.9 Sample yield distribution from Risk Profiler tool with ten years of data.

As we saw in Chapter 10, complicated methods can be used to estimate CDFs and PDFs. Plugging the ten data points of Figure 14.6 into the Risk Profiler tool yields the result shown in Figure 14.9.

We could use Latin hypercube sampling to pull data points from the CDF distribution shown in Figure 14.9 and, plugging those data points into the budget, create a corresponding total gross margin distribution. However, the user is cautioned that this process is still dependent upon the accuracy of the data used to produce it. The decision maker is the final judge and should exercise his or her authority in whether or not this result paints an accurate picture of expectations.

In the Risk Navigator Stochastic Crop Budget Generator introduced in this chapter, we took a more simplistic view in generating distributions for the stochastic elements of the budget and based it on only the identified high, low, and most likely values to occur. Using 30 for the most likely yield outcome, 14 for the low, and 38 for the high, we can produce a distribution function for yield similar to Figure 14.9. Under the assumption that all other elements of the budget are constant, the Stochastic Crop Budget Generator generates a distribution of gross margins as depicted in Figure 14.10.

14.5 MANAGEMENT STRATEGIES

Production risk management strategies involve a plan to address variable output or to mitigate possible negative effects on output quantities. For crop producers, crop yield insurance is an example of a risk management tool that can be used to decrease the

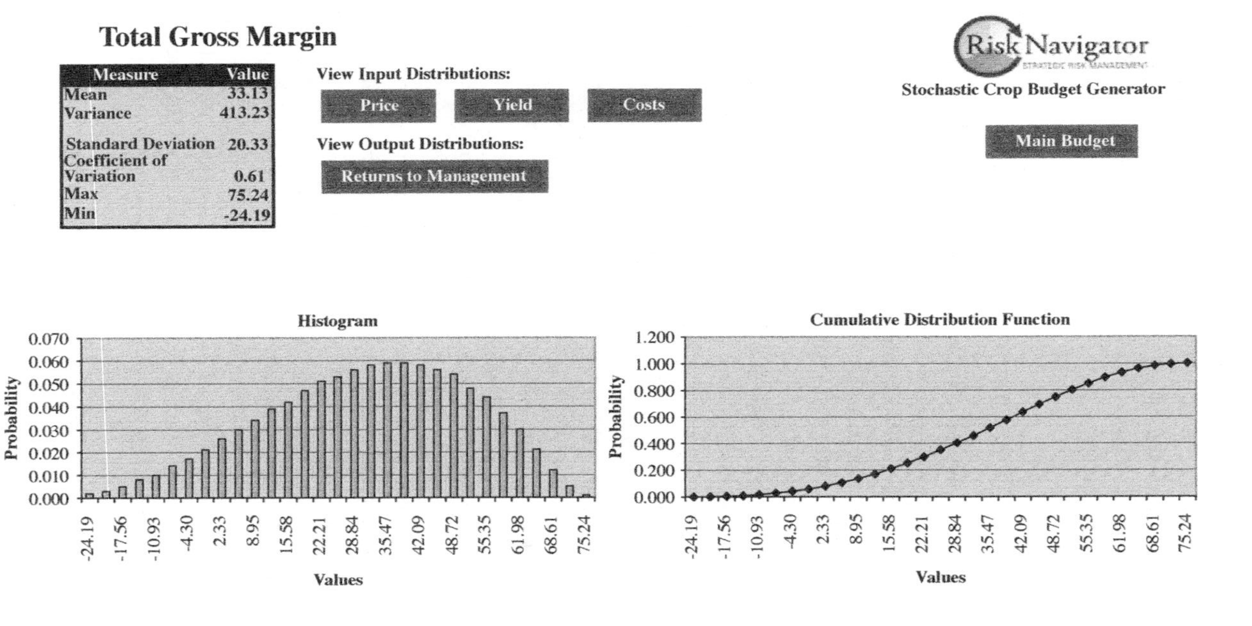

FIGURE 14.10 A gross margin distribution from the Stochastic Crop Budget Generator.

effect an extremely poor crop yield can have on the operation. The insurance concept can be implemented in a number of different ways to address production risks and, as explained below, it doesn't always involve a written policy contract with an outside party. Ultimately, the producer has many tools and strategies at his or her disposal. Many of these were discussed in Chapter 9. Some involve turning to outside parties for help, but others can be handled entirely within the operation. Consideration of enterprise selection and production practices plays a key role in producing a solid production risk management strategy.

14.5.1 Reducing Risk within the Business Operation

Reducing risk within the operation can be done in a number of different ways. One of the ways is to simply select low-risk production enterprises. Given a choice between a high-risk enterprise and a low-risk enterprise, the prudent choice may be the latter, even though it may mean accepting significantly lower profits. The trade-off is the decision maker's choice to make and should be done with an open mind, considering the overall health of the operation.

Another key way to reduce risk within the operation is to diversify. Crop yields tend to be positively correlated because of the weather; however, diversifying crops can reduce risk considerably in some situations. Crop diversification can reduce production risk from an input standpoint by spreading out the timing of demand for those inputs, especially labor and machinery. Diversifying between crops and livestock is also a tried and true method to reduce risk in an operation, though it is usually more of an approach to address market risk than to address production risk. A hail storm may pose a tremendous threat to a dryland wheat producer on the plains, but it is only a minor threat to a cattle producer.

Diversification also can disburse the agricultural operation over a wide region. If a natural land formation that has a tremendous effect on weather patterns runs through a geographical area, it may be wise to exploit that in land purchase and lease decisions. The extra travel will add to expenses, but it may be a welcome trade-off to know that one storm is very unlikely to wipe out your entire year's worth of crops.

Using low-risk production practices can directly reduce production risks for both livestock and crop producers. Vaccinating livestock is a good example of a production practice that will reduce risk at a minimal cost. In some cases, the risk of disease may be low, but if it were contracted by the herd, its effect might be devastating. Many times, having piece of mind is well worth the cost of the vaccine. For crop producers, low-risk production practices include using irrigation systems, pesticides, and reliable equipment.

Finally, maintaining flexibility is vital to reducing production risk in any operation. When faced with a decision, it is important for management to not only consider the short-term impact but also the long-term consequences associated with each alternative. Crop selection is a big decision, especially when it comes to maintaining cropping rotations or dealing with a multiple-year forage crop situation. This is also true for related input decisions such as machinery purchases. We are all guilty at times of being too short-sighted. The disciplined decision maker must force himself to think things through completely to make sure that chosen alternatives do not reduce flexibility to the point that it severely increases future risk factors.

14.5.2 Transferring Risk Outside the Operation

Insurance is the most obvious example of how to transfer risk outside the business operation. For example, crop yield insurance is simply a matter of transferring some of the risk in regard to poor production yields to an outside party, the insurance provider. In exchange for this transfer, the producer agrees to pay the insurance provider a premium. In a good production year, the producer is agreeing to accept a lower income equal to the net return without insurance minus the cost of the insurance. This lowered income potential is accepted as a trade-off for the added income in poor production years that results from the insurance indemnity payment being greater than the insurance premium. The effect of crop insurance on the producer's production risk situation is to truncate the ends of the probability density function for profit. This truncation reduces the risk by lowering the traditional statistical measures such as standard deviation and variance; however, it also lowers the mean return because of the premium.

We would be remiss if we didn't talk about two of the major reasons crop insurance is such a popular risk management tool. One reason is that the premiums charged for most U.S. crops coverable by crop insurance are subsidized by the government. Crop insurance is a complicated subject and is growing more complicated every day with the advent of more and more insurance products.

The insurance company charges the producer a premium based upon the expected losses that could be incurred. If the premium is exactly equal to the expected losses, the insurance premium is said to be actuarially fair. But because the insurance company needs to pay operating expenses and expects to be compensated for taking on the risk, it needs to charge premiums considerably higher than the actuarially fair cost to stay in business. The government closely regulates the fees and premiums that crop insurance companies can charge for insurance products; however, the government also wants to encourage producers to be proactive in addressing production risk and to not rely upon government emergency funding to bail them out in poor production years. For this reason, crop insurance premiums are subsidized at various rates, depending on the crop, to bring the premium charged to producers down to a level that encourages high participation rates.

The second reason crop insurance is a popular risk management tool with today's agricultural producers is because more and more lending institutions are requiring crop insurance to be in place as a loan condition. When land and equipment prices increase, the operating debt for most producers naturally increases, as well. Servicing this increased debt is a worry to the lenders in the context of risky agricultural production settings. Crop insurance is a tool to insure that a base level of cash flow is maintained to service debt and cover expenses.

The final tool to strategically transfer risk outside the operation is the negotiation of favorable land lease agreements. A cash lease agreement leaves all of the production risk on the producer's shoulders. To reduce this burden, the producer can negotiate a crop share agreement in lieu of cash rent. Then, the production risk is shared between the landlord and the producer. The effects of a poor crop are offset by a reduced rental cost. In some cases, the producer can negotiate a cost share agreement for some of the inputs into the rental contract, further reducing the producer's exposure to production risk while transferring that exposure to the landlord. In good years, the landlord is rewarded for taking on this risk by sharing in the bumper crop yields.

14.5.3 Building Capacity to Bear Risk

Building capacity within the operation to bear risk is a risk management strategy that most producers are familiar with. It could be described simply by the old adage of "saving for a rainy day" or, in the case of agricultural production, perhaps saving for a drought. This strategy utilizes the concept of self-insurance, as it is all handled internally.

Building capacity to bear risk involves adopting practices such as investing in extra machinery capacity and maintaining resource reserves. A relatively common practice among producers who buy a new piece of equipment is to hang on to their old equipment, even though it's been replaced. Their logic usually goes something like this, "It is worth more to me in reserve than I could get for it on trade-in." This is simply a case of saying that the risk-bearing capacity increase that the additional machinery capacity provides is worth more than the cash offered for parting with the old piece of machinery.

Technically, extra machinery capacity is a specialized case of maintaining resource reserves. It is usually thought of separately because it calls for a conscientious decision to invest in that extra capacity. Other examples of maintaining resource reserves include understocking pastures to maintain a forage reserve, holding on to excess feed inventories, and keeping a spare-parts inventory.

14.6 SUMMARY

Production risk is both an easy and a difficult topic to address. It is easy because in its simplest sense one can focus on output yields and the variances in those yields as a summary statistic of everything that is going on in the production realm. Obviously, it can become a very difficult topic when one digs into the details behind those production numbers. Taking the time to understand the input and output risk involved with your operation is an important step in finding ways to address production risk. Insurance products are a way to transfer production risk outside of the operation. Internally, building capacity to bear risk and adopting risk-reducing production practices are two high-impact strategies any producer can use. The stochastic budget is a tool that can be utilized to analyze the effects of risk on the operation and the effectiveness of different strategies to address it.

The most important thing a producer can do to address production risks is to seek additional information. This includes paying attention to weather patterns and forecasts, as well as being up to date on the latest production practices. It takes time and energy to remain current, but it can pay big dividends when it comes to sustaining a profitable operation.

14.7 THE STOCHASTIC CROP BUDGET GENERATOR

The Stochastic Crop Budget Generator is a tool designed on our Web site to help the producer learn about and assess the impact that production risk can have on his or her crop enterprise. The tool's budget is set up for a dryland wheat enterprise with stochastic elements allowed for output price, output yield, and total operating costs. The tool can be adapted for any crop that fits the template. For each of these stochastic values, the user can enter in a maximum, minimum, and most likely value in the box to the right of the budget on the main page (see Figure 14.11). The navigation button below this input box will take the user to the various distributions the tool is designed to produce. Notice that the columns of the input box for the stochastic information are color coded with the corresponding spot where the values occur in the crop budget.

For example, the user may feel that $3 is the minimum price that wheat will bring this year, with $6 as the high and $4.60 as the most likely price received. If the user inputs that information into the budget generator for the stochastic element for crop price and clicks the price button to view input distributions, the distribution in Figure 14.12 will be generated. Notice that because of the input values for the stochastic elements, this distribution is nearly symmetric. Of course that doesn't have to be the case. A peek at the crop yield distribution (minimum: 14; maximum: 38; most likely: 30) and the total operating costs (minimum: 90; maximum: 113; most likely: 98.79) reveal examples that are much less symmetric in nature.

Once the user has the stochastic elements for output price, output yield, and total operating costs set where they want them, they have two choices for output distributions. Total gross margin is the result of combining the effect of the three stochastic elements. Returns to management and risk takes into account fixed costs and factor payments to land ownership. Both of these are essentially the same distribution, differing only in the form of a scalar shift adjusting for the deterministic values associated with fixed costs and factor payments to land. Figure 14.13 shows the total gross margin distribution that is the result of the stochastic information input in Figure 14.11.

Many of the values and text in the crop budget are open for editing to model the user's own situation; however, the stochastic elements determine the output distribution for the gross margin, regardless of what the budget says goes into them. The user is cautioned not to get too carried away with the cost details that make up the budget other than to the extent that they help get a handle on the values to input for the stochastic element for total operating costs. The property and ownership costs and the factor payments to land are deterministic in the tool, and are therefore open to editing by the user to fit their situation for modeling the returns to management.

	Unit	Price or Cost/Unit	Quantity	Value or Cost per Acre
Gross Receipts from Production				
Hard Red Winter Wheat	BU	4.60	30	138.00
Crop Insurance Indemnity				
Total Receipts				**138.00**
Direct Costs				
Operating Preharvest				
Seed	Dols	6.39	1.00	6.39
Fertilizer	Dols	31.76	1.00	31.76
Herbicide	Dols	6.37	1.00	6.37
Custom Application	Dols	4.00	1.00	4.00
Crop Insurance	Dols			7.85
Fuel	Dols			10.84
Repair & Maintenance	Dols			8.24
Labor	Dols			2.23
Interest Expense	Dols	7%	38.84	2.72
Total Preharvest				**80.40**
Operating Harvest				
Fuel	Dols			4.83
Repair & Maintenance	Dols			7.88
Labor	Dols			1.48
Hauling	BU	0.14	30	4.20
Total Harvest				**18.39**
Total Operating Costs				**98.79**
Total Gross Margin				**39.21**
Property and Ownership Costs				
Machinery Ownership Costs	Dols			42.78
General Farm Overhead	Dols			10.00
Real Estate Taxes	Dols			1.80
Total Property and Ownership Costs				**54.58**
Net Receipts Before Factor Payments				**−15.37**
Factor Payments				
Land @ 4.00%	Dols			32.00
Return to Management and Risk				**−47.37**

Risk Navigator
STRATEGIC RISK MANAGEMENT

Stochastic Crop Budget Generator

Jay Parsons
jay.parsons@OptimalAg.com

Stochastic Elements			
	Crop Price per Unit	Crop Yield per Acre	Total Operating Costs/Acre
Minimum	3.00	14	90.00
Most Likely	4.60	30	98.79
Maximum	6.00	38	113.00

View Input Distributions

Price | Yield | Costs

View Output Distributions

Gross Margin | Returns to Management

FIGURE 14.11 Stochastic Crop Budget Generator main page.

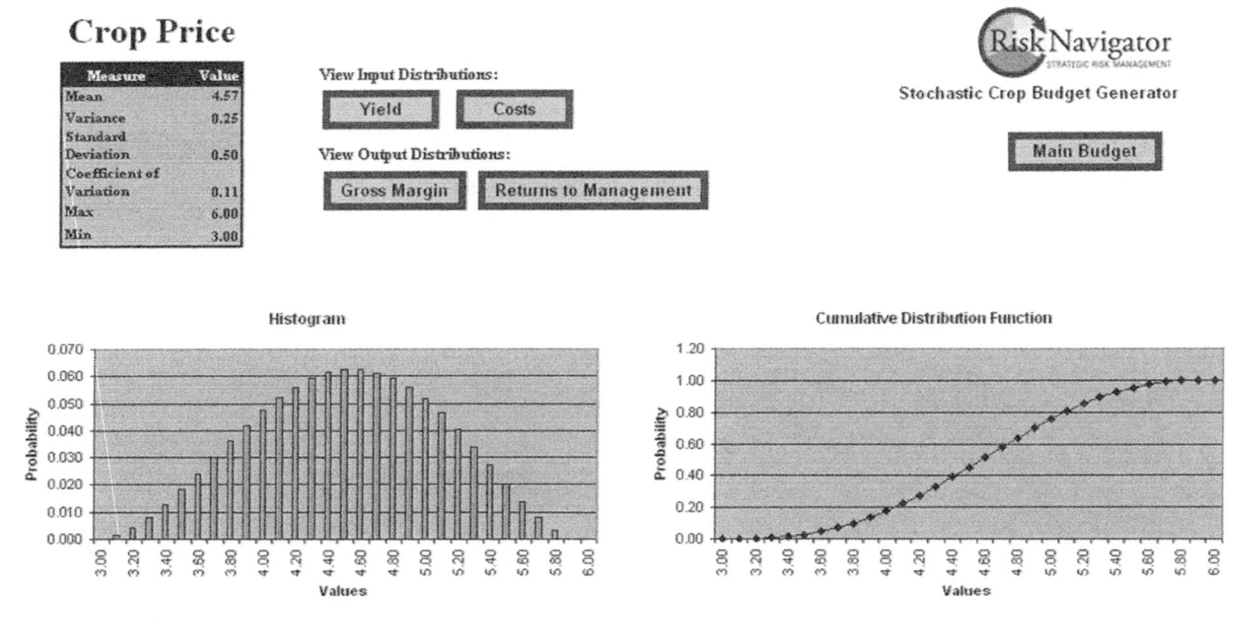

FIGURE 14.12 Stochastic Crop Budget Generator sample crop price distribution.

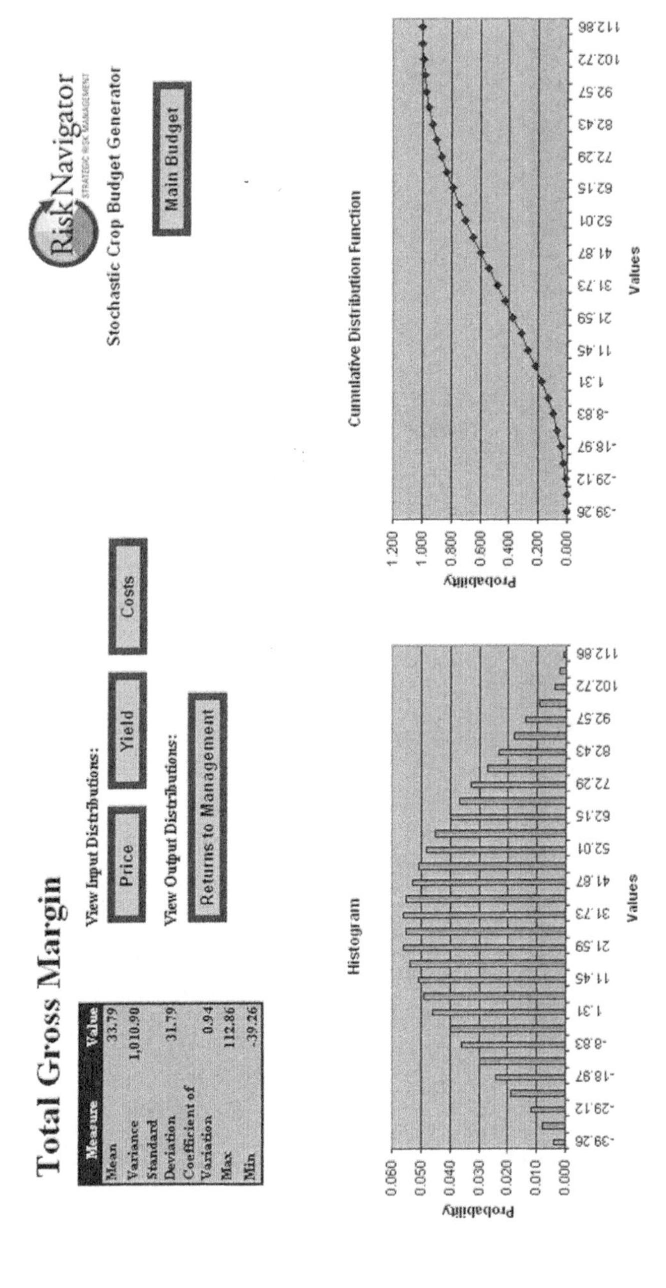

FIGURE 14.13 Stochastic Crop Budget Generator sample total gross margin distribution.

REFERENCES

Chambers, R. G., and J. Quiggin. 2000. *Uncertainty, production, choice, and agency: The state-contingent approach.* Cambridge: Cambridge University Press.

Dinar, A., and J. Letey. 1991. Agricultural water marketing, allocative efficiency, and drainage reduction. *Journal of Environmental Economics and Management* 20: 210–223.

Dinar, A., J. Letey, and K. C. Knapp. 1985. Economic evaluation of salinity, drainage and non-uniformity of infiltrated irrigation water. *Agricultural Water Management* 10: 221–233.

Feinerman, E., Y. Shani, and E. Bresler. 1989. Economic optimisation of sprinkler irrigation considering uncertainty of spatial water distribution. *Australian Journal of Agricultural Economics* 33(2): 88–107.

Hanna, H. M., M. L. White, T. S. Colvin, and J. L. Baker. 2002. Anhydrous ammonia distribution during field application. *Applied Engineering in Agriculture* 18(4): 443–451.

Knapp, K. C. 1992. Irrigation management and investment under saline, limited drainage conditions 1. Model formulation. *Water Resources Research* 28(12): 3085–3090.

Robison, L. J., and P. J. Barry. 1987. *The competitive firm's response to risk.* New York: MacMillan.

15 Understanding Your Financial Statements and Ratios for Risk Management

Duane Griffith

CONTENTS

We discussed financial analysis and introduced a condensed set of financial statements to help learn the concepts of financial analysis in Chapter 5. Other chapters covered use of risk management tools to reduce or eliminate the five basic types of risk. Each of these tools requires a clear understanding of how the tool helps control a particular risk and how each tool is implemented, but the results of implementing a particular risk management strategy will usually have some type of financial implication. We will continue to explore financial analysis concepts and the impact of implementing some of the risk management strategies discussed in this chapter.

Financial analysis for an operation measures the cumulative effect of managing, or not managing, various individual types of risk (Boehlje, Dobbins, and Miller, 2000). The outcome of selecting and using an individual risk management strategy, such as price risk management with options, is eventually incorporated into financial statements. Financial statements can rarely identify the success or failure of one particular risk management effort.

15.1 EWS FARMS FINANCIAL POSITION AND PERFORMANCE

The four basic financial statements for EWS Farms introduced in Chapter 5 are shown again in Figure 15.1. Details about EWS Farms were provided in Chapter 3, including characteristics about the operation's size, enterprise mix, the production capabilities of the resource base, climate, soil type, and number of potential families that may want to be involved. Examples in Chapter 5 focused on implementing family goals and the impacts this would have on family withdrawals to illustrate potential types of human resource risk. These transactions affected the cash flow statement and balance sheets but did not affect the income statement. Examples presented in this chapter will continue to analyze financial implications of implementing some of EWS Farms' goals and selected management strategies covered in Chapters 6 through 12. This chapter covers the tactical planning stage of the Strategic Risk Management (SRM) process with respect to financial planning.

15.2 FINANCIAL RISKS

Previous examples using the condensed form of the financial statements helped make it clear that the typical practice of preparing only a balance sheet and a projected cash flow for agricultural operations is not adequate. Long-run survival of agriculture operations depend on a risk management plan that incorporates all sources of risks with adequate financial analysis capabilities to evaluate alternative risk management strategies and implement tactical financial plans. Financial analysis is the beginning of the ten-step SRM process. *What's your plan?*

The discussion in Chapter 3 of goals and objectives were detailed in Chapter 7. One of these goals is using conservation tillage methods to assure continued health of the land resource base for future generations. Chapter 7 discusses the development of a plan to purchase machinery and equipment for the operation that will allow EWS Farms to meet these goals. Making detailed financial plans years in advance is very difficult and may be rather impractical. However, the financial implications for

Percent Crop Revenue	100%	Percent Livestock Revenue	100%
Percent Cost of Production - Crops	100%	Percent cost of Production - Livestock	100%

Percent Gov. Payments 100%

Balance Sheet

	Assets Beginning	Assets Ending		Liabilities Beginning	Liabilities Ending
Cash on Hand	1,500	7,591	Accounts Payable (Exp)	2,000	2,000
Crops Held for Feed (Exp)	8,400	8,400	Accrued Interest (Exp)	12,540	11,889
Crops Held for Sale (Inc)	200,000	200,000	Current Principal	11,908	12,559
Market Livestock (Inc)	0	0	Other Current Liability (Exp)	10,000	10,000
Other Current Assets (Inc)	15,000	15,000	Short Term Notes Payable (Exp)	0	0
Cash Invt Growing Crops (Exp)	0	0	Other Current Liab. (Not Adj.)	0	0
Supplies&Prepaid Exp. (Exp)	10,000	10,000	Def. Tax on Current Assets	0	0
Total Current Assets	234,900	240,991	Operating Loan Carryover	0	0
Non-Current Assets			**Total Current Liab.**	36,448	36,448
Mach. & Equipment	325,000	292,500	**Non-Current Liabilities**		
Breeding Livestock	0	0	Prin. on T.D. & C.L.	255,399	242,840
Real Estate (Land, Bldgs, Impr)	1,040,000	1,038,000	Total Business Liab.	291,847	279,288
Total Business Assets	1,599,900	1,571,491	**Business Net Worth**	1,308,053	1,292,204
			Change in Equity From Beginning to End of Year		(15,849)

Cash Flow Statement

		Inflows			OutFlows
Crop Sales & Net Insurance Payments		319,920	Operating Expenses	No Interest >	278,020
Mrkt & Cull Livestock Sales		0	Other Cash Business Expense		0
Government Payments		38,505	Cash Int. Exp. - T.D. & C.L.*		12,540
Other Cash Business Income		0	Cash Int. Exp. - Operating	7%	4,865
Operating Loan Proceeds	50.0%	139,010	Loan Prin. Payments - T.D. & C.L.		11,908
Loan Proceeds Capital Assets		0	Breeding Livestock Asset Purchases		0
Non-Business Inflows/Revenue			Mach & Equip & Real Estate Purchase		0
Other Nonfarm Inflows, net of taxes		0	Owner withdrawals		45,000
Other Nonfarm Inflows, net of taxes		0	Cash Taxes Paid (Income & SS)		0
Total Cash Inflows		$497,434	Other Cash Outflows (Not Expenses)		0
				Subtotal	$352,333
			Operating Loan Prin. Payments		$139,010
			Total Cash Outflows		$491,343
			Annual Net Cash Flow (never < zero)		7,591

**T.D. = Term Debt, C.L. = Capital Lease*

Income Statement - Accrual Adj.

	Income
Cash Income (adj. for cull lvstk sales)	$358,424
Non-Cash Income Adjustments	0
Non-Cash Income (Raised Brdg Lvstk)	0
Capital Gain/Loss on Breeding Lvstk (Net)	0
Gross Revenue	$358,424
	Expense
Cash Expense (Excluding Interest)	278,020
Non-Cash Feed Inventory Adjustment	0
Other Non-Cash Non-Interest Expense	0
Depreciation (Land, Bldgs, Equip.)	34,500
Total Operating Expense	312,520
Cash Int. Exp. - T.D. & C.L.	12,540
Cash Int. Exp. - Operating	4,865
Non-Cash Interest Expense	(651)
Total Expense	$329,274
Net Business Income From Operations	29,151
Net Business Income	29,151
Income & SS Taxes (Cash & Non-Cash)	0
Net Income	$29,151

Statement of Owner Equity

Beginning Net Worth (Cost/Mrkt)		1,308,053
Net Income	+	29,151
Non-Business Cash Inflows	+	0
Owner Withdrawals (Cash)	-	45,000
Asset Valuation Change or Cont./Distrib.	+/-	$0
Calculated Ending Net Worth	=	1,292,204
Reported Ending Net Worth (Cost/Mrkt)		1,292,204
Discrepancy		0

FIGURE 15.1 Condensed version of the four basic financial statements in RDFinancial.

an operation of implementing future plans, even years in advance, should always be explored. Early indications of potential adverse effects of implementing a particular plan may save a lot of precious time and effort.

Let's further explore the flow of information between the financial statements with transactions that affect the income statement by evaluating the purchase of machinery and equipment necessary to help improve EWS Farms' conservation plans or replace existing machinery used to carry out those plans. The series of transactions we are about to use relate to the purchase of a piece of equipment at the end of the year. Often, purchasing equipment at year end is a result of managing potential income tax liabilities. This series of transactions will also illustrate potential negative financial impacts on an operation due to this type of management decision. As a reference point, we will start with the financial status shown in Figure 15.1. This series of transactions will initially be shown with the tax estimator turned off so it is easy to follow the numbers that change on all the financial statements.

For example purposes we will use an equipment purchase of $100,000 of farm machinery. It does not matter what it is, just that it is a business asset. The equipment will be 90% financed for a seven-year period with an interest rate of 8%. To help illustrate the interaction of information on the financial statements, this transaction is separated into three steps.

The first step shows the purchase with cash outflows of $100,000 and an increased ending balance for Non-Current assets for Mach. & Equipment on the ending balance sheet of $100,000.

Figure 15.2 shows that the $100,000 purchase price has been added to the outflow side of the cash flow statement (Mach & Equip & Real Estate Purchase) and to the ending balance sheet for Mach. & Equipment under Non-Current assets. Use Figure 15.1 to verify these changes. The Cash Flow Statement now shows a zero Annual Net Cash Flow and the balance sheet shows a zero ending Cash on Hand. The Balance Sheet also shows a $92,409 Operating Loan Carryover on the liabilities side. This is the difference in positive cash flow shown in Figure 15.1 and the $100,000 purchase price. Net Worth did not change and the statement of owner equity (SOE) and Income Statement show no change. The first portion of this transaction is simply trading two types of assets, cash for a capital asset. No income or expense is involved so the Income Statement and the SOE are not affected.

The second portion of the transaction adds the 90% financing arrangement. RDFinancial calculates the principal and interest payment for the amount financed and adds that information to the principal and interest due on the new loan in the Current and Non-Current Liabilities section on the ending Balance Sheet. One aspect of this transaction is different than most cash basis tax payers would encounter. The Income Statement in RDFinancial is accrual adjusted so the interest portion of the payment for the new machinery is included on the Income Statement as an expense for the period the Income Statement covers. A cash basis tax payer would not count this expense until it was actually paid, sometime in the following year.

Figure 15.3 shows the financial implications of adding the financed portion of the purchase price. The inflow of cash from receiving loan proceeds is listed on the inflows side of the Cash Flow Statement. This inflow, not income, reduces the operating loan carryover to $2,409. Ending Cash on hand is still zero, which indicates

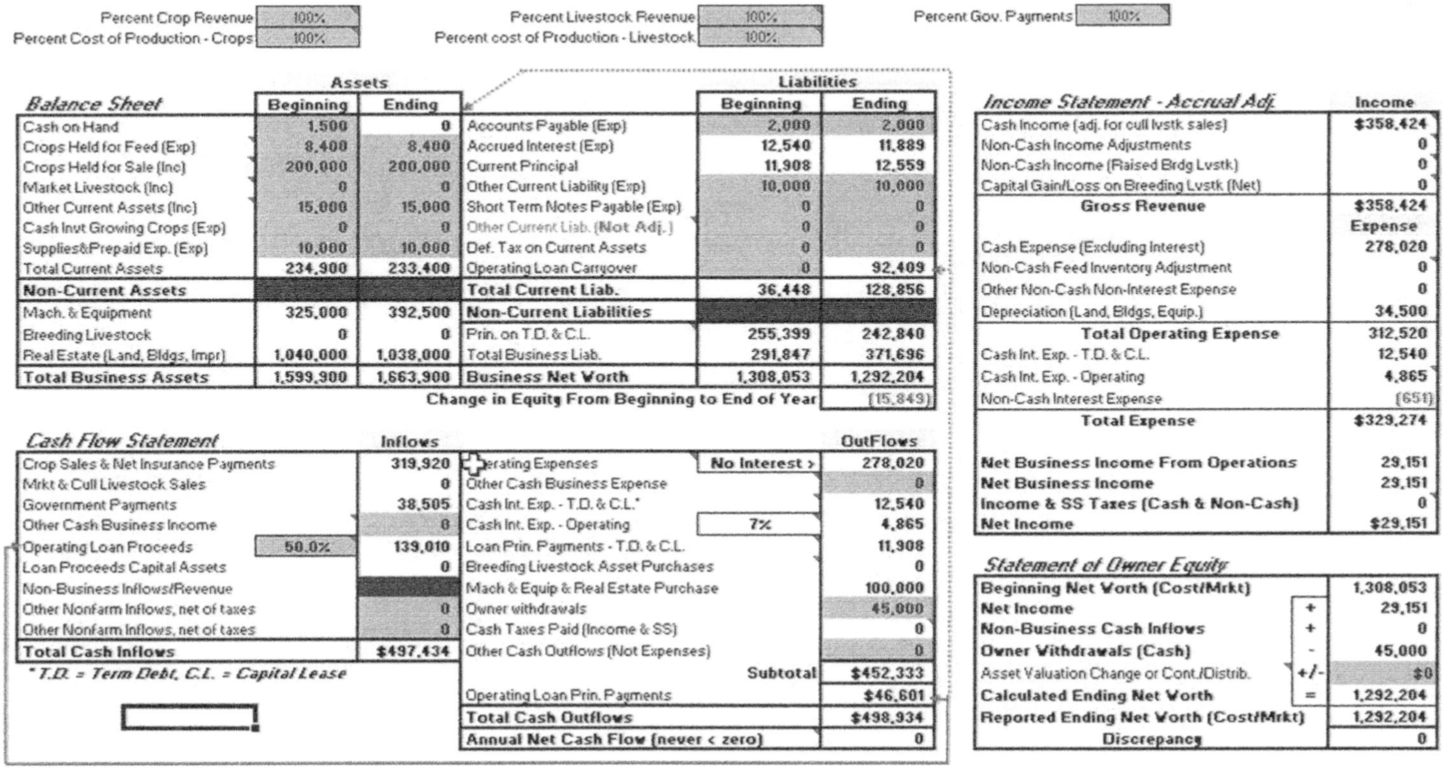

Percent Crop Revenue	100%
Percent Cost of Production - Crops	100%
Percent Livestock Revenue	100%
Percent cost of Production - Livestock	100%
Percent Gov. Payments	100%

Balance Sheet

Assets	Beginning	Ending
Cash on Hand	1,500	0
Crops Held for Feed (Exp)	8,400	8,400
Crops Held for Sale (Inc)	200,000	200,000
Market Livestock (Inc)	0	0
Other Current Assets (Inc)	15,000	15,000
Cash Invt Growing Crops (Exp)	0	0
Supplies&Prepaid Exp. (Exp)	10,000	10,000
Total Current Assets	234,900	233,400
Non-Current Assets		
Mach. & Equipment	325,000	392,500
Breeding Livestock	0	0
Real Estate (Land, Bldgs, Impr)	1,040,000	1,038,000
Total Business Assets	1,599,900	1,663,900

Liabilities	Beginning	Ending
Accounts Payable (Exp)	2,000	2,000
Accrued Interest (Exp)	12,540	11,889
Current Principal	11,908	12,559
Other Current Liability (Exp)	10,000	10,000
Short Term Notes Payable (Exp)	0	0
Other Current Liab. (Not Adj.)	0	0
Def. Tax on Current Assets	0	0
Operating Loan Carryover	0	92,409
Total Current Liab.	36,448	128,856
Non-Current Liabilities		
Prin. on T.D. & C.L.	255,399	242,840
Total Business Liab.	291,847	371,696
Business Net Worth	1,308,053	1,292,204
Change in Equity From Beginning to End of Year		(15,849)

Cash Flow Statement

		Inflows
Crop Sales & Net Insurance Payments		319,920
Mrkt & Cull Livestock Sales		0
Government Payments		38,505
Other Cash Business Income		0
Operating Loan Proceeds	50.0%	139,010
Loan Proceeds Capital Assets		0
Non-Business Inflows/Revenue		
Other Nonfarm Inflows, net of taxes		0
Other Nonfarm Inflows, net of taxes		0
Total Cash Inflows		$497,434

T.D. = Term Debt, C.L. = Capital Lease

		OutFlows
Operating Expenses	No Interest >	278,020
Other Cash Business Expense		0
Cash Int. Exp. - T.D. & C.L.*		12,540
Cash Int. Exp. - Operating	7%	4,865
Loan Prin. Payments - T.D. & C.L.		11,908
Breeding Livestock Asset Purchases		0
Mach & Equip & Real Estate Purchase		100,000
Owner withdrawals		45,000
Cash Taxes Paid (Income & SS)		0
Other Cash Outflows (Not Expenses)		0
	Subtotal	$452,333
Operating Loan Prin. Payments		$46,601
Total Cash Outflows		$498,934
Annual Net Cash Flow (never < zero)		0

Income Statement - Accrual Adj.	Income
Cash Income (adj. for cull lvstk sales)	$358,424
Non-Cash Income Adjustments	0
Non-Cash Income (Raised Brdg Lvstk)	0
Capital Gain/Loss on Breeding Lvstk (Net)	0
Gross Revenue	$358,424
	Expense
Cash Expense (Excluding Interest)	278,020
Non-Cash Feed Inventory Adjustment	0
Other Non-Cash Non-Interest Expense	0
Depreciation (Land, Bldgs, Equip.)	34,500
Total Operating Expense	312,520
Cash Int. Exp. - T.D. & C.L.	12,540
Cash Int. Exp. - Operating	4,865
Non-Cash Interest Expense	(651)
Total Expense	$329,274
Net Business Income From Operations	29,151
Net Business Income	29,151
Income & SS Taxes (Cash & Non-Cash)	0
Net Income	$29,151

Statement of Owner Equity

Beginning Net Worth (Cost/Mrkt)		1,308,053
Net Income	+	29,151
Non-Business Cash Inflows	+	0
Owner Withdrawals (Cash)	-	45,000
Asset Valuation Change or Cont./Distrib.	+/-	$0
Calculated Ending Net Worth	=	1,292,204
Reported Ending Net Worth (Cost/Mrkt)		1,292,204
Discrepancy		0

FIGURE 15.2 Financial statements showing the first step in the purchase of $100,000 of machinery.

Percent Crop Revenue	100%	Percent Livestock Revenue	100%	Percent Gov. Payments 100%
Percent Cost of Production - Crops	100%	Percent cost of Production - Livestock	100%	

Balance Sheet

	Assets Beginning	Assets Ending		Liabilities Beginning	Liabilities Ending
Cash on Hand	1,500	0	Accounts Payable (Exp)	2,000	2,000
Crops Held for Feed (Exp)	8,400	8,400	Accrued Interest (Exp)	12,540	18,189
Crops Held for Sale (Inc)	200,000	200,000	Current Principal	11,908	22,959
Market Livestock (Inc)	0	0	Other Current Liability (Exp)	10,000	10,000
Other Current Assets (Inc)	15,000	15,000	Short Term Notes Payable (Exp)	0	0
Cash Invt Growing Crops (Exp)	0	0	Other Current Liab. (Not Adj.)	0	0
Supplies&Prepaid Exp. (Exp)	10,000	10,000	Def. Tax on Current Assets	0	0
Total Current Assets	234,900	233,400	Operating Loan Carryover	0	2,409
Non-Current Assets			**Total Current Liab.**	36,448	55,556
Mach. & Equipment	325,000	392,500	**Non-Current Liabilities**		
Breeding Livestock	0	0	Prin. on T.D. & C.L.	255,399	322,440
Real Estate (Land, Bldgs, Impr)	1,040,000	1,038,000	Total Business Liab.	291,847	377,996
Total Business Assets	1,599,900	1,663,900	**Business Net Worth**	1,308,053	1,285,904
			Change in Equity From Beginning to End of Year		(22,149)

Income Statement - Accrual Adj.

	Income
Cash Income (adj. for cull lvstk sales)	$358,424
Non-Cash Income Adjustments	0
Non-Cash Income (Raised Brdg Lvstk)	0
Capital Gain/Loss on Breeding Lvstk (Net)	0
Gross Revenue	$358,424
	Expense
Cash Expense (Excluding Interest)	278,020
Non-Cash Feed Inventory Adjustment	0
Other Non-Cash Non-Interest Expense	0
Depreciation (Land, Bldgs, Equip.)	34,500
Total Operating Expense	312,520
Cash Int. Exp. - T.D. & C.L.	12,540
Cash Int. Exp. - Operating	4,865
Non-Cash Interest Expense	5,649
Total Expense	$335,574
Net Business Income From Operations	22,851
Net Business Income	22,851
Income & SS Taxes (Cash & Non-Cash)	0
Net Income	$22,851

Cash Flow Statement

		Inflows			OutFlows
Crop Sales & Net Insurance Payments		319,920	Operating Expenses	No Interest >	278,020
Mrkt & Cull Livestock Sales		0	Other Cash Business Expense		0
Government Payments		38,505	Cash Int. Exp. - T.D. & C.L.*		12,540
Other Cash Business Income		0	Cash Int. Exp. - Operating	7%	4,865
Operating Loan Proceeds	50.0%	139,010	Loan Prin. Payments - T.D. & C.L.		11,908
Loan Proceeds Capital Assets		90,000	Breeding Livestock Asset Purchases		0
Non-Business Inflows/Revenue			Mach & Equip & Real Estate Purchase		100,000
Other Nonfarm Inflows, net of taxes		0	Owner withdrawals		45,000
Other Nonfarm Inflows, net of taxes		0	Cash Taxes Paid (Income & SS)		0
Total Cash Inflows		$587,434	Other Cash Outflows (Not Expenses)		0
				Subtotal	$452,333
			Operating Loan Prin. Payments		$136,601
			Total Cash Outflows		$588,934
			Annual Net Cash Flow (never < zero)		0

T.D. = Term Debt, C.L. = Capital Lease

Statement of Owner Equity

Beginning Net Worth (Cost/Mrkt)		1,308,053
Net Income	+	22,851
Non-Business Cash Inflows	+	0
Owner Withdrawals (Cash)	-	45,000
Asset Valuation Change or Cont./Distrib.	+/-	$0
Calculated Ending Net Worth	=	1,285,904
Reported Ending Net Worth (Cost/Mrkt)		1,285,904
Discrepancy		0

FIGURE 15.3 Financial statements showing the impact of the 90% financing arrangement of the $100,000 machinery purchase.

the lender financed more than the $90,000 loan amount. Net Business Income has declined by $6,300 from $29,151 to $22,851. This decline is due to the interest expense calculated for the loan, $6,300, showing up as an accrual adjustment to the Income Statement under Non-Cash Interest Expense, Figure 15.3. Figure 15.1 shows the value of this line on the Income Statement as a negative $651. The value shown in Figure 15.3 for Non-Cash Interest Expense is $5,649, (–$651 + $6,300). From previous examples we know that when Net Income declines, any positive change in net worth also declines or turns negative, if all other things remain the same. Net Worth also declines by $6,300, as shown on the ending Balance Sheet. The Change in Equity from Beginning to End of Year calculation now shows a negative $22,149 = (–$15,849 – $6,300).

The third step to complete this transaction is including an allowable depreciation expense for tax purposes. In this example, the depreciation expense is $10,000 and is included on the income statement and also affects the ending balance for Mach. & Equipment on the Balance Sheet. Since depreciation is not a cash expense, the Cash Flow Statement is not affected and the ending Cash on Hand does not change. Depreciation listed on the income statement increased by $10,000 from $34,500, Figure 15.3, to $44,500 in Figure 15.4. Net Income declines from the $22,851 in Figure 15.3, to $12,851 in Figure 15.4. Total decline in Net Income is now $16,300, $10,000 of depreciation plus $6,300 of interest.

The equipment purchase example is used to help illustrate a critical concept when measuring business financial position and performance. The timing of this purchase was such that the equipment was not used during the year and therefore no additional expense, such as repairs, were recorded and the equipment did not have an opportunity to help generate added revenue or reduce expenses. Our example simply illustrates the implication of a stated goal for EWS Farms. Depending on the actual timing of this purchase in the future, the purchase may also serve a dual purpose, to reduce taxes. The financial management concept we wish to emphasize is: *you cannot buy equity with the purchase of a capital asset.* Net worth declined by $16,300 after the equipment purchase ($10,000 of depreciation plus $6,300 of interest). The financial statements are telling us that growth in equity must be earned. The purchase of this piece of machinery decreased Net Income for the current year and will continue to have a negative impact on Net Income until the depreciation and interest expense for this piece of equipment disappear. The purchase of this equipment fundamentally changed the financial health of this operation.

Isolating this transaction at year end helps evaluate the impact of a capital asset purchase and also shows the financial impact the capital asset must provide the operation in order to financially justify the purchase. Earned growth in equity is dependent on Net Income. When this machine is put to work the following year, it must generate an increase in net income of $16,300 just to maintain the net worth prior to its purchase. From a strict financial health perspective, EWS Farms must be able to argue that expenses (repair, cash labor, fuel. etc.) will be reduced and/or that income to the operation will be increased by at least $16,300. Income may increase if the new machine allows better timing for planting or harvesting thereby increasing yields for one or more enterprises. If this argument cannot be made, the financial health of the operation has declined. Of course, this does not mean you should never

Percent Crop Revenue	100%	Percent Livestock Revenue	100%	Percent Gov. Payments	100%
Percent Cost of Production - Crops	100%	Percent cost of Production - Livestock	100%		

Balance Sheet

	Assets Beginning	Assets Ending		Liabilities Beginning	Liabilities Ending
Cash on Hand	1,500	0	Accounts Payable (Exp)	2,000	2,000
Crops Held for Feed (Exp)	8,400	8,400	Accrued Interest (Exp)	12,540	18,189
Crops Held for Sale (Inc)	200,000	200,000	Current Principal	11,908	22,959
Market Livestock (Inc)	0	0	Other Current Liability (Exp)	10,000	10,000
Other Current Assets (Inc)	15,000	15,000	Short Term Notes Payable (Exp)	0	0
Cash Invt Growing Crops (Exp)	0	0	Other Current Liab. (Not Adj.)	0	0
Supplies&Prepaid Exp. (Exp)	10,000	10,000	Def. Tax on Current Assets	0	0
Total Current Assets	234,900	233,400	Operating Loan Carryover	0	2,409
Non-Current Assets			**Total Current Liab.**	36,448	55,556
Mach. & Equipment	325,000	382,500	**Non-Current Liabilities**		
Breeding Livestock	0	0	Prin. on T.D. & C.L.	255,399	322,440
Real Estate (Land, Bldgs, Impr)	1,040,000	1,038,000	Total Business Liab.	291,847	377,996
Total Business Assets	1,599,900	1,653,900	**Business Net Worth**	1,308,053	1,275,904
			Change in Equity From Beginning to End of Year		(32,149)

Cash Flow Statement

		Inflows			OutFlows
Crop Sales & Net Insurance Payments		319,920	Operating Expenses	No Interest >	278,020
Mrkt & Cull Livestock Sales		0	Other Cash Business Expense		0
Government Payments		38,505	Cash Int. Exp. - T.D. & C.L.*		12,540
Other Cash Business Income		0	Cash Int. Exp. - Operating	7%	4,865
Operating Loan Proceeds	50.0%	139,010	Loan Prin. Payments - T.D. & C.L.		11,908
Loan Proceeds Capital Assets		90,000	Breeding Livestock Asset Purchases		0
Non-Business Inflows/Revenue			Mach & Equip & Real Estate Purchase		100,000
Other Nonfarm Inflows, net of taxes		0	Owner withdrawals		45,000
Other Nonfarm Inflows, net of taxes		0	Cash Taxes Paid (Income & SS)		0
Total Cash Inflows		$587,434	Other Cash Outflows (Not Expenses)		0
				Subtotal	$452,333
			Operating Loan Prin. Payments		$136,601
			Total Cash Outflows		$588,934
			Annual Net Cash Flow (never < zero)		0

**T.D. = Term Debt, C.L. = Capital Lease*

Income Statement - Accrual Adj.

	Income
Cash Income (adj. for cull lvstk sales)	$358,424
Non-Cash Income Adjustments	0
Non-Cash Income (Raised Brdg Lvstk)	0
Capital Gain/Loss on Breeding Lvstk (Net)	0
Gross Revenue	$358,424
	Expense
Cash Expense (Excluding Interest)	278,020
Non-Cash Feed Inventory Adjustment	0
Other Non-Cash Non-Interest Expense	0
Depreciation (Land, Bldgs, Equip.)	44,500
Total Operating Expense	322,520
Cash Int. Exp. - T.D. & C.L.	12,540
Cash Int. Exp. - Operating	4,865
Non-Cash Interest Expense	5,649
Total Expense	$345,574
Net Business Income From Operations	12,851
Net Business Income	12,851
Income & SS Taxes (Cash & Non-Cash)	0
Net Income	$12,851

Statement of Owner Equity

Beginning Net Worth (Cost/Mrkt)		1,308,053
Net Income	+	12,851
Non-Business Cash Inflows	+	0
Owner Withdrawals (Cash)	-	45,000
Asset Valuation Change or Cont./Distrib.	+/-	$0
Calculated Ending Net Worth	=	1,275,904
Reported Ending Net Worth (Cost/Mrkt)		1,275,904
Discrepancy		0

FIGURE 15.4 Financial statements showing the impact of claiming depreciation in the year of purchase.

purchase machinery. It may be absolutely necessary to replace aging equipment due to its unreliability and cost of operation. The decision to purchase capital assets should be based on sound financial analysis, not just tax management. If a producer is not preparing all of the financial statements suggested, the impact of purchasing any type of capital assets can easily be missed. The assumption that the equipment was purchased at the end of the year simplified the example, but the basic concept and discussion holds for a purchase made earlier in the year. Analysis of the financial impact would simply be more dynamic. Detailed capital budgeting analysis procedures, typically Excel spreadsheets, are available through land grant universities. These types of procedures include complete life cycle analysis of potential income and expenses of the capital item under consideration.

What is the result of our financial analysis if we ignore the income statement and use only a cash flow and balance sheet to analyze the $100,000 machinery purchase? Making the assumption that in aggregate the example reflects the real world, the cash flow has change from a positive $7,591 to zero with an operating loan carryover on the ending balance sheet of $2,409. The ending cash balance on the balance sheet is also zero. The ending balance sheet shows Net Worth has declined to $1,275,904 from $1,292,204, as seen in Figure 15.1. The decline is equal to $16,300. There are a tremendous number of timing issues here that we are purposefully ignoring. These include whether the new equipment listed on the ending balance sheet at $100,000 or $90,000. It is likely the machinery was listed at full market value as the accountant would not have time to adjust the inventory value of equipment for depreciation purposes. A cash basis taxpayer would not have incurred any interest expense in our example and it is not shown in the cash flow statement. If the only cash flow prepared is a projection for next year rather than a historic cash flow, the impact on cash flow may also be missed. The management of this operation may never know the implications to the financial health of the operation as a result of purchasing this capital asset or have a measure of the size of the impact, $16,300, if an Accrual Adjusted Income Statement is not prepared and used to help reconcile the complete set of financial statements.

Let's follow this theme of limited financial information via a cash flow and balance sheet versus a more complete set of financials. It may be possible that EWS Farms could see a little bit of a cash flow crunch coming after the equipment purchase and decided they needed to sell some crop from inventory to help meet their cash flow needs. Let's assume EWS Farms sells $30,000 of the $200,000 crop inventory shown on the Balance Sheets under Crops Held for Sale (Inc) in the Current Assets section, as shown in Figure 15.4. Figure 15.5 shows the financial results of this transaction in addition to all of the transactions reflected in Figure 15.4. Two entries were made on the financial statements. The first is to reduce the dollar value of Crops Held for Sale on the ending Balance Sheet to $170,000. The second is to show the additional cash inflow on the cash flow statement, $30,000 under Other Cash Business Income, from the sale of the stored grain. The Cash Flow Statement now shows a positive $27,591 and this amount is also listed as the Cash on Hand on the ending Balance Sheet. However, Net Worth, Net Income, the calculated change in Net Worth, and the SOE have not changed from those shown in Figure 15.4.

The increase in cash income generated by the sale of stored grain is shown on the Income Statement, but it is offset by a negative accrual adjustment, also shown on the

Income Statement under Non-Cash Income Adjustments. The Matching Principle (Financial Accounting Standards Board [FASB]; Farm Financial Standards Council [FFSC], 1997) also provides the logic behind this procedure. If the grain was in inventory at the beginning of the period (listed on the beginning Balance Sheet) then it must have been produced in a prior period. The sale of the grain in this period should not be attributed to business performance in the current analysis period.

The Accrual Adjusted Income Statement shows a more conservative, accurate evaluation of business performance than a Cash Flow or cash basis Income Statement in this instance. A cash basis Income Statement would also show additional cash sales of $30,000 without any matching cash expenses. The larger net cash income would indicate that the business, on a cash basis, preformed well. Cash-only analysis adds to the pool of misinformation used to make future management decisions and can lead to serious long-term financial consequences.

15.3 FINANCIAL RATIOS AND MEASURES: THE SWEET SIXTEEN

The above examples were intended to illustrate the potential benefits of financial tactical planning, the interaction of the four basic financial statements, and show why all four statements are required. Information directly from the financial statements is very useful, but it would also be useful to be able to compare the performance of an individual operation with benchmarks for similar operations. Virtually all financial institutions and farm business management associations use ratios and other financial measures to evaluate individual operations as well as comparative analysis among operations (Langemeier, 2004; Ellinger, 1997a,b; Miller, Boehlje, Dobbins, 2001).

Figure 15.6 is an illustration of the financial ratios using Figure 15.1 as a starting point for comparison purposes. The cumulative effects of the transactions shown in Figures 15.4 and 15.5 result in the ratio values shown in the last column of Figure 15.6.

The impacts of the transactions incorporated into Figure 15.6 have had a significant impact on some of the ratios for EWS Farms. We encourage the reader to explore transactions that affect the various financial statements and the resulting impact on the financial ratios. Figure 5.9 in Chapter 5 shows the ratio values for EWS Farms using the financial statement values shown in Figure 15.1.

Ratios can aid in summarizing masses of financial data into meaningful management information. Table 15.1 uses the current ratio and working capital, both of which use the same information in their calculations, to show the difference between ratios and number values.

The current ratio retains the same value for every set of sample values for current assets and liabilities, while working capital changes values for each set of asset and liability values. The nature of the working capital calculation indicates that you should also know farm size to determine what working capital really means for any individual operation. It is theoretically possible that all of the values in Table 15.1 came from the same size operation, but not very likely. This simple example illustrates that even standardized ratios and measures should be used with full knowledge of how to interpret them and knowing the limitations that apply for each measure.

Percent Crop Revenue	100%	Percent Livestock Revenue	100%	Percent Gov. Payments	100%
Percent Cost of Production - Crops	100%	Percent cost of Production - Livestock	100%		

Balance Sheet

	Assets Beginning	Assets Ending		Liabilities Beginning	Liabilities Ending
Cash on Hand	1,500	27,591	Accounts Payable (Exp)	2,000	2,000
Crops Held for Feed (Exp)	8,400	8,400	Accrued Interest (Exp)	12,540	18,189
Crops Held for Sale (Inc)	200,000	170,000	Current Principal	11,908	22,959
Market Livestock (Inc)	0	0	Other Current Liability (Exp)	10,000	10,000
Other Current Assets (Inc)	15,000	15,000	Short Term Notes Payable (Exp)	0	0
Cash Invt Growing Crops (Exp)	0	0	Other Current Liab. (Not Adj.)	0	0
Supplies&Prepaid Exp. (Exp)	10,000	10,000	Def. Tax on Current Assets	0	0
Total Current Assets	234,900	230,991	Operating Loan Carryover	0	0
Non-Current Assets			**Total Current Liab.**	36,448	53,148
Mach. & Equipment	325,000	382,500	**Non-Current Liabilities**		
Breeding Livestock	0	0	Prin. on T.D. & C.L.	255,399	322,440
Real Estate (Land, Bldgs, Impr)	1,040,000	1,038,000	Total Business Liab.	291,847	375,588
Total Business Assets	1,599,900	1,651,491	**Business Net Worth**	1,308,053	1,275,904
			Change in Equity From Beginning to End of Year		(32,149)

Cash Flow Statement

		Inflows			OutFlows
Crop Sales & Net Insurance Payments		319,920	Operating Expenses	No Interest >	278,020
Mrkt & Cull Livestock Sales		0	Other Cash Business Expense		0
Government Payments		38,505	Cash Int. Exp. - T.D. & C.L.*		12,540
Other Cash Business Income		30,000	Cash Int. Exp. - Operating	7%	4,865
Operating Loan Proceeds	50.0%	139,010	Loan Prin. Payments - T.D. & C.L.		11,908
Loan Proceeds Capital Assets		90,000	Breeding Livestock Asset Purchases		0
Non-Business Inflows/Revenue			Mach & Equip & Real Estate Purchase		100,000
Other Nonfarm Inflows, net of taxes		0	Owner withdrawals		45,000
Other Nonfarm Inflows, net of taxes		0	Cash Taxes Paid (Income & SS)		0
Total Cash Inflows		$617,434	Other Cash Outflows (Not Expenses)		0
				Subtotal	$452,333
			Operating Loan Prin. Payments		$139,010
			Total Cash Outflows		$591,343
			Annual Net Cash Flow (never < zero)		27,591

**T.D. = Term Debt, C.L. = Capital Lease*

Income Statement - Accrual Adj.

	Income
Cash Income (adj. for cull lvstk sales)	$388,424
Non-Cash Income Adjustments	(30,000)
Non-Cash Income (Raised Brdg Lvstk)	0
Capital Gain/Loss on Breeding Lvstk (Net)	0
Gross Revenue	$358,424
	Expense
Cash Expense (Excluding Interest)	278,020
Non-Cash Feed Inventory Adjustment	0
Other Non-Cash Non-Interest Expense	0
Depreciation (Land, Bldgs, Equip.)	44,500
Total Operating Expense	322,520
Cash Int. Exp. - T.D. & C.L.	12,540
Cash Int. Exp. - Operating	4,865
Non-Cash Interest Expense	5,649
Total Expense	$345,574
Net Business Income From Operations	12,851
Net Business Income	12,851
Income & SS Taxes (Cash & Non-Cash)	0
Net Income	$12,851

Statement of Owner Equity

Beginning Net Worth (Cost/Mrkt)		1,308,053
Net Income	+	12,851
Non-Business Cash Inflows	+	0
Owner Withdrawals (Cash)	-	45,000
Asset Valuation Change or Cont./Distrib.	+/-	$0
Calculated Ending Net Worth	=	1,275,904
Reported Ending Net Worth (Cost/Mrkt)		1,275,904
Discrepancy		0

FIGURE 15.5 Financial statements showing the impact of selling $30,000 of Crops Held for Sale. Includes transactions shown in Figure 15.4.

	Beginning	Ending	Fig. 15.4 & 15.5 Transactions Ending
Liquidity			
Current Ratio	6.44	6.61	4.35
Working Capital	$198,452	$204,544	$177,844
Solvency			
Debt/Asset Ratio	18.24%	0.178	0.227
Equity/Asset Ratio	81.76%	0.822	0.773
Debt/Equity Ratio	0.22	0.216	0.294
Profitability		Ending	Ending
Rate of Return on Business Assets		1.32%	0.67%
Rate of Return on Business Equity		0.32%	–0.94%
Operating Profit Margin Ratio		0.06	0.03
Net Business Income		$29,151	$12,851
Repayment Capacity			
Term Debt and Capital Lease Coverage Ratio		1.28	1.02
Capital Replacement and Term Debt Repayment Margin		$6,742	$442
Financial Efficiency			
Asset Turnover Ratio		0.23	0.22
Operating Expense Ratio		0.78	0.78
Depreciation Expense Ratio		0.10	0.12
Interest Expense Ratio		0.05	0.06
Net Farm Income From Operations Ratio		0.08	0.04
Check Sum		100.00%	100.00%

FIGURE 15.6 Ratio value comparison before and after the transaction shown in Figures 15.4 and 15.5.

TABLE 15.1
Example Calculations for the Current Ratio and Working Capital

Current Assets	Current Liabilities	Current Ratio	Working Capital
$10,000	$7,500	1.333	$2,500
$50,000	$37,500	1.333	$12,500
$100,000	$75,000	1.333	$25,000
$200,000	$150,000	1.333	$50,000
$500,000	$375,000	1.333	$125,000
$1,000,000	$750,000	1.333	$250,000

Notes: Current Ratio = Current Asset divided by Current Liabilities and Working Capital = Current Assets minus Current Liabilities.

Ratios and measures along with the procedures to calculate these measures are provided by the accounting profession for most major industries, and can be found at the FASB Web site (http://www.fasb.org). However, they do not provide standardized procedures for the agricultural industry. Lending institutions for agriculture have proliferated literally hundreds of ratios and measures now used for agricultural financial analysis.

After the farm financial crisis of the mid-1980s, the Farm Financial Standards Council (FFSC) was formed to address this problem and standardize data requirements and formulas for calculating financial ratios. Chapter 5 discussed these standards and procedures. The objective of this standardization was to increase the quality of the financial data used for analysis and reduce the number of poor lending decisions and farm and ranch failures resulting from using poor financial information. A portion of this effort resulted in the promotion of a selected set of ratios and measures called the Sweet Sixteen.

There is no single measure that tells the entire story about an operation. The sixteen ratios and measures recommended by the FFSC are by no means an exhaustive list. Other ratios and measures can be calculated if quality data is available and a particular ratio provides more or better information than one of the Sweet Sixteen.

The initial recommendations of the FFSC were made in 1997. Adoption of these recommendations has been slow, to say the least. There is still little or no standardization in the way financial statements are prepared, what set of ratios and measures are calculated from the financial statements, or in the formulas used to calculate the financial ratios. With that as a disclaimer, benchmarks have been developed for agricultural financial analysis ratios. Figure 15.7 lists benchmarks that have been developed over time through empirical research on financial data sets for agricultural operations. Financial benchmarks for agricultural operations are almost always derived by averaging the actual performance data from a large group of smaller operations. Figure 15.7 also includes the Sweet Sixteen ratios from EWS Farms for comparison purposes. The benchmarks show ranges of green, yellow, and red, a typical procedure used to visually provide a good to poor financial analysis indication. Examination of Figure 15.7 shows the results vary depending on the way the benchmarks were calculated, the type of farms and ranches included in the data sets used to derive the benchmarks, and in part due to the interpretation as to the value necessary to meet a green, yellow, red system of financial indicators.

A single set of benchmarks cannot be applied across all types of agricultural operations. A dryland grain operation harvesting its production once a year will have very different ratios than a dairy or confinement hog operation. These benchmarks should only be used as guides, not as absolute evaluation criteria, and benchmark comparisons should only be made using results from similar operations.

A very brief interpretation and discussion of each of the sixteen ratios and measures is given below. More detail is available from the references at the end of this chapter. While you read about each ratio or measure, use the Sweet Sixteen Ratio Analyzer tool to view the impact on the financial ratios from changes in selected financial data. Figure 15.8 shows the ratio analyzer page for the liquidity measures for EWS Farms. The user is able to enter their own values for beginning and end-

Generally Better If	Financial Statement Ratios and Measures	EWS Case Farm		Green = Sound Financially Yellow = Caution Red = Take Immediate Action					
				University of Vermont Extension[1]			Purdue University[2]		
	Liquidity	Beginning	Ending	Green	Yellow	Red	Green	Yellow	Red
Larger	Current Ratio	6.44	6.61	>2	1 to 2	<1	>1.50	1.00 to 1.50	<1.00
Larger	Working Capital	$198.452	$204.544				Compare to Business Expense; Depends on Scope of operation		
	Solvency								
Smaller	Debt/Asset Ratio	18.24%	17.77%	<30%	30% to 60%	<60%	<20%	20% to 60%	>60%
	Debt to Asset (mostly rented/leased land)						<30%	30% to 70%	>70%
Larger	Equity/Asset Ratio	81.76%	82.23%	>70%	40% to 70%	<40%	>80%	40% to 80%	<40%
	Equity to Asset (mostly rented/leased land)						>70%	30% to 70%	<30%
Smaller	Debt/Equity Ratio	0.22	0.22	<43%	43% to 150%	>150%	<25%	25% to 150%	>150%
	Debt to Equity (mostly rented/leased land)						<42%	42% to 230%	>230%
	Profitability								
Larger	Rate of Return on Business Assets		1.32%	>5%	1% to 5%	<1%	>5%	1% to 5%	<1%
	Rate of Return on Business Assets (mostly leased land)						>11%	3% to 11%	<3%
Larger	Rate of Return on Business Equity		32.00%	>10%	5% to 10%	<5%	Compare to other non-farm investments; look trends		
Larger	Operating Profit Margin Ratio		5.83%	>35%	20% to 35%	<20%	>25%	10% to 25%	<10%
Larger	Net Business Income		$29.151						
	Repayment Capacity								
Larger	Term Debt and Capital Lease Coverage Ratio		1.28	>135%	110% to 135%	<110%	>150%	110% to 150%	<110%
Larger	Capital Replacement and Term Debt Repayment Margin		$6,742						
	Financial Efficiency								
Larger	Asset Turnover Ratio		0.2260	>40%	20% to 40%	<20%	Depends on type of operation & owned versus lease		
Smaller	Operating Expense Ratio		77.57%	<60%	60% to 80%	>80%	<55%	55% to 65%	>65%
	Operating Expense Ratio (mostly leased land)						<65%	65% to 75%	>75%
Smaller	Depreciation Expense Ratio		9.63%	<10%	10% to 20%	>20%	Compare to capital replacement and term debt repament margin		
Smaller	Interest Expense Ratio		4.67%	<10%	10% to 20%	>20%	<10%	10% to 20%	>20%
Larger	Net Farm Income From Operations Ratio		8.13%	>20%	10% to 20%	<10%	Look at trends; varies with cyclical nature of agriculture prices & incomes		

[1] Developed by: Rick Wackernagel, Dennis Kauppila, and Glenn Rogers. University of Vermont Extension, 1998
[2] Purdue University: http://fbfm.ace.uiuc.edu/cooperators/PDF/fin%20 char/financialbenchmarks.pdf: Modified from Daivd Kohl
[3] Purdue University. *Interpreting Financial Performance Measures*. William Edwards, 1998
[4] University of Minnesota. Summary FinBin report. Average of All farms in Database from 1993 to 2004.

FIGURE 15.7 Benchmark ratios and measures derived from studies of empirical data.

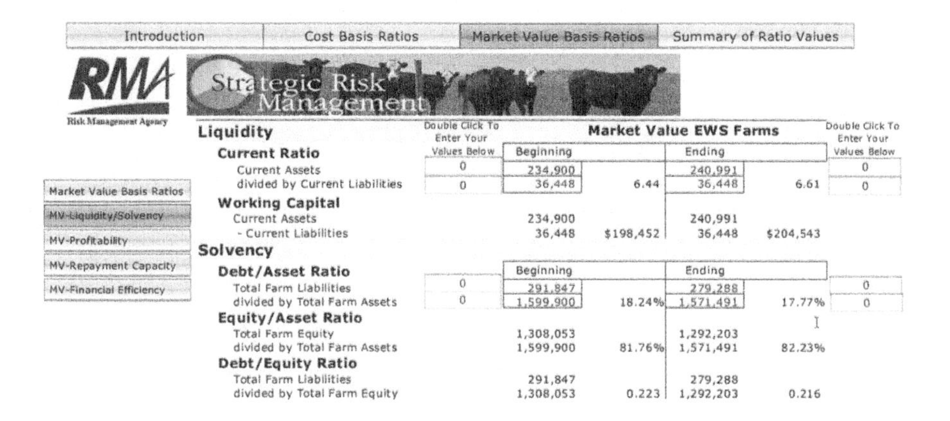

Introduction
Cost Basis Ratios
Market Value Basis Ratios
Summary of Ratio Values
RMA
Risk Management Agency
Strategic Risk Management
Market Value Basis Ratios
MV-Liquidity/Solvency
MV-Profitability
MV-Repayment Capacity
MV-Financial Efficiency
Liquidity
Double Click To Enter Your Values Below
Market Value EWS Farms
Double Click To Enter Your Values Below
Beginning
Ending
Current Ratio
Current Assets 0 234,900 240,991 0
divided by Current Liabilities 0 36,448 6.44 36,448 6.61 0
Working Capital
Current Assets 234,900 240,991
- Current Liabilities 36,448 $198,452 36,448 $204,543
Solvency
Debt/Asset Ratio
Beginning
Ending
Total Farm Liabilities 0 291,847 279,288 0
divided by Total Farm Assets 0 1,599,900 18.24% 1,571,491 17.77% 0
Equity/Asset Ratio
Total Farm Equity 1,308,053 1,292,203
divided by Total Farm Assets 1,599,900 81.76% 1,571,491 82.23%
Debt/Equity Ratio
Total Farm Liabilities 291,847 279,288
divided by Total Farm Equity 1,308,053 0.223 1,292,203 0.216

FIGURE 15.8 Sweet Sixteen Ratio Analyzer Tool

ing data necessary to calculate all of the ratios for their own operation by stepping through the menus provided in this tool.

15.3.1 Liquidity Measures

Liquidity is measured by the Current Ratio and a calculation called Working Capital. Liquidity measures the ability of a farm business to meet financial obligations as they come due in the ordinary course of business, without disrupting the normal operations of the business.

15.3.1.1 Current Ratio

The benchmark data listed in Figure 15.7 shows the low end value of 1.0 with the high end value ranging from greater than 1.50 to greater than 2.00. The value of the Current Ratio for the case farm calculated from the beginning Balance Sheet is 6.44 and 6.61 from the ending Balance Sheet. The current ratio is well within the green range and shows improvement during the year. For the case farm, the ending value indicates there is $6.61 of current assets for each $1.00 of current liability at year end.

15.3.1.2 Working Capital

Working capital for the case farm also improved during the year, ending at $204,544, and is large enough to meet all of the current liabilities coming due in the next 12 months. The necessary value of working capital will depend on the type and size of the operation and that is why there are no benchmarks listed for working capital in Figure 15.7. Working capital can also be evaluated by comparing it with total operating expense for the year. EWS Farms working capital is not large enough to cover current liabilities and projected cash operating needs for the coming year. The income statement for EWS Farms, Figure 15.1, lists total operating expense as $312,520. Both the producer and lender can use this value and the working capital value to help plan lines of credit for the coming year.

15.3.2 Solvency Measures

There are three solvency measures in the Sweet Sixteen, the debt-to-asset, debt-to-equity, and equity-to-asset ratios. Solvency measures the amount of borrowed capital, debt, leasing commitments, and other expense obligations used by a business relative to the amount of owner equity invested in the business. Debt capital is interest bearing and/or has a date by which it must be paid. Therefore, solvency measures provide

- an indication of the firm's ability to repay all financial obligations if all assets were sold (for the valuation indicated on the balance sheet), and
- an indication of the ability to continue operations as a viable business after a financial adversity, such as several years of drought, which can result in increased debt and reduced equity.

Solvency measures are used as a measure of bankruptcy. A firm is bankrupt if its total liabilities are greater than its total assets. Solvency measures are also used by lenders as a measure of their risk in advancing money for operating loans or the purchase of capital assets. Solvency measures are significantly affected by including or excluding contingent tax liabilities.

15.3.2.1 Debt-to-Asset Ratio

The ending case farm value for the debt-to-asset ratio, 17.77, is in the green range on all of the benchmarks listed. The debt-to-asset ratio compares total farm debt obligations owed against the value of total farm assets. This ratio expresses the proportion of total farm assets owed to creditors. The ratio value shows the creditors' claims against the assets of a business. A larger ratio value indicates greater financial risk exposure for the farm business.

15.3.2.2 Equity-to-Asset Ratio

EWS Farms value from the ending Balance Sheet is 82.23%. This value is also in the green range on all the benchmark criteria listed. As a check sum for accuracy, the debt-to-asset and equity-to-asset ratios should always add to 100%. This is due to the accounting equation: Net Worth (Equity) = Total Assets − Total Liabilities. Alternatively, Total Assets + Total Liabilities = Net Worth. The higher the value of the ratio, the more total capital has been supplied by the owner(s).

15.3.2.3 Debt-to-Equity Ratio

The case farm value from the ending Balance Sheet is 0.22. An interpretation of this value is that for every dollar of equity, there is $.22 of debt on the operation. This value is also in the green range on all the benchmark criteria listed in Figure 15.7. Larger ratio values indicate more total capital is been supplied by the creditors and less by the owner(s).

As discussed in Chapter 5, these three ratios can be calculated using either the cost or market value approach to value farm assets. If the market value approach is used to value farm assets, then deferred taxes that may be incurred on asset sales should be included as liabilities as they may have very large impacts on these ratio values.

These ratios are most meaningful for comparisons between farms when the market value approach is used to value farm assets. However, due to the impact of fluctuations in market values of farm assets, the ratios are most meaningful for comparisons between accounting periods for an individual farm operation when the cost approach is used to value farm assets.

15.3.3 Profitability Measures

Profitability measures the extent to which a business generates a profit from the use of land, labor, management, and capital. Three ratio values and one dollar value, Net Business (Farm) Income, are used to measure profitability. Producers seldom prepare an Income Statement, which provides the Net Business Income measure of profitability. Net Farm Income, listed on the Schedule F tax return, is typically substituted for Net Business Income from an Income Statement. This substitution casts considerable suspicion on the value of the Profitability Measures calculated using a tax schedule net income value.

15.3.3.1 Rate of Return on Farm Assets (ROA)

The case farm value for this ratio is 1.32%. This value falls in the yellow range for the benchmarks listed in Figure 15.7. This ratio measures the rate of return on farm assets and is often used as a key measure of profitability. Larger ratio values indicate more profitable farming operations.

15.3.3.2 Rate of Return on Farm Equity (ROE)

The case farm value for ROE is 0.32% and falls in the red category of the benchmarks provided. This ratio measures the rate of return on equity capital employed in the farm business. Larger values indicate the equity capital is being used effectively to generate higher returns to the owner's investment in the business. This ratio can be used to compare with the rate of return on alternative investments if the equity was pulled out of the farm business and invested elsewhere. Care must be taken that the after-tax dollar values of equity used in another investment are calculated correctly.

15.3.3.3 Operating Profit Margin Ratio

The case farm value for this ratio is 5.83%, which puts it in the red category on the benchmark criteria listed. This ratio measures profitability in terms of return per dollar of gross revenue. A farm business has two ways to increase profits—by increasing the profit per unit produced or by increasing the volume of production, if the business is profitable. If the asset turnover ratio is multiplied by the operating profit margin ratio, the result is the rate of return on assets.

15.3.3.4 Net Business (Farm) Income (NFI)

Net farm income for the case farm is $29,151. There are no benchmarks for this measure. Like the Working Capital value, it is necessary to consider many other factors. Empirical data is available showing net farm income when averaged across all farms in a data set, but caution should be used to make sure the type and size of the operations in the data set are similar to the farm operation being analyzed.

NFI is the return to three factors of production—unpaid labor and management and owner equity (capital). Net farm income can be allocated to each of these factors of production for comparison purposes with alternative investments. For example, if the farm owner could get a job off the farm earning a salary of $30,000 per year, then the return to management and owner equity is a negative $849 ($30,000 – $29,151). Alternatively, if the owner would be happy with a 5% return on his equity, $65,403 (0.05 × 1,308,053) using beginning net worth on the case farm, the residual value is a negative $36,252 as a return to labor and management. Note that deferred tax liabilities are not listed on the Balance Sheets used in this example and would affect these calculations.

15.3.4 Repayment Capacity Ratios and Measures

Repayment capacity measures the ability of a borrower to repay term farm debt from farm and nonfarm income. Principal payments on term loans must come from net income (with depreciation added back) after owner withdrawals, income, and Social Security taxes.

15.3.4.1 Term Debt and Capital Lease Coverage Ratio

The case farm value for this ratio is 1.28, meaning that for every dollar of scheduled term debt and capital lease payments required there is $1.28 available to make those payments. This is a yellow rating for the operation using the benchmark criteria listed in Figure 15.7. This ratio provides a measure of the ability of the borrower to cover all term debt and capital lease payments. Larger ratio values indicate an increased ability to cover these payments.

15.3.4.2 Capital Replacement and Term Debt Repayment Margin

This is not a ratio but a dollar value calculation. The case farm value is $6,742. No benchmark data is listed for this value indicating that the dollar value alone, while informative, is not enough to establish a benchmark.

This measure enables borrowers and lenders to evaluate the ability of the operation to generate funds necessary to repay debts with maturity dates longer than one year and to replace capital assets. It also enables users to evaluate the ability to acquire capital or service additional debt and to evaluate the risk margin for capital replacement and debt service. This measure assumes that credit obtained for current-year operating expenses will be repaid in one year as a result of the normal operation of the business. Unpaid operating debt from a prior period should exclude lines of credit and debt for livestock purchased in that period for sale in the current period (if part of the normal course of business).

15.3.5 Financial Efficiency Ratios

Just as the name suggests, financial efficiency measures the efficiency with which a business uses its assets to generate gross revenues and the effectiveness of production, purchasing, pricing, financing, and marketing decisions. Financial efficiency ratios are essential a division of the expenses incurred on an operation into three categories:

- Operating expenses
- Depreciation (capital equipment replacement) expense
- Interest (financing) expense

Collectively, these three ratios are the portion of every dollar generated used to cover the particular expense. The net farm income from operations ratio is the portion of every dollar generated that is left after covering the three expense categories. As a cross-check for accuracy, the three operating expense ratios and the net farm income from operations ratio should sum to 100%.

15.3.5.1 Asset Turnover Ratio

EWS Farms value for this ratio is 0.2260 and one of the benchmark criteria listed in Figure 15.7 indicates this is in the yellow range. Another indicates that you cannot determine adequate benchmarks using averages for different types and sizes of farms. An example of this would be comparing a dryland grain farm that has a once-a-year production cycle to a dairy operation that has a daily production cycle.

The asset turnover ratio is a measure of how efficiently farm assets are being used to generate revenue. A farm business has two ways to increase profits:

- Increase the profit per unit produced
- Increase the volume of production (if the business is profitable)

A relationship exists among the rate of return on farm assets, the asset turnover ratio, and the operating profit margin ratio. If the asset turnover ratio is multiplied by the operating profit margin ratio, the result is the rate of return on farm assets. Consequently, the same asset valuation approach should be used to calculate the asset turnover ratio and the rate of return on farm assets. The higher the ratio, the more efficiently assets are being used to generate revenue.

15.3.5.2 Operating Expense Ratios

EWS Farms value for the Operating Expense Ratio is 77.57% indicating that approximately 78% of every dollar of income generated on the operation is used to pay operating expenses, fuel, repairs, fertilizer, cash labor, etc. Another 9.63% of every dollar is used to cover the depreciation expense for this operation. Remember, depreciation is a noncash expense. Interest expense on this operation requires another 4.67% of every dollar of revenue generated. For this particular year, 8.13% of each dollar of income generated went to Net Farm Income from Operations. As with other case farm ratio values, the benchmark criteria used results in different ratings for this set of case farm ratio values. These four ratios reflect the relationship of expense and income categories to gross revenues. The sum of the first three expresses total farm expenses per dollar of gross revenue and the sum of all four ratios should equal 100%.

15.4 CREDIT SCORING MODELS

As discussed in Chapter 5, credit scoring models are an attempt to incorporate the combined effect of all the financial information available into one number. This number is then ranked using a predetermined scale to get an objective evaluation of whether or not a business is credit worthy. These models can be very useful to the extent they correctly incorporate all relevant information about a particular business operation. A simple credit scoring model based on the Sweet Sixteen is included in the EWS Farms RDFinancial files. However, this model does not include character attributes. Find out if your lender uses a credit scoring model on the financial information you provide and what formation is included in the model. Your financial tactical planning may get a substantial boost if you are aware of all the tools your lender uses to evaluate your credit worthiness.

15.5 SUMMARY

Financial analysis provides the ability to evaluate the implications of a wide array of risk management decisions. For the analysis to be accurate, all four of the recommended financial statements must be prepared and reconciled. This process helps assure financial information used to prepare the ratios and measures is valid and that the values for the Sweet Sixteen are correct. Preparation of all four of the financial statements can be a tedious process, but will provide the information necessary for sound financial strategic and tactical planning.

REFERENCES

Boehlje, M., C. Dobbins, and A. Miller. 2000. *Farm business management for the 21st century.* Purdue Extension, ID-237, pp. 1–11, West Lafayette, IN: Purdue Extension, Purdue University, August.

Ellinger, P. 1997a. Ratio analysis: Look among the numbers. *Doane's AgLender*, November issue.

Ellinger, P. 1997b. Ratio analysis: Look behind the numbers, *Doane's AgLender*, July issue.

Farm Financial Standards Council (FFSC). 1997. *Financial guidelines for agricultural producers: Recommendations of the Farm Financial Standards Council*, http://www.ffsc.org/Order_Form.pdf.

Financial Accounting Standards Board (FASB), http://www.fasb.org.

Langemeier, M. R. 2004. *Financial ratios used in financial management*, MF-270. Kansas State University Agricultural Experiment Station and Cooperative Extension Service, October.

Miller, A., M. Boehlje, and C. Dobbins. 2001. *Key financial performance measures for farm general managers.* Purdue Extension, ID-243. West Lafayette, IN: Purdue University.

FURTHER READING

Ellinger, P. 1998. *Ratio Analysis: Comparative Analysis: Guidelines for Liquidity and Solvency Measures*, Ag Lender, March.

16 Human and Institutional Risk

John P. Hewlett and Dana Hoag

CONTENTS

Our lives improve only when we take chances—and the first and most difficult risk we can take is to be honest with ourselves.

—Walter Anderson

Human and institutional risks represent the last two of the five sources of identified risk. These risks are serious threats to most organizations and enterprises but are seldom given the attention they should receive. As such, these sources of risk may well represent the risks that most seriously threaten the long-term survival of today's farm businesses. This is especially true where human and institutional risks are not as easily measured or managed when compared to production, marketing, or financial risks.

The term *human resource management* usually brings employee management to mind. Human resources certainly include employees; however, human resources are much more than that. Human resources, in the broad sense used here, are the people involved in an organization either directly or indirectly. This will include the owner and the owner's immediate family and heirs. If hired management is involved, it includes the managers and their immediate families, even if those immediate family members are not directly involved in the operation. The list also includes full-time and part-time hired employees, contractors, and other service providers. Beyond that, it includes employees' immediate families, as they may influence the quality and quantity of the service provided to the organization. Indirect or secondary service providers such as bankers, insurance sales representatives, ditch riders, and U.S. Forest Service agents may also be human resources involved in an operation.

In contrast, institutional resources cover two broad aspects of business management: policy and legal dimensions. Policy dimensions include federal crop insurance, loan programs, price supports, grazing policies, tax waivers, interest support programs, international trade, foreign subsidies, global competition, and other assistance policies and programs. The legal dimensions include structural issues (forms of business ownership), estate transfer issues, powers of attorney, contract obligations, tort liability, statutory obligations, and food safety.

Nearly all dimensions of institutional resources influence or have consequences for other business resources. For example, the financial decision to borrow capital creates a contractual obligation to repay the funds. This transaction affects both financial and institutional resources. Institutional resources also can change independently. Tax codes, government regulations, support programs, and so forth, can change independently from the conditions faced by a farm or ranch manager. In this way, the institutional resources represent a source of uncertainty for the operator.

Three important human and institutional sources of risk are:

- Government intervention: Federal, state, and local governments increasingly, and sometimes unpredictably, exercise control over farms and ranches, from labor issues to the environment. At the same time, federal policy makers are reducing massive programs, like price and income supports, in favor of the market. Reforms in farm support place more of the risk management burden on the shoulders of agricultural managers.
- Market complexity: Increased global competition, the changing structure of agriculture, new marketing alternatives, technological advances, and increasingly available information individually and collectively increase the pressures for managers to make sound business decisions.
- Risk connections: Decisions in one area of an agricultural enterprise increasingly affect other aspects of the operation. For example, marketing risk management easily translates into increasing pressures to better manage financial resources when lenders exercise increasing scrutiny and set higher standards for financial performance. (Adapted from U.S. Department of Agriculture, Risk Management Agency [USDA-RMA], 1997.)

Effective risk management involves anticipating any possible difficulties and making plans to reduce potential impacts (Singer, 2002). This applies to both human and institutional issues. As found in Chapter 8, agricultural managers did not rate human or institutional sources of risk very high when asked to rank their leading sources of risk. Given how profoundly these sources of risk can affect a business when things go wrong, this is surprising. The findings of Tversky and Kahneman (1992), well-known researchers in the psychology of human attitudes toward risk and uncertainty, may help to explain this phenomenon. The eight factors affecting a person's psychology toward risk (see text box) work against addressing human or institutional risks relative to the other types of risk. In addition, many agricultural operations are managed as family owned and operated or closely held organizations, where human resources and associated risks are more challenging to manage.

16.1 HUMAN RESOURCE RISK

The risk in human resources stems from the fact that managers are uncertain about whether or not people involved in the system will deliver as they agreed. Human resource uncertainty can come from many sources or combinations of sources, including sickness, disease, injury, death, relationship problems, divorce, lack of proper training, failure to recognize the importance of a task, poor communication, lack of qualifications, carelessness, inconsistent performance, and poor work ethic.

Experienced managers recognize that not all employees have equal abilities. Individuals vary by depth of experience, tenure with an organization, attitude, physical capability, analytical skills, credentials, willingness to put in extra hours, desire to be promoted, etc. These variations across individuals make employees less substitutable for each other; therefore, where one or more individuals are impaired or unable to complete a task, the cost of replacement may be greater than the cost of the original resource. This serves to further increase the uncertainty associated with human resources and the level of services they provide.

Observations about the Psychology of Risk

People tend to ignore the fact that runs of good and bad luck tend to regress to the mean over time.

Emotion can damage the ability to decide rationally.

Humans often do not possess all information necessary to decide in an economically rational manner.

Human choice is often based on inadequate sampling (one's own experience is not representative).

Humans tend to be loss averse more than risk adverse.

Humans tend to overestimate low-probability/high-drama risks and underestimate high-frequency/low-drama risk.

The manner in which questions about a risk are framed can influence human attitudes about that risk.

Obtaining more information on certain risks tends to promote a willingness to take those risks. (Tversky and Kahneman, 1992)

The uncertainty associated with the human and institutional dimensions of an operation brings about many risks. Sometimes the consequences of these risks can be estimated, thereby making selection of a risk management strategy easier. At other times, consequences are completely discounted (not considered or ignored) by managers, leaving them and their business activities completely exposed to risk.

Determining the correct response to identified risks is not always easy. Notably, agricultural managers tend to know more about traditional management practices, for example, production management, marketing management, and so forth. In one of the three studies about risk sources in Chapter 8, producers ranked liability insurance and government farm programs high, but these were not even ranked in the other two studies. Where human resources and the institutional dimensions of most agricultural operations are often not as aggressively managed as other dimensions of the business, a more in-depth look at the alternatives available may reveal options previously unknown to many managers.

16.1.1 Sources of Human Resource Risk

Human resources are unique in that they cannot be stored for later use. Labor is necessary to utilize almost all other resources on a farm or ranch (Jones, 2002). In addition, human resources are adaptable in the face of risk. In other words, people are capable of developing strategies or systems for reducing risk, possibly even benefiting from the exposure.

Human resource risk has many facets in the typical business. Characterizing human resource risk involves both the quantity and quality of the resource. Human resource quantity will change over time due to any number of events such as:

- Limited hours of work due to employee illness. Where a sick leave policy does not allow for employee recuperation away from the job site, this can also lead to additional employees becoming ill.
- Death or illness of employee family members.
- Failure to start work on time or leaving work early.
- Injury from job-related sources or from personal activities off the job.
- Lack of an employer policy regarding employee absence from work or having a policy that is too liberal to achieve the employer's goals.
- Childbirth or adoption and associated childcare responsibilities.
- Problems with children at school requiring employee attention, and insurance plans and the level of corresponding medical care, which may lead to prolonged employee absence from the workplace.
- Lack of access to adequate childcare services.
- Changes in marital status of the employee or his family members.
- Lack of substitute labor.
- Changes in employment status of the employee's family members.
- Changes in physical capability due to advancing age or lack of physical conditioning.
- Inability to perform job duties due to substance abuse (either on or off the job) or fatigue.
- Quality of human resource performance may vary for many reasons, some of which may include one or more of the following:

- Preoccupation with nonjob concerns, such as family member illness, marital strife, or other relationship difficulties.
- Lack of an employee reward system that emphasizes output quality.
- Failure to provide a healthy work environment, i.e., adequate access to hydration, latrine facilities, rest breaks, etc.
- Failure to adequately understand employer goals for the work assigned.
- Variation in work output due to advancing age or lack of physical conditioning.
- Inability to maintain focus due to substance abuse (either on or off the job) or fatigue.
- Lack of initiative due to poor supervisor attitude or management style.
- Any other factors that reduce the employee's sense of job satisfaction, such as lack of a system for advancement, failure to reward higher levels of employee performance by increases in pay or nonmonetary benefits, etc.

Employees are affected differently by the same risk. What constitutes an adequate incentive system for one worker my not motivate another. Further, some individuals will not perform well in a particular job no matter what the incentive or working environment. This is true where an employee's personality and likes and dislikes are not matched with the work to be accomplished. Of course this is not necessarily the responsibility of the employer; however, since the best work is performed by happy, healthy employees, the employer should match employees with the work they prefer whenever possible.

Organizational structure can be a source of institutional risk. Autocratic management styles are often a source of strife in a small business. Ambiguous assignments, not knowing who supervises who and in what circumstances, lack of a clear outline of job roles and responsibilities or expectations for work hours or overtime are also institutional shortcomings. This would be particularly true in a family business that is just beginning to add nonfamily laborers or in an organization experiencing growth.

A similar source of risk arises from managing (or not managing) management. In an agricultural business that employs multiple managers, a system should exist to provide feedback to management in general. Given that management sets the pace and provides direction for the rest of the business, some type of evaluation mechanism should be in place to assess how well this is being accomplished. Main street businesses often use an outside audit. Something similar could be arranged for agricultural firms. The risk of going off-course from poorly managed employees is a real risk to businesses looking to be successful over the long term.

16.1.2 Assessing Human Resource Risk

While we have looked at many human resource risks, we have not yet considered how we might evaluate the extent of those risks. That is to say, not all sources of human resource risk are equal in their potential to influence business performance.

Human Resource Risk Assessment

Source of Risk (describe)	Loss Frequency Score Low=1, Med=2, High=3	Loss Severity Score Low=1, Med=2, High=3	Human Resource Risk Index (LFS X LSS)
______	______	______	______
______	______	______	______
______	______	______	______
______	______	______	______
______	______	______	______
______	______	______	______

FIGURE 16.1 Human Resource Risk Assessment sheet (Risk Navigator Tool).

In order to determine which sources represent the greater threat, a system for comparing one source against another is necessary.

Human resource risk can be measured by two metrics: loss frequency and loss severity. Loss frequency is assessed by estimating the number of times business loss occurs from human resource events. For example, the number of days where production quotas are missed due to poor employee health would represent an estimate of loss frequency. Another example would be the estimated hours of work missed due to on-the-job accidents.

Loss severity is a measure of the extent of loss. For example, an estimate of the revenue lost due to employee absenteeism would represent loss severity due to human resource risk. Another estimate might be to calculate the cost to the business of employees arriving late or leaving early. Although each represents a cost to the business, it is unlikely that the two sources of human resource risk are equal.

The Risk Navigator Strategic Risk Management (SRM) tool called the Human Resource Risk Assessment sheet applies the techniques of assessing loss frequency and loss severity (Figure 16.1). This worksheet allows the user to describe the sources of human resource risk that are of greatest concern. A relative score—high, medium, or low—is assigned to loss frequency and loss severity for each of the sources listed. An index number is calculated from these two scores, allowing the manager to get a basic idea of which human resource risks represent the greatest threat to the business. This tool can help the manager decide which sources of human resource risk require controls, which might need attention in the future, and which represent little threat to continuing business activity.

Loss frequency and loss severity are helpful in estimating business losses resulting from the action or inaction of employees or service providers. Yet there is another type of human resource risk derived from the persons most integrally involved in operating the business, and this is called *direct human resource risk*. The death or exit of a key individual, especially in sole proprietorships and partnerships, represents a whole new source of risk. In cases where financing for the business has been extended based on a history with a key individual, such credit could be withdrawn when the individual is no longer involved in the operation. Because they almost always cause disruption to the business, direct human resource risks are particularly important for agricultural operations to consider in any risk management planning.

A final source of human resource risk exists for agricultural businesses that are also family businesses. Such a structure can and does offer strengths that other business forms do not share; however, the risks are greater. Although the sources of human resource risk noted above are all present in family businesses, these businesses are faced with an entirely different set of risks as well—risks to family relationships. This becomes especially true where family members are involved in supervisor–employee roles. It is one thing to manage human resource risks where arms-length relationships exist, but quite another if the manager is forced to consider firing Cousin Jim or promoting Nephew Jack over Aunt Mable. Such relationships make already difficult decisions more difficult and make it all the more important that risks be carefully considered. (One resource family businesses may find helpful can be found at Enterprising Rural Families: Making It Work, a Web site and course dedicated to helping family businesses become more successful. The site offers course materials organized in part to provide rural families with the tools and skills to deal with immediate challenges and build long-term resilience. http://eRuralFamilies.org,)

16.1.3 Tools for Managing Human Resource Risk

Risk management is usually focused on limiting the negative consequences of adverse events. This is no different when considering the controls available for managing human resource risk, with one exception. Human resources can be improved and shaped to operate in a manner that may either eliminate certain risks altogether or diminish their consequences should they occur. Information is the currency to facilitate risk reduction in this area. For example, training employees to follow certain safety practices or to ensure that safety shields are in place are two ways of reducing the uncertainty of workplace accidents.

A number of tools exist for managing human resource risks in a business; however, to discuss those tools in depth is beyond the scope of this text. A number of outside resources are available to assist with this and have been noted in the references at the end of this chapter. Of those listed, *Agricultural Help Wanted: Guidelines for Managing Agricultural Labor* (Rosenberg et al., 2002), is one of the most recent and most helpful in developing a sound risk management strategy for human resources. Others worth noting include: *Managing People on Your Farm* (Canadian Farm Business Management Council, 2002), *Managing the Multi-Generational Family Farm* (Owen and Howard, 1997), and *Farming with Neighbours: Preventing, Managing and Resolving Conflicts over Farming Practices—A Guide for Canadian Farmers* (Carter and Owen, 2002).

Human Resource Risk Controls

Strategic business planning
Estate and ownership succession planning
Modern personnel practices and human resource management
Health, disability, and life insurance
Prenuptial agreements and marriage counseling
Using management consultants (Musser and Patrick, 2002)

A basic introduction to the tools of human resource management should include a discussion of organizational structure, staffing the farm business, managing employee performance, the employee handbook, and family business issues. While other tools are certainly available, these address issues at the core of most agricultural businesses, regardless of their size or focus.

16.1.3.1 Organizational Structure

Structure is necessary for a well-functioning business. Lines of authority, as well as the other roles and responsibilities for individuals in the business, should be clearly delineated. Organizational charts outlining who reports to whom and lists of job duties can help in communicating this information to the people involved. The structure, however, may change over time with the addition of new employees, promotions, use of contractors and service providers, and through changes in the organization itself. As a result, information about the organization and its structure should be periodically reviewed and updated as necessary. This is especially true during times of rapid growth or when bringing nonfamily members into a previously family-only business.

16.1.3.2 Staffing the Farm Business

Staffing the business includes a broad range of activities from describing what jobs need to be performed to orient new employees. Specific tools that can be helpful with staffing include a draft job description, an interview outline and a process for orienting employees. Job descriptions are critical to obtaining the type and level of labor resources desired. Although this is standard practice in most of the business world, it is often neglected by agricultural businesses. One important dimension to completing a job description is thinking through exactly what duties the job will include and the expected time allocation for accomplishment. Getting this down on paper for everyone involved can be helpful in advertising the position, selecting a candidate, orienting the new employee, and evaluating employee performance. It is definitely worth the effort.

A consistent and routine employee orientation has at least two substantial benefits. First, it gives management the chance to clearly outline job duties and responsibilities and gives a new employee the chance to ask questions about anything that is unclear. Second, it provides an employee the greatest chance for success in the new position. When the employee is successful, the business as a whole is more likely to succeed. Although the process can take time away from the work at hand, getting a new employee comfortable with the job and discussing any questions about the position make the chances of a positive and ongoing relationship more likely.

16.1.3.3 Managing Employee Performance

Once the employee is off to a good start in a new position through proper orientation, the next critical step is to ensure he or she maintains a high level of performance over time. Satisfied employees tend to maintain higher levels of performance than those who are dissatisfied with their positions. Periodically evaluating an employee provides an opportunity to give feedback on the positive and negative aspects of his or her performance, and an opportunity to learn from the employee how the position might be improved. Again, this process will proceed more smoothly when based upon a formal, written job description.

A formal and consistently applied process for evaluating employee performance is another tool for managing human resource risk. Such a process helps to ensure that evaluations are completed similarly for both new and longer-tenured employees. In addition, using a formal instrument to facilitate the process both provides a written record and helps to cut down on communication problems associated with verbal-only evaluations.

16.1.3.4 Employee Handbook

A document organizing all policies and guidelines regarding employee management can be invaluable to the success of a business. Taking the time to assemble such a document can help to reduce uncertainty not only on the part of management and supervisors, but also on the part of employees. Reduced uncertainty leads to reduced human resource risk. Furthermore, having such a handbook available can help to reduce legal risks, as well.

Such a handbook need not be complicated but should address the following headings: employment status and records, employee benefits, timekeeping/payroll, work conditions and hours, leaves of absence, and employee conduct and discipline.

16.1.3.5 Family Business Issues

As noted earlier, family businesses are unique. While such businesses tend to be managed in a much less formal manner than others, the issues they face are no less complex. In fact, family businesses may well face more serious and difficult issues than many main street enterprises. Specific tools family business should consider for reducing human resource risks include a family business charter, a code of conduct, and a succession plan.

A family charter is a decision-making tool that sets out the values that are important to the family and the rules for resolving problems in a united and peaceful way. It is not intended to be a legal document (as is a Shareholders' Agreement) but a reference point that clearly sets out the criteria for the goals, management philosophy, shared ownership, working relationships, family relationships, and succession of the family business.

The code of conduct is a written statement that clearly points out what is important in conducting business affairs. It reminds members of their commitment to and interaction with each other. In addition, it reaffirms what is important in conducting family business affairs, both within the team and with others outside the business. It must be designed to suit a family's specific needs and situation (Hewlett, 2006).

A succession plan is a written document outlining how a family business will be passed from one generation to the next. It will address income, ownership, management, and information about how the process will be carried out. Open, honest communication in drafting the plan and throughout its implementation is critical. Furthermore, it is important to recognize that succession is a process, not an event, and that it involves many other lives outside of the principle parties involved. Well-planned and executed, a succession plan can impart an element of immortality to a successful family business. Poorly executed, succession can mean the destruction of many lives (Sharp, 2007; Tranel, 2007).

16.2 INSTITUTIONAL RISK

Where institutional and legal resources are important to the success and functioning of most agricultural businesses, uncertainty about how these resources will operate in the future introduces risk to the business environment. In addition, although controls are available for managing risks from the policy and legal arenas, these areas are subject to changing interpretation and political environment. As a result, contracts or government programs put in place to protect against certain adverse outcomes can be done away with or offer little protection with changes in interpretation or by legislative action. Therefore, this area of business uncertainty (perhaps more than other areas) must be constantly monitored and assessed to maintain some basic level of risk protection.

16.2.1 Business Structure

How a business is structured and its form of ownership can be a source of risk for the individuals involved in an operation. The various modes of business ownership, ranging from sole proprietorship to corporation, provide different levels of risk protection and liability exposure. Additionally, the level of income and property taxation varies by form of ownership and from state to state.

The form of business ownership also has many implications for planning estate transfers. In short, a professional should be consulted when considering the pros and cons associated with business ownership within a given state. Alternative scenarios should be worked out with someone familiar with the relevant laws and regulations, as well as with how agricultural business is normally conducted. Only then can the individuals involved make an informed decision regarding the level of risk they find acceptable and the necessary financial commitment to maintaining the chosen business organization.

16.2.2 Contractual Arrangements

Contractual arrangements are made more frequently in day-to-day management of an agricultural business than many realize. These arrangements span the gambit, ranging from financial agreements with lenders, to lease arrangements with neighbors, to work agreements with laborers, to life insurance contracts. In fact, legally binding agreements between two parties do not even have to be written down. Certain verbal agreements can be just as binding as a written contract signed by both parties.

In essence, any arrangement that spells out the promises made by each party in return for some compensation or benefit constitutes a contract. Such an agreement may also describe what should happen in the case of nonperformance; however, this is not strictly required. Again, how contracts are formed and the regulations regarding their enforcement vary from state to state. In order to minimize this type of risk, a professional legal consultant should be retained to review or draft formal legal agreements between parties before they are put into place.

16.2.3 Tort Liability

Tort liability comes about when there is a risk of harm to someone outside the operation. If a salesperson came onto the farm or ranch and was injured in some way due to negligence on the part of the farm or ranch owner, it would be considered a tort liability. Or if the farm or ranch owner knowingly discharged a pollutant from the operation that caused harm to people in the community, for example, smoke from burning straw, he or she may be accused of tort liability.

Often this type of risk is covered by a general farm liability policy; however, where the laws governing this type of liability vary from state to state, a qualified professional should be consulted to ensure adequate protection against this source of risk.

16.2.4 Statutory Compliance

Statutory compliance risk arises from failure to follow regulations governing the operation of the business. The regulations facing today's agricultural operations seem to be constantly increasing. In addition, they span many different aspects of the operation, from nondiscrimination in employee management to federal crop program compliance issues.

Many businesses now provide assistance to agricultural managers who manage this type of risk. Tax preparation services, pesticide and herbicide application companies, employee location services, etc., all provide services to agriculturalists willing to pay. In addition, they provide a mechanism for spreading some of the risks of statutory compliance. Keep in mind that although the sheer number of regulations governing how you run your farm or ranch may appear too intimidating to tackle, ignorance and the resulting noncompliance can result in severe fines or other penalties.

16.2.5 Public Policy

Government policies can influence agricultural risks in areas from health and safety regulations, to macroeconomic and environmental policies, to labor laws and trade rules. Compacts with the World Trade Organization or the North American Free Trade

Agreement, for example, have created, curtailed, and eliminated international markets, which account for 20% of U.S. coarse grains and more than 40% of cotton. Trade uncertainty has been very high over bovine spongiform encephalopathy (BSE), more commonly known as mad cow disease, in cattle. Canada, for example, lost around half of its income in cattle sales when BSE was discovered in their cattle population.

Macroeconomic policies can have sweeping effects on agricultural markets, too. Fleisher (1990) suggests that agriculture is more susceptible to macroeconomic policies, such as interest rates and foreign exchange rates, than are most other sectors. One reason is that there is more price volatility in agriculture—agricultural markets are quick to respond to shocks, but agricultural production reacts slowly. Gardner (2002) found that policies to reduce inflation keep real interest rates high and farm prices low. Efforts by the Federal Reserve to reduce inflation in the 1970s led to farm failures in 1979. According to Gardner, high real interest rates in the early 1980s caused one of the biggest crashes in U.S. agricultural history.

The federal farm bill is probably the program that affects agriculture the most. A new farm bill is passed about every five years or so. The last bill passed was called the Food, Conservation, and Energy Act of 2008. Price and income support payments now account for up to half of net farm income and significantly influence crop prices, whether they are supported by the program or not. In 2007, just 10% of participating farmers received over 60% of the money paid out (Environmental Working Group, 2007). It would seem that losing or reducing these supports would have huge impacts on farm income. For example, a $0.15 per-bushel decrease in corn price translates into a revenue loss of about $25 per acre on EWS Farms. This would reduce net farm income by about 50% and land value by about $300 to $600 per acre. Therefore, loss of deficiency payments could be devastating.

Federal price and income support programs are unquestionably an important source of risk to crop and livestock producers. However, research on the matter is mixed (Gardner, 2002). The Economic Research Service (Harwood et al., 1999) found that eliminating deficiency payments makes risk worse for most producers. However, they cite other studies that found that 29% of corn acres and 26% of wheat acres were in counties that were destabilized by deficiency payments. Another study found that the changes made to the 1996 farm bill (Federal Agricultural Improvement and Reform [FAIR]) did not lead to a permanent significant increase in the volatility of farm prices or revenues (Lence and Hayes, 2002). Some people suggest that variability even encourages farmers to try different production methods, making them more efficient.

All of these studies about farm policy risk are usually too little too late to help producers manage their risks. The 2008 farm bill offers producers yet another option to lure them away from traditional price and income support programs. The Average Crop Revenue Election (ACRE) program can be selected in place of traditional programs like direct, marketing loan, and countercyclical programs. Producers will have a difficult time deciding whether to take the ACRE option, because there are many complicated details to consider. Many are unknown because they are newly offered. Fortunately, many extension programs develop evaluation tools for programs like ACRE. One comprehensive program is available from the University of Illinois at Urbana-Champaign at www.farmdoc.uiuc.edu.

The point of policy risk is that changing government programs can be difficult for producers to track, evaluate, and participate in.

Legal and Environmental Risk Controls

Liability insurance
Nutrient management plans
Retaining legal counsel
Good neighbor relations
Road vehicle maintenance (Musser and Patrick, 2002)

16.3 SUMMARY COMMENTS

While producers don't appear to think of them as such, human and institutional risks can be as important as or more important than any of the other types of risk. Musser and Patrick suggest five ways to control risk (see boxed text), but there are many others to consider because human and institutional risks are multidimensional. Because human and institutional risks are unique to every operation, conventional risk management tools might not work. Nevertheless, taking time to think about what your risks are and applying commonsense tools could make the difference between a failed farm and a successful one.

REFERENCES

Canadian Farm Business Management Council. 2002. *Managing people on your farm*. Ontario: Canadian Farm Business Management Council.

Carter, J., and L. Owen. 2002. *Farming with neighbours: Preventing, managing and resolving conflicts over farming practices—A guide for Canadian farmers*, 2nd ed. Ontario: Canadian Farm Business Management Council.

Environmental Working Group. EWG Farm Bill 2007 Policy Analysis Database. Accessed June 9, 2009 from http://farm.ewg.org/sites/farmbill2007/.

Fleisher, B. 1990. *Agricultural risk management*. Boulder, CO: Lynne Rienner Publishers.

Gardner, B. 2002. Risk created by policy in agriculture. In *A comprehensive assessment of the role of risk in U.S. agriculture*, edited by R. Just and R. Pope, pp. 489–510. Boston: Kluwer Academic Publishers.

Harwood, J., R. Heifner, K. Coble, J. Perry, and A. Somwaru. 1999. *Managing risk in farming: Concepts, research, and analysis*. Washington, DC: U.S. Department of Agriculture Economic Research Service, Market and Trade Economics Division and Resource Economics Division. Rep. 774.

Hewlett, J. P. 2006. What is a family business charter and why would a family business need one? *Enterprising Rural Families* online newsletter. II-7. Accessed June 2, 2009 from http://eRuralFamilies.org.

Jones, R. 2002. Fundamentals of strategic and tactical business planning. A document prepared for the Management, Analysis, and Strategic Thinking (MAST) Program at Kansas State University.

Lence, S. H., and D. J. Hayes. 2002. U.S. farm policy and the volatility of commodity prices and farm revenues. *American Journal of Agricultural Economics* 84(2): 335–351.

Musser, W., and G. Patrick. 2002. How much does risk really matter to farmers? In *A comprehensive assessment of the role of risk in U.S. agriculture*, edited by R. Just and R. Pope, pp. 537–556. Boston: Kluwer Academic Publishers.

Owen, L., and W. Howard. 1997. *Managing the multi-generational family farm*. Ontario: Canadian Farm Business Management Council.

Rosenberg, H. R., R. Carkner, J. P. Hewlett, L. Owen, T. Teegerstrom, J. E. Tranel, and R. R. Weigel. 2002. *Agricultural help wanted: Guidelines for managing agricultural labor*. Western Farm Management Committee, University of Wyoming. Accessed June 2, 2009 from http://AgHelpWanted.org.

Sharp, R. L., J. P. Hewlett, and J. E. Tranel. 2007. *A Lasting Legacy Course 1*. Accessed June 2, 2009 from http://RightRisk.org

Singer, D. 2002. *Managing farm business risk*. Ontario: Canadian Farm Business Management Council.

Tranel, J. E., J. P. Hewlett, and R. L. Sharp. 2007. *A Lasting Legacy Course 2*. Accessed June 2, 2009 from http://RightRisk.org

Tversky, A., and D. Kahneman. 1992. Advances in prospect theory: Cumulative representation of uncertainty. *Journal of Risk and Uncertainty* 5(4): 297–323.

U.S. Department of Agriculture, Risk Management Agency (USDA-RMA). 1997. *Introduction to risk management: Understanding agricultural risks: Production, marketing, financial, legal, and human resources*. Washington, DC: U.S. Department of Agriculture, Risk Management Agency.

FURTHER READING

Williams, C. A, M. L. Smith, and P. C. Young. 1998. *Risk management and insurance*, 8th ed. New York: Irwin McGraw-Hill.

17 Ag Survivor

Dana Hoag and Jay Parsons

Ag Survivor is a simulation program developed by the RightRisk Education Team (www.RightRisk.org) to teach risk concepts and management strategies to agricultural producers in an experiential learning environment. This hands-on educational program lets producers test firsthand whether they are better off implementing newly learned risk management tools and strategies, like those learned in the Strategic Risk Management (SRM) program. This chapter has been included to encourage you to go to the Ag Survivor Web site and play the simulations as a way of solidifying some of the concepts that you have learned here. The simulations play like real life. If you take your time, you can use the scenario guides that come with each simulation to plot out your risk management strategy before playing. Build a payoff matrix. Practice using different decision rules. You may print the scenario guides straight from the Web site. An example scenario guide for EWS Farms is provided here to get you started. We wait until chapter 18 to show you how to use the Ag Survivor scenarios with SRM tools.

Ag Survivor can be played directly on the Web site (www.AgSurvivor.com) or at one of the RightRisk workshops (schedule shown at www.RightRisk.org). Realistic Ag Survivor scenarios present complicated risk management subject matter in an easily understood format. This is done by fully engaging users in a hands-on farm or ranch simulation. Participants use one of our eight different Ag Survivor scenarios that test everything from marketing to feed inventories.

We have worked hard to make the outcomes as realistic as possible. As a result, many of the scenarios take place in the Intermountain West, where we had access to the actual data required to develop the range of values and associated probabilities. However, we believe that producers from all regions of the country will be able to practice the risk management strategies using these scenarios. The following is a summary of these 10 scenarios, including a description of the different producer groups, risks, and lessons learned:

- **King Family Ranch:** This is a high mountain ranch typical of Colorado and Wyoming, with cattle and hay production. Participants manage hay inventory, pricing calves, and vaccinations.
- **Wheatfields:** In this scenario, participants manage 3,000 acres of wheat and 100 head of cattle. The decisions include managing hay inventory and pricing wheat and cattle.
- **Lazy U Ranch:** The Lazy U is a sheep and hay ranch reflective of the weather and risks associated with the Rocky Mountains in Colorado. Participants manage hay inventory, decide whether to supplement feed, or to hire extra labor during lambing, and how to price wool and lambs. Users also must decide whether to acquire more grazing land when rainfall is tight.

- **Public Lands:** This scenario was developed to represent cow/calf/hay operations along the Arizona and Utah border that rely heavily on grazing public lands. Managers oversee 650 mother cows and make decisions about feed, forward pricing yearlings, retained ownership, buying and selling cow/calf pairs, and leasing additional grazing resources.
- **Bar BQ Ranch:** The Bar BQ Ranch is much like the King Family ranch, but the decisions are centered on grazing intensity and revenue insurance. Participants in this simulation make decisions about feed and stocking rates, as well as whether they wish to buy Livestock Risk Protection Insurance.
- **EWS Farms:** EWS Farms is an irrigated and dryland farming operation, typical of northeastern Colorado. It is loosely based on the book case example, EWS Farms. The farm includes 500 irrigated acres of corn and 2,000 dryland acres (including 750 acres of wheat). The decisions include forward contracting and hedging wheat and corn.
- **Mountain View Farms:** Mountain View Farms is a dryland operation in southeastern Idaho that consists of 300 acres of contracted malt barley, 500 acres of feed barley, and 1,200 acres of winter wheat. Mountain View Farms also has a small, 100-head cattle operation. Decisions include forward pricing wheat, production of feed barley and calves, cross-hedging barley, and buying crop insurance for both wheat and barley.
- **Big Horn Basin Farms:** This farm grows 800 acres of crops, including 265 acres of malting barley, 250 acres of sugar beets, 105 acres of alfalfa, and 180 acres of corn in the Big Horn Basin of Wyoming. Risk management decisions include fertilization, crop insurance, the option to replant beets, and selling options for barley.
- **Oasis Ranch:** This cow/calf operation runs 800 cows on federal and private rangeland. The ranch is dealing with drought and whether it should purchase rainfall-based insurance.
- **High Plains Ranch:** This Rocky Mountain ranch runs 500 mother beef cows on 16,200 acres of public and private grazing lands. Ranchers also raise 350 acres of hay each year for winter feed.

Through friendly team competition, and interaction with trained RightRisk instructors, participants at workshops are able to experience a unique, interactive learning environment conducive to producing long-term growth in decision-making skills. RightRisk workshop participants are put in the role of a farm or ranch manager and asked to make decisions for the operation over a one- or multiyear time span in a simulated environment. This creates an energetic and interactive group learning experience, with many teachable moments. The discussions that take place within management teams as decisions are being made add tremendous value to the workshop experience.

Likewise, the discussions that take place between management teams as they compare their team performances create some interesting and lively conversations. Workshop participants are typically highly engaged and eager to repeat the experience.

Ag Survivor scenarios use real probabilities and impacts to depict risks. With this information, participants make risk management decisions for the operation as it progresses through several decision-making periods. In each period, a click of the button determines the random outcomes and moves the management team forward in time with updated prices, yield estimates, and inventories, among other information.

By the end of the simulation, each team or player will have progressed through one or more production years with the representative farm or ranch. Along the way, each management team will have experienced the same prices, weather, and other factors as the other management teams, but will have distinguished themselves by their unique set of input decisions. This provides the basis for a lively, slightly competitive conversation about the best outcomes.

Ag Survivor provides a platform for the participants to use several different measures to determine the best outcomes. For example, in a lot of ways, a single run through the simulation time period represents a combination of decision-making strategy and the luck of the draw. While a single simulation provides a good starting point, with the click of a button, the Ag Survivor software can rerun the model 100 times using random draws. This removes the element of luck. Outcomes can change with multiple iterations, and the results can add considerable depth to the conversation. Output from these 100 repeated runs include graphical measures such as bar graphs, and statistical measures such as mean, high, low, variance, and other factors as shown in Figure 17.1, which help the user to differentiate their luck from a single run from the overall worth of their strategy.

The results from Figure 17.1 are based on both random draws of the states of nature and the player's management decisions, which provides a measure of how well players manage risk. The histogram is a risk profile; how well you did will

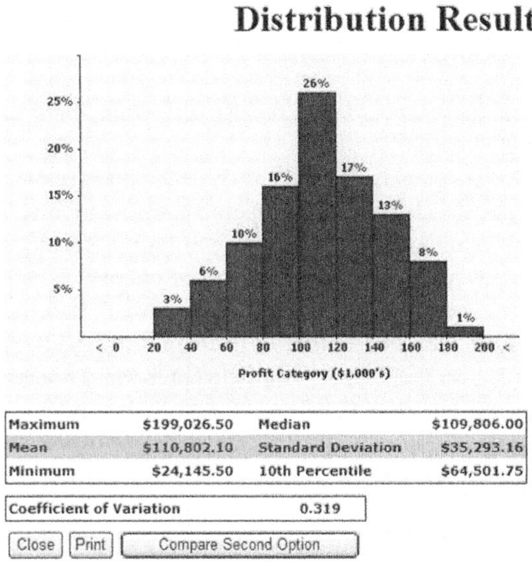

Distribution Results

Profit Category ($1,000's)	< 0	20	40	60	80	100	120	140	160	180	200 <
Percent		3%	6%	10%	16%	26%	17%	13%	8%	1%	

Maximum	$199,026.50	Median	$109,806.00
Mean	$110,802.10	Standard Deviation	$35,293.16
Minimum	$24,145.50	10th Percentile	$64,501.75

Coefficient of Variation	0.319

Close | Print | Compare Second Option

Decisions	
Year 1: Nov. 1 - Feb. 1	
Alternative Feeding	Yes
Buy or Sell Hay	
Year 1: Feb. 1 - June 1	
Add Labor	No
Sell Wool at Shearing	No
Year 1: June 1 - Sept. 1	
Forward Price Weaned Lambs to Feedlot	0
Acquire additional grazing	No
Sell Wool from Storage	No
Year 1: Sept. 1 - Nov. 1	
Acquire Fall Grazing Resource	No
Buy or Sell Hay	
Sell Wool from Storage	No
Retained Ownership of Lambs	0
Year 2: Nov. 1 - Feb. 1	
Alternative Feeding	No
Buy or Sell Hay	
Year 2: Feb. 1 - June 1	
Add Labor	No
Sell Wool at Shearing	No
Year 2: June 1 - Sept. 1	
Forward Price Weaned Lambs to Feedlot	0
Acquire additional grazing	No
Sell Wool from Storage	No
Year 2: Sept. 1 - Nov. 1	
Acquire Fall grazing resource	No
Buy or Sell Hay	
Sell Wool from Storage	No

FIGURE 17.1 Graphical result of 100 runs in the Lazy U Ranch Ag Survivor scenario.

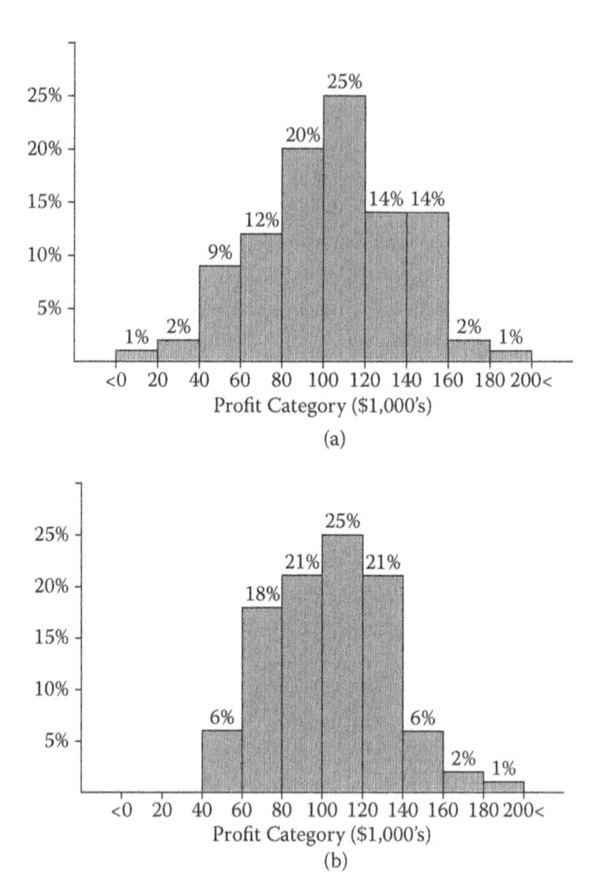

FIGURE 17.2 A comparison of two distributions for different playing styles: (a) no decisions; (b) forward price lambs.

depend on your risk personality. That is, you could look for the best mean, the best low, the smallest variance, or the highest high.

Players can also use Ag Survivor to quickly learn how different management options would have affected the results. By clicking on Compare Second Option you can change any decision made in the simulation, without having to replay the entire game. This provides a trial-and-error way to test how management options change distributions, as presented in Chapters 2 and 9. The new histogram reveals how the changes made an impact on risk. For example, contrast the two histograms, A and B, in Figure 17.2. Compare the change in the histogram when lambs are not forward priced, scenario A, to scenario B where they are. Note that the distribution with forward pricing is shifting to the right. There is a higher mean, higher low, and the same high with the new distribution. In this case, forward pricing is a good risk management tool.

To get the hang of the simulation exercises, it is best to first attend an Ag Survivor workshop. However, anyone can play any of the scenarios at any time by simply going to the Web site. The RightRisk Education Team encourages people to use the different scenarios, and playing the different scenarios provides good risk management skill practice. We do caution people, however, that these scenarios reflect the "typical," and therefore may not represent a producer's specific situation. Ag Survivor should only be used as a starting point to learn lessons and never be used as a decision tool. This book and other more detailed sources should be used for such detailed and personalized planning.

Each scenario has a corresponding scenario guide that contains supporting documentation. Scenario guides are also available on the Web site. We provide the EWS Farms scenario guide since it is related to the book's case farm.

How Much Risk Is Right for You?

Scenario Guide (January 2007) #SG-05-05

EWS FARMS

A. Sprague, J. Pritchett, J. Parsons, D. Hoag, and J. Deering

EWS Farms is representative of an irrigated and dryland farming operation in northeastern Colorado. Irrigated corn, dryland corn, and dryland wheat are the three enterprises on this farm. Production practices, costs of production, market prices, production yields, and other information are based on data from the region in order to provide a realistic setting. The probabilities of risk events were also calculated using actual data where available. Slight modifications were sometimes made to maintain the workability and realism of the game.

The farm includes 500 irrigated acres and 2,000 dryland acres raising 750 acres of dryland wheat and 500 acres of dryland corn annually in a three-year rotation. Irrigated corn is grown on the remaining 500 irrigated acres. Production costs for the three enterprises include direct cash costs to the operation excluding factor payments to land, generally calculated as a percentage return to land value. These costs include the cash labor, maintenance, and replacement costs associated with a typical operation of this size. Average expected yields for the three enterprises are 35 bu/acre (dryland wheat), 50 bu/acre (dryland corn), and 200 bu/acre (irrigated corn) resulting in a total expected yield of 26,250 bushels of wheat and 125,000 bushels of corn to be marketed each year.

Each year, the farm chooses from three options to market their grain crops: (1) forward contract (corn/wheat) to the elevator for harvest delivery, (2) hedge (corn/wheat) against the (December corn/September wheat) futures contracts for harvest settlement, (3) sell all grain inventory at the harvest cash price.

Dryland Wheat Production

Crop Acres	750
Average Annual Yield	35 Bushels per Acre
Production Costs	$96.77 per Acre
Average Market Price	$2.98 per Bushel
Average Yearly Production	26,250 Bushels
Annual Government Payment	$13,677

Dryland Corn Production

Crop Acres	500
Average Annual Yield	50 Bushels per Acre
Production Costs	$113.50 per Acre
Average Market Price	$2.25 per Bushel
Average Yearly Production	25,000 Bushels
Annual Government Payment	$8,017

Irrigated Corn Production

Crop Acres	500
Average Annual Yield	200 Bushels per Acre
Production Costs	$352.00 per Acre
Average Market Price	$2.25 per Bushel
Average Yearly Production	100,000 Bushels
Annual Government Payment	$16,811

Taking all of the above information into account, the farm expects to sell 125,000 bushels of corn and 26,250 bushels of wheat each year. Total revenues would equal net sales of $318,920 (after subtracting the landlord's share on leased ground) and $39,505 in government payments. Total farm operating expenses would total $329,274 leaving a total return to land of $29,151.

Expected Annual Net Farm Income

Expected Revenues		**Expected Expenses**	
Wheat	26,250 Bushels = $78,225	Wheat	750 Acres = $72,578
Dryland Corn	25,000 Bushels = $56,250	Dryland Corn	500 Acres = $56,750
Irrigated Corn	100,000 Bushels = $225,000	Irrigated Corn	500 Acres = $176,000
Lease Payments	–$39,555	Cost Share	–$32,760
Gov't Payments	$39,505	Other Costs	$56,706
Annual Total:	$358,425	Annual Total:	$329,274

Return to Land = $29,151

DECISIONS

Period 1	Risk and Probability of Occurrence	Impact
	Ending Stocks Report High Medium Low	• Wheat prices will decrease with a higher than expected ending stocks report. • Wheat prices will stay relatively unchanged with an ending stocks report at or near normal or expected. • Wheat prices will increase with a lower than expected ending stocks report.
	Risk Management Strategy Decisions	
	Decision 1: Forward Contract Wheat You have the opportunity to forward contract all or part of your expected production of winter wheat at the posted contract price for harvest delivery. Keep in mind that actual production may differ from expected. Contracts not filled by actual production will be settled by buying grain at harvest at the cash market price. **Decision 2: Hedge Wheat** A hedge may be placed against the posted September Kansas City Wheat Futures price in 5,000 bushel increments. A "round-turn" commission of $50 per contract will be charged to your account for this transaction. Basis at harvest (Cash – Futures) may be stronger or weaker than expected causing the realized price to differ from the expected market price. Variation in actual production from expected may cause you to be overhedged in poor production years.	
Period 2	**Risk and Probability of Occurrence**	**Impact**
	Wheat Seedings Report High Medium Low	• Wheat prices will decrease with a higher than expected reported planted acreage. • Wheat prices will remain relatively unchanged with an average wheat seedings report. • Wheat prices will dramatically increase with a significantly smaller than expected planted acreage report.
	Prospective Plantings High Medium Low	• Corn prices will decrease with a higher than expected prospective plantings report. • Corn prices will remain relatively steady with an average prospective plantings report. • Corn prices will dramatically increase with a significantly smaller than expected prospective plantings report.
	Risk Management Strategy Decisions	
	Decision 3: Forward Contract Wheat You have the opportunity to forward contract all or part of your expected production of winter wheat at the posted contract price for harvest delivery. Keep in mind that actual production may differ from expected. Contracts not filled by actual production will be settled by buying grain at harvest at the cash market price.	

	Risk Management Strategy Decisions	
	Decision 4: Hedge Wheat A hedge may be placed against the posted September Kansas City Wheat Futures price in 5,000 bushel increments. A "round-turn" commission of $50 per contract will be charged to your account for this transaction. Basis at harvest (Cash – Futures) may be stronger or weaker than expected causing the realized price to differ from the expected market price. Variation in actual production from expected may cause you to be overhedged in poor production years.	
	Decision 5: Forward Contract Corn You have the opportunity to forward contract all or part of your expected production of corn at the posted contract price for harvest delivery. Keep in mind that actual production may differ from expected. Contracts not filled by actual production will be settled by buying grain at harvest at the cash market price.	
	Decision 6: Hedge Corn A hedge may be placed against the posted December Corn futures contract in 5,000 bushel increments. A "round-turn" commission of $50 per contract will be charged to your account for this transaction. Basis at harvest (Cash – Futures) may be stronger or weaker than expected causing the realized price to differ from the expected market price. Variation in actual production from expected may cause you to be overhedged in poor production years.	
Period 3	**Risk and Probability of Occurrence**	**Impact**
	Crop Progress Report Excellent Good Poor	• Wheat prices will decrease with a better than expected crop progress report. • Wheat prices will remain relatively unchanged with a crop progress report in line with expectations. • Wheat prices will dramatically increase with a poorer than expected crop progress report.
	Cattle on Feed Far below Expectations In Line with Expectations Much Greater than Expected	• Corn prices will decrease with a higher than expected cattle on feed report because of decreased corn demand. • Corn prices will remain relatively steady with an average cattle on feed report. • Corn prices will dramatically increase with a significantly smaller than expected cattle on feed report.
	Risk Management Strategy Decisions	
	Decision 7: Forward Contract Wheat You have the opportunity to forward contract all or part of your expected production of winter wheat at the posted contract price for harvest delivery. Keep in mind that actual production may differ from expected. Contracts not filled by actual production will be settled by buying grain at harvest at the cash market price.	

<table>
<tr><td></td><td colspan="2">Risk Management Strategy Decisions</td></tr>
<tr><td></td><td colspan="2">Decision 8: Hedge Wheat
A hedge may be placed against the posted September Kansas City Wheat Futures price in 5,000 bushel increments. A "round-turn" commission of $50 per contract will be charged to your account for this transaction. Basis at harvest (Cash – Futures) may be stronger or weaker than expected causing the realized price to differ from the expected market price. Variation in actual production from expected may cause you to be overhedged in poor production years.</td></tr>
<tr><td></td><td colspan="2">Decision 9: Forward Contract Corn
You have the opportunity to forward contract all or part of your expected production of corn at the posted contract price for harvest delivery. Keep in mind that actual production may differ from expected. Contracts not filled by actual production will be settled by buying grain at harvest at the cash market price.</td></tr>
<tr><td></td><td colspan="2">Decision 10: Hedge Corn
A hedge may be placed against the posted December Corn futures contract in 5,000 bushel increments. A "round-turn" commission of $50 per contract will be charged to your account for this transaction. Basis at harvest (Cash – Futures) may be stronger or weaker than expected causing the realized price to differ from the expected market price. Variation in actual production from expected may cause you to be overhedged in poor production years.</td></tr>
<tr><td>Period 4</td><td>Risk and Probability of Occurrence</td><td>Impact</td></tr>
<tr><td></td><td>Crop Progress Report
Excellent
Good
Poor</td><td>• Corn prices will decrease with a better than expected crop progress report.
• Corn prices will remain relatively unchanged with a crop progress report in line with expectations.
• Corn prices will dramatically increase with a poorer than expected crop progress report.</td></tr>
<tr><td></td><td colspan="2">Risk Management Strategy Decisions</td></tr>
<tr><td></td><td colspan="2">Decision 11: Forward Contract Corn
You have the opportunity to forward contract all or part of your expected production of corn at the posted contract price for harvest delivery. Keep in mind that actual production may differ from expected. Contracts not filled by actual production will be settled by buying grain at harvest at the cash market price.</td></tr>
<tr><td></td><td colspan="2">Decision 12: Hedge Corn
A hedge may be placed against the posted December Corn futures contract in 5,000 bushel increments. A "round-turn" commission of $50 per contract will be charged to your account for this transaction. Basis at harvest (Cash – Futures) may be stronger or weaker than expected causing the realized price to differ from the expected market price. Variation in actual production from expected may cause you to be overhedged in poor production years.</td></tr>
<tr><td></td><td colspan="2">Game End</td></tr>
</table>

RightRisk™ is an innovative risk research and education program. It uses real-world farm and ranch settings and agricultural economics to help you understand and explore risk management decisions and evaluate the effects of those decisions. You will learn about your personal risk management style and build your decision-making skills.

RightRisk is not only a simulation model. You will have ongoing access to agricultural economists with expertise in risk management. The RightRisk Education Team consists of a team of researchers and extension specialists from eight Western states including Arizona, Colorado, Idaho, Montana, Nevada, Utah, Washington, and Wyoming.

For more information about RightRisk, please visit our Web site. There you can learn more about RightRisk, about risk and managing risks, how to contact resource people, and where and when up-coming RightRisk meetings will be held. Also, you can play RightRisk online!

18 Building Customized Risk Management Plans

Dana Hoag and Duane Griffith

CONTENTS

18.1 OBJECTIVES

The ten-step Risk Navigator Strategic Risk Management (SRM) process was designed to organize comprehensive, diverse, and complicated information into a simple format to manage risk. As you master these steps, you will start to find the SRM process more and more limiting. It won't allow you to do exactly what you want. Fortunately, there are very good software programs available to take that next step into customizing your risk management program.

Microsoft Excel™ spreadsheets will provide the basic platform for our economic analyses. You can mix and match SRM tools and your own customized Excel programs to develop a more meaningful overall plan. Spreadsheets can be simple or relatively complex and provide analysis of a particular activity or estimate results of a particular action. The vast majority of spreadsheets are deterministic. The term *deterministic* is used to indicate that when one value is changed, the spreadsheet makes the appropriate calculations and displays the new results. The new value is treated, at least with the spreadsheet, as if it is absolutely correct. For example, what is the effect on Returns over Variable Costs (ROVC) if wheat yield is changed from 40 bushels to 45 bushels? Replacing 40 with 45 bushels in the appropriate cell in a spreadsheet makes all necessary calculations and displays a new value for the ROVC calculation. Producers can explore the boundaries that various input values (fertilizer, repairs, fuel, seed, crop insurance, chemicals, labor, etc.) have on the ROVC, but every calculation is based on tedious trial and error, and results are not weighted by how likely they are to occur.

There are no models that give "the right answer," but more complex models can be used to enhance our understanding of the real world. Some of these models can include the variability of historical characteristics taken from past measures of historical data. Including risk turns deterministic models into stochastic models. The objective in this chapter is to introduce the reader to a variety of techniques and software programs that are useful for customizing risk management to include the stochastic nature inherent in agriculture. Specifically, we introduce the reader to Excel and two add-in programs specifically designed to make Excel capable of advanced risk management, @Risk™ and Simetar©. These programs can be used for everything from developing customized probabilities to full-blown simulations that use stochastic processes.

18.2 HOW DO YOU GO ABOUT BUILDING A STOCHASTIC RISK MODEL?

Using simulation to model financial risk usually results in some kind of a probability distribution for a particular outcome, such as net business income. The resulting probability distribution of net business income is a picture of the cumulative effect of the risk inherent in the production process. Stochastic models include the variability in the production process, to the best of our ability to build the risks into any model. Let's consider an example of how one might customize information.

A producer may enter the 45 bushels knowing full well that it is not a certainty. Alternatively, he could look at historical records if he wanted to determine the probability of getting a 45-bushel yield. The probability might be only 30%, for example. Rather than relying solely on a deterministic approach, the producer could incorporate a probability distribution into his estimation of ROVC. This approach can incorporate real-world factors (yields, rainfall, temperatures, availability of irrigation water, etc.) and enhance our understanding about the possible outcomes of model estimates. Rather than one answer, each measured outcome will have a distribution of possible answers based on the distribution of possible input values (yield in this case) for any given set of variables.

Figure 18.1 shows a distribution of dryland corn yields for Phillips County Colorado, where EWS Farms is located, using the historic data in Table 18.1. Historic yields are shown as a histogram and the fitted distribution is the bell-shaped curve overlain on the histogram. Using the historic data, the mean yield for dryland corn is 45.03 bushels, as shown in Figure 18.1. This information adds a great deal of understanding to the way we interpret our desired results, such as net business income.

Table 18.1 provides a sample of historic data for dryland corn yields. However, suppose you are interested in understanding and explaining when and why yields will be high or low. For this reason we have also included the Palmer Drought Severity Index (PDSI) in Table 18.1. PDSI is a measure of several factors that when combined provide a measure of how good or bad the weather was for a particular period of time. Inspecting the table values we can see that large negative

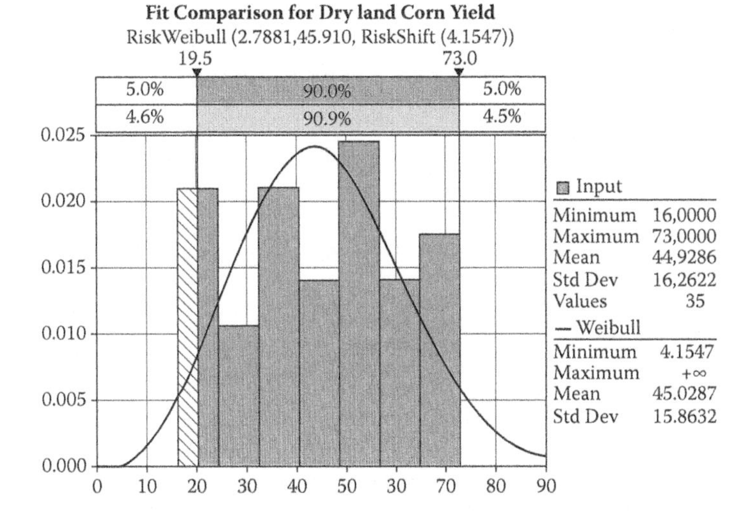

FIGURE 18.1 Corn yield distribution using the @Risk Fit Distribution Tool.

TABLE 18.1
Historic Values for Dryland Corn Yields and the Palmer Drought Severity Index (PDSI) in August in Weather Region 3 in Phillips County Colorado

Year	Dryland Corn Grain Yield	PDSI August	Year	Dryland Corn Grain Yield	PDSI August
1970	22	−1.7	1987	56	1.58
1971	30	−1.59	1988	57	−0.48
1972	29.5	0.48	1989	65	0.93
1973	34	3.17	1990	65	0.97
1974	21	−2.26	1991	53.5	0.86
1975	16	−0.97	1992	73	3.54
1976	19.5	−3.62	1993	52	2.52
1977	44	−1.61	1994	49.5	1.3
1978	19.5	−3.19	1995	39.5	5.65
1979	56	2.02	1996	63.5	3.41
1980	32	2.67	1997	64	1.52
1981	43	2.94	1998	67.5	1.44
1982	55	2.23	1999	73	1.67
1983	41.5	3.9	2000	34.5	−2.63
1984	37.5	2.65	2001	57.5	−1.72
1985	40	−0.76	2002	24	−4.95
1986	45	−2.27	2003	36	−2.23
			2004	56.5	−2.5
Correlation Coefficient, Dryland Corn Yields and PDSI for August					0.46478198

TABLE 18.2
Correlations for Multiple Crop Yields and Months for the PDSI for Phillips County Colorado, Using Monthly PDSI Values

Crop Name	Mar	Apr	May	Jun	Jul	Aug	Sep	Oct
WW Dryland on Fallow	0.4099	0.3583	**0.5151**	0.4271	0.4719	0.4123	0.4496	0.4463
WW Dryland Continuous	0.3139	0.2619	0.4038	0.3992	0.4458	0.3343	0.3795	**0.4820**
Corn Grain Dryland	0.1404	0.0518	0.0486	0.1859	0.3360	**0.4648**	0.4387	0.3539

PDSI values are associated with low corn yields. But what is the exact relationship between yields and the PDSI? Is PDSI the right measure to associate with corn yields or is there a better measure? Is August the correct month to select as having significant impact on corn yields or is another month a better indicator? Would the cumulative value of the PDSI over some period of time be a better measure than a single month? All these questions and more should be addressed when building stochastic models.

Table 18.2 explores the relationship between the PDSI and yields. Historic yields for several crops were correlated with values for several months of the PDSI. In this simple example, the highest correlation value for dryland corn is the month of August and this value is displayed in Table 18.1. Initially this suggests that the most important month for good weather for a corn crop grown at this location under dryland conditions is August. Other crops have different values for the correlation coefficient and they occur in different months.

In addition to developing corn yield PDSI correlation coefficients, we can fit a distribution using @Risk for the August PDSI as shown in Figure 18.2. These are just two examples about how information can and often needs to be customized to fit every unique situation.

18.3 ADVANCED MODELING SOFTWARE

Many programs are available to build simulation models. @Risk, used to generate Figures 18.1 and 18.2, is one of these programs. @Risk is a Microsoft Excel add-in. Add-ins like @Risk and Simetar© allow users to introduce stochastic analysis as the next step in model building. While the authors find these programs useful, this brief review is not intended as an endorsement of any of these commercially available software packages.

18.3.1 @Risk by Palisade Corporation (http://www.palisade.com/)

The @Risk add-in for Excel provides powerful simulation tools for risk analysis. While better measurements of expected results are obtained, the level of difficulty in collecting and processing the additional information necessary to build the advanced

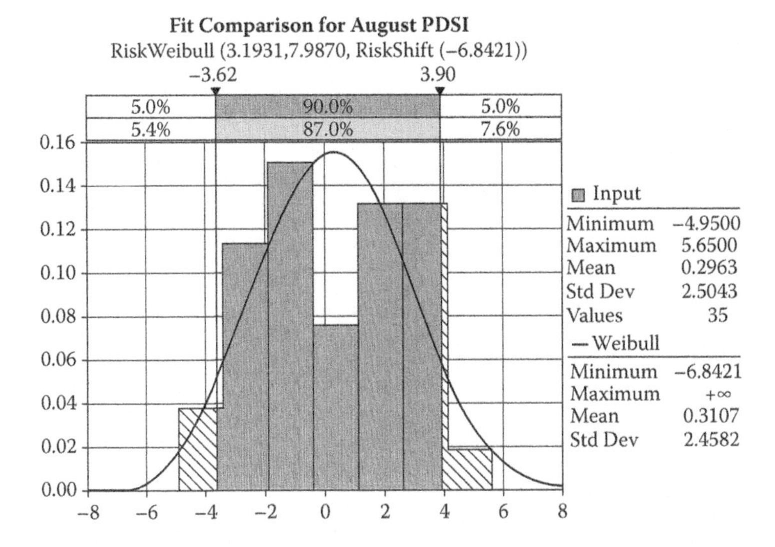

FIGURE 18.2 Fitted distribution for the August PDSI in northeastern Colorado, using Table 18.1 data.

models can increase substantially. Many tools and functions are available to enhance your risk management analysis. The type and number of tools and functions depends on the version of @Risk.

Crop yields as they relate to weather are a simple example. @Risk provides many functions that help build stochastic models and these functions can be applied to most historic data that could be used to build a stochastic model. The type of data and its characteristics will determine how it must be incorporated into a model, but @Risk provides several tools to manage raw data for this purpose.

The tools include fitting probability distributions to raw data, the ability to run simulations, stress testing, building correlation matrixes, and many others. After using the tools to develop model parameters, the variability of the raw data can be incorporated into the model using functions supplied by the @Risk software. These functions are similar to typical Excel functions such as Lookup, If Then, Min, Max, and Average, to name a few. These functions are entered in a manner similar to typical Excel functions with the distinct difference that they do not generate a single value for a particular cell. The value for any cell that uses that value is then also variable. @Risk has a tool that allows you to simulate the results of using many different input values a selected set of output values. This tool accumulates the selected output values in a probability distribution. This probability distribution for the selected result(s), based on using probabilities for historic data values, is then used to evaluate the overall result(s) the model was designed to analyze.

One of the many @Risk functions generated using the Fit Distribution tool in @ Risk is:

=RiskWeibull(3.1931,7.987,RiskShift(-6.8421),RiskName("Dataset #1"))

This function can then be entered into an Excel cell for use with simulations like that found in Figure 18.2. During a simulation, a random draw of a PDSI value from the RiskWeibull distribution is used with the correlation coefficient to generate a random yield. Associating expected yields with a detriment variable is only one possible step toward a stochastic model.

@Risk has been around a long time and is probably the leader in the industry. More information about the program can be obtained at the @Risk Web site along with a lot of other interesting information. Also at the Web site are links and references to several books and articles that have been written about using the software; these case studies provide an excellent opportunity for learning more about the program.

18.3.2 Simetar© (http://simetar.com)

Simetar©, which stands for Simulation & Econometrics to Analysis Risk, is an alternative to @Risk. While it is lesser known and has been around a shorter time, Simetar is an advanced program with many of the same, and some different, features. According to their Web site, more than 250 functions in Simetar can be categorized into seven groups: (1) simulating random variables, (2) statistical analyses and tests, (3) graphical analysis, (4) ranking risky alternatives, (5) data manipulation and analysis, (6) econometric modeling, and (7) forecasting. A complete description of the program, including links to dozens of examples, opportunities for training, and links to purchase the software, can be found at the Web site.

One particularly attractive feature of Simetar is that the Web site contains many examples that are oriented toward agriculture. One of those examples, "Best Demo," is shown in Figures 18.3a and 18.3b. The point of customizing your own plan is to develop information the way you want it to be. Therefore, you need to look at a lot of examples. In this case, the Simetar team has developed five result tables showing different risk measures from the first table that shows 50 data points collected for each of five different management plans, A through E. In this example, management plans are organized by mean in Table S2, standard deviation in S3, and so on.

18.4 IMPLEMENTING A SIMPLE STOCHASTIC MODEL

EWS Farms is targeting the productive capacity of their resource base to internally generate enough net income for family activities. One possible option is changing the enterprise mix. New crops included in the operation may be higher in value but they also may have more variability in prices due to volatile markets and fluctuating yields. Scoring big, occasionally, with high-valued crops may be offset by lower average net income. Diversification of crop mix is one strategy suggested to lower production and financial risks. However, from a financial/economic perspective, lowering production risks should not come at the expense of lowering average net income for the operation, unless the variability of net income is reduced accordingly. Another option is leasing more farmland.

Figure 18.4 shows the results of an @Risk simulation and for EWS Farms before and after adding 1,500 acres of irrigated crop and 1,500 acres of pasture

	A	B	C	D	E	F
1	Table S1. The raw data for 5 strategies, scenarios, or investments that need to be ranked.					
2						
3		A	B	C	D	E
4	1	22.74	37.30	18.65	10.58	39.56
5	2	16.95	21.85	10.93	9.37	29.91
6	3	23.41	39.09	19.55	10.69	40.68
7	4	17.58	23.55	11.78	9.48	30.97
	:	:		:	:	
51	48	19.09	27.58	13.79	9.79	33.49
52	49	17.08	22.21	11.10	9.39	30.13
53	50	23.82	40.19	20.09	10.75	41.37
54	Average	19.90	29.73	14.86	9.97	34.83
55	Std Dev	2.90	7.74	3.87	0.55	4.84
56	Min	14.33	14.87	7.44	9.08	25.55
57	Max	27.70	50.54	25.27	10.99	47.84
58						
59						
60	Table S2. Rank scenarios based on the MEAN (Xbar).					
61	Scenario	Mean	Xbar Rank			
62	A	19.90	−2			
63	B	29.73	−3			
64	C	14.86	−1			
65	D	9.97	0			
66	E	34.83	−4			
67						
68	Table S3. Rank scenarios based on the STANDARD DEVIATION (SD).					
69						
70	Scenario	Mean	SD Rank			
71	A	2.90	2			
72	B	7.74	5			
73	C	3.87	3			
74	D	0.55	1			
75	E	4.84	4			

FIGURE 18.3A Risk information computed by Simetar© in Best Demo (altered for presentation purposes).

	H	I	J	K	L	M
65	**Table S4. Rank scenarios based on the Mean Variance (MV).**					
66	**Scenario**	**Variance**	**Mean**	**MV Rank**		
67	A	8.422	19.90	1		
68						
69	B	59.893	29.73	5		
70	C	14.973	14.86	3		
71	D	0.298	9.97	4		
72	E	23.396	34.83	2		
73						
74						
75						
76	**Table S5. Rank scenarios based on the MINIMUM & MAXIMUM**					
77	**Scenario**	**Minimum**	**Min Rank**		**Maximum**	**Max Rank**
78	A	14.32714	3		27.70283	3
79	B	14.87238	2		50.54087	1
80	C	7.436189	5		25.27044	4
81	D	9.076193	4		10.98937	5
82	E	25.54524	1		47.83805	2
83						
84						
85						
86	**Table S6. Rank scenarios based on the Absolute**					
87	**COEFFICIENT OF VARIATION (CV).**					
88	**Scenario**	**Minimum**	**Std Dev**	**CV**	**CV Rank**	
89	A	19.90	2.902144	14.58553	3	
90	B	29.73	7.739049	26.03423	4	
91	C	14.86	3.869525	26.03424	5	
92	D	9.97	0.545843	5.473157	1	
93	E	34.83	4.836907	13.88758	2	

FIGURE 18.3B Risk information computed by Simetar© in Best Demo (altered for presentation purposes).

with leased land. For EWS Farms to lease extra land, they must also invest in more machinery and equipment. Irrigated corn and wheat are grown on the leased acres. Simulation indicates that the standard deviation of net farm income (risk) increased from $14,397 to $42,479, something that should be avoided. However, the mean net farm income increased from $35,372 to $194,698, shifting the distribution of net farm income to the right substantially. If the leased ground is added, there is only a 2.3% chance of getting a net farm income of less than $110,000. Without the leased ground, there is zero percent chance of getting a net farm income above $110,000. This appears to be an attractive strategy to change the profitability of the operation through on-farm activities to provide for additional family members.

Off-farm activities may also be considered as a way to relieve the expected financial stress on the operation. Surveys indicate that on average, only 12.2% of surveyed

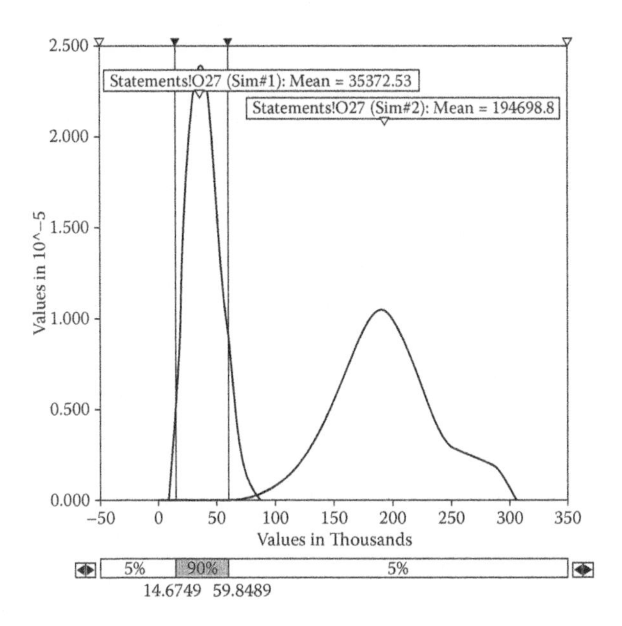

FIGURE 18.4 Simulation showing net business income from the Income Statement, with and without expansion of leased acres.

producers indicated off-farm investments were effective, while 14.7% indicated off-farm employment was effective (Coble, Knight, Patrick, and Baquet, 1999).

An off-farm strategy to reduce or eliminate financial stress focuses on the *cash flow* problems of the combined business and family. This strategy is fundamentally different than trying to change the productive capacity of the operation and increase *net farm income*. For a variety of reasons, this strategy may be a preferred, or the only alternative, for some producers. Financial information in Figure 5.5, tells us exactly what type of off-farm job to look for. Figure 5.5 indicates an Owner Withdrawal of $85,000. Increasing this by $15,000 (adjusting for the affects of long-term inflation) and using an Owner Withdrawal of $100,000, the Operating Loan Carryover increases to $47,409. Therefore, one or more of the children and or their spouses must get a job off-farm that will pay $47,409 = (32,409 + 15,000).

Figure 18.5 is a simulation showing, as you would expect, that net farm income does not change comparing before- and after-simulation results for the addition of inflows from off-farm sources—the distributions overlap perfectly. Off-farm inflows are not included on the Income Statement. Another simulation, not shown, indicates that annual net cash flow for the combined family and business will always be zero without added off-farm cash inflows of $40,000 and will have a mean value of $5,995 with $40,000 of added cash inflow from off-farm sources.

The off-farm inflow is entered on the inflow side of the Cash Flow Statement under the heading of Other Nonfarm Inflows and reduces the $47,409 Operating Loan Carryover to $7,409. The Cash Flow Statement, statement of owner equity (SOE), and the ending Balance Sheet are all affected by this change. Annual Net Cash Flow and the ending Cash on Hand is $0. The SOE now shows Non-Business

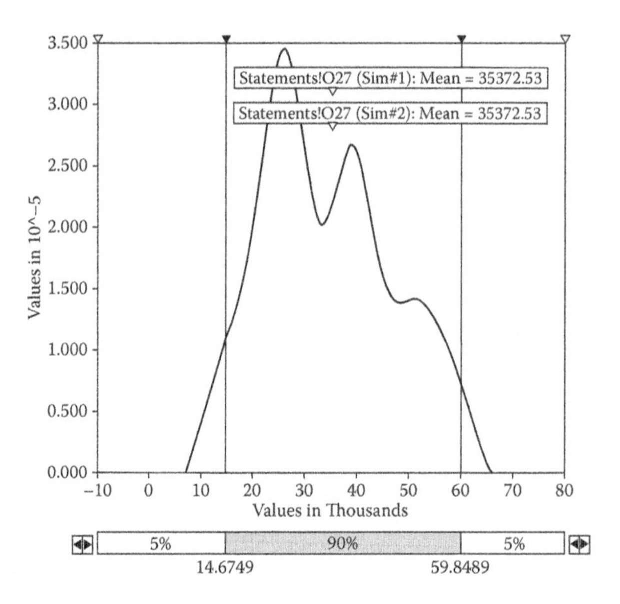

FIGURE 18.5 Simulation of net income with and without off-farm inflows and no expansion.

Cash Inflows and calculated and reported ending net worth are identical, maintaining a zero discrepancy. Net Farm Income does not change. While off-farm wages and salaries are income to the wage earner, they are not income to the farm business, they are simply inflows. The negative change in net worth reported on the Balance Sheet, –$30,849, is calculated as ($29,151 + $40,000 – $100,000). Another off-farm job of $7,409 or greater will eliminate the remaining operating loan carryover and provide a positive cash flow for the combined business and family financial statements.

18.5 RISK NAVIGATOR SRM LITE©

SRM Lite© was developed for fast, easy, and intuitive risk comparisons. It involves filling out two worksheets: the risk payoff worksheet and the graphical payoff worksheet. You have to use a bit of ingenuity to organize your information, but this Risk Navigator tool makes it easy to compare risks.

18.5.1 Using Ag Survivor

You can use Ag Survivor scenarios to test, apply, and evaluate different risk management rules because the program allows users to compare the results of changes quickly and easily. Several Fact Sheets to learn about decision analysis tools like those discussed in Chapter 11 are available at the RightRisk Web site, such as a safety first guide by Bastian and Hewlett (2004). One of those Fact Sheets shows how to use the scenarios in SRM Lite© (Hoag and Parsons, 2009). Let's see how it works.

The distribution for making no decisions in the Mountain View Farms barley scenario is presented in Figure 18.6. Note that the mean is about $243,000, with a minimum of (–$37,685) and a maximum of $392,173. Suppose I wanted to know whether to purchase crop insurance for malt barley. There are two opportunities to purchase insurance, as shown in the box titled "Decisions" in Figure 18.6. Mountain View Farms considers two types of insurance, multiperil crop insurance (MPCI) and revenue assurance (RA). In each case the optional malt barley supplement, plan B, is also purchased. Consult the scenario guide for details about this scenario or Hoag and Parsons for details.

As with many of the Ag Survivor scenarios, users can click Compare Second Option to recompute and redisplay the histogram with the new insurance options. The results of having MPCI are shown in Figure 18.7 and the results for RA are shown in Figure 18.8. There is a lot of information to be gained just by looking at how the histograms changed. But let's be a little more organized by putting the information into SRM Lite©.

Record the information from each of the three screens (Figures 18.5, 18.7, and 18.8) into the top portion of the risk payoff worksheet, as shown in Table 18.3. Next,

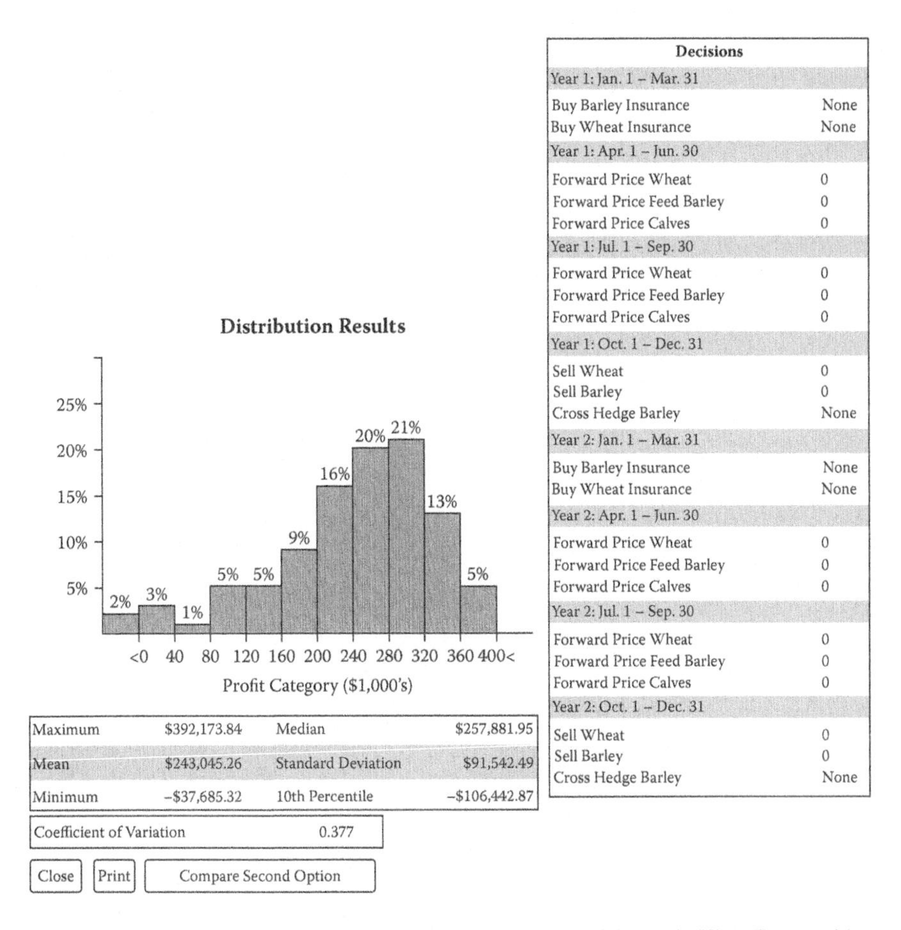

FIGURE 18.6 Payoff histogram from Ag Survivor Scenario, Mountain View Farms with no crop insurance. (From http://www.AgSurvivor.com.)

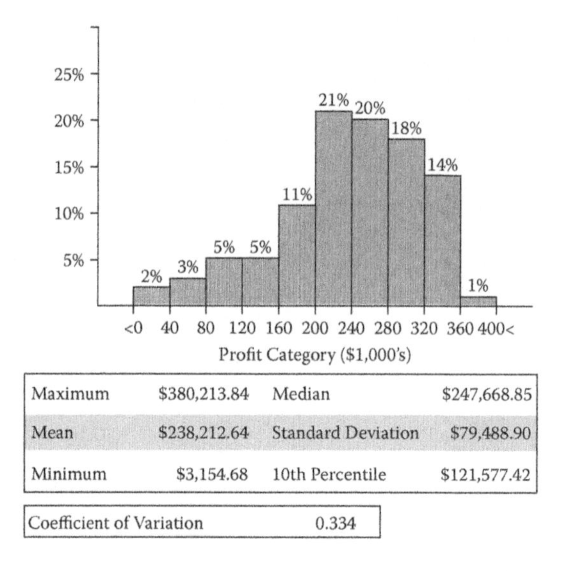

Maximum	$380,213.84	Median	$247,668.85
Mean	$238,212.64	Standard Deviation	$79,488.90
Minimum	$3,154.68	10th Percentile	$121,577.42

Coefficient of Variation	0.334

FIGURE 18.7 Payoff histogram from Ag Survivor Scenario, Mountain View Farms, with multiperil crop insurance and malt barley supplemental insurance. (From http://www.AgSurvivor.com.)

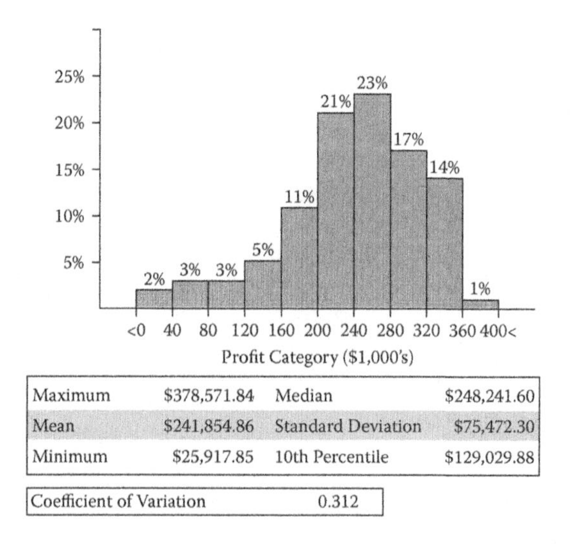

Maximum	$378,571.84	Median	$248,241.60
Mean	$241,854.86	Standard Deviation	$75,472.30
Minimum	$25,917.85	10th Percentile	$129,029.88

Coefficient of Variation	0.312

FIGURE 18.8 Payoff histogram from Ag Survivor Scenario, Mountain View Farms, with multiperil crop insurance and malt barley supplemental insurance. (From http:www.AgSurvivor.com.)

TABLE 18.3
Risk Payoff Worksheet for SRM Lite©, Hoag, 2009

Action Values	Management Action 1: No Insurance	Management Action 2: MPCI	Management Action 3: RA	Management Action 4
Maximum	392	380	378	
Minimum	−37	3	26	
Mean/Expected Value	243	238	242	
Standard Deviation	92	79	75	
Coefficient of Variation	.38	.33	.31	

Risk Management Techniques

Enter one value per row, using information from above

Maximum Expected Value	√		
Maximin			√
Maximax	√		
Minimize Variation (SD)			√
Minimize Relative Variation (CV)			√

Source: Dana Hoag and Jay Parsons. "Using Strategic Risk Management Lite (SRM Lite©): An Example on Mountain View Farms." LG_01_009, RightRisk Education Team Lesson Guide, LG_01_09, http://www.RightRisk.org, February 2009.

apply the risk management criterion in the bottom half of the table by checking the box that optimizes the criterion. In this case, no insurance maximizes expected value and the maximax criterion. Revenue assurance does best at reducing risk. MPCI is never preferred.

A graphical look at this same information can help. Enter the high, low, and average returns into the three rows below the graphical payoff worksheet as shown in Figure 18.9. Row one should contain information about the "no insurance" scenario, row two is about MPCI, and row three has the results for RA insurance. Draw a triangle on top of these values by connecting the high and low to the ends of the triangle and making the point of the triangle align with the mean. The height of the triangle is not important for now.

Now look at the three graphs and compare the aspects that you are interested in. For example, you can raise the minimum outcome from −$37,000 to about $3,000 by adopting MPCI. The cost of doing so is a drop of $5,000 in your mean and a lower high outcome. Would you give up an average of $5,000 per year to raise that worst outcome to $3,000? Revenue insurance presents a much better option. You can raise the minimum to $26,000 and it only costs you $1,000 thousand in lost mean. Choosing RA is not a slam dunk; some people would give up the $1,000 and accept a lower maximum, while others would not. The graphical analysis simply helps make it easy to compare.

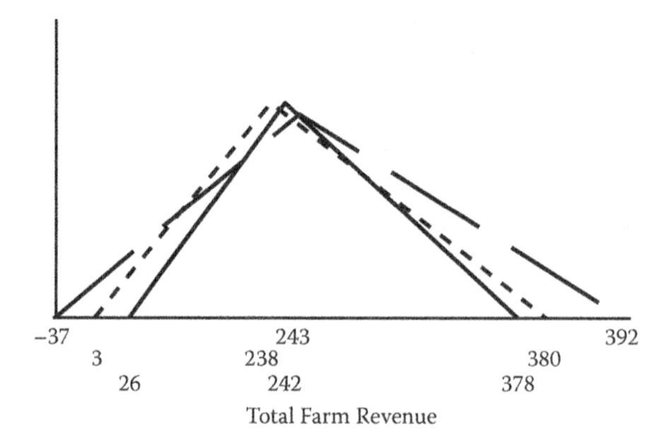

FIGURE 18.9 Graphical Payoff Worksheet for SRM Lite (From Hoag and Parsons, 2009).

18.5.2 Using Farm Data and Expert Opinion

Let's repeat the example used above with information that comes from a farm instead of a synthetic computer scenario; we create such an example based on the Mountain View Farms scenario in order to make our comparisons consistent (Parsons and Hoag, 2009). Suppose that barley yields in the last five years were 20, 37, 58, 44, and 59 bushels per acre. Using these yields, a person might attach a 40% chance of getting 58 bushels (since roughly 58 bushels was achieved 2 out of 5 years). Furthermore, you might decide that there is a 40% chance of 40 bushels per acre and 20% chance of getting 20 bushels. You also decide that you will get one of three prices, $2.55, $2.30, or $2.20 based on past experience. Combining these three prices with three yields gives me 9 possible outcomes, as shown in Table 18.4.

For simplicity, let's assume these 9 possibilities cover all outcomes. The last two columns are created when we combine the yield probability with an assumed 33% chance of each price. We assign a 33% probability on prices because we have no information about how likely one price is to occur over another. There is a 13%

TABLE 18.4
Yield and Price Possibilities for Barley Production on Mountain View Farms

Outcome	Yield	Probability	Price	Probability	Yield*Price	Probability
1	58	40%	2.55	33%	147.90	13%
2	58	40%	2.30	33%	133.40	13%
3	58	40%	2.20	33%	127.60	13%
4	40	40%	2.55	33%	102.00	13%
5	40	40%	2.30	33%	92.00	13%
6	40	40%	2.20	33%	88.00	13%
7	20	20%	2.55	33%	51.00	7%
8	20	20%	2.30	33%	46.00	7%
9	20	20%	2.20	33%	44.00	7%

TABLE 18.5
Crop Insurance Indemnities for MPCI and RA on Mountain View Farms

Outcome	Multiperil Crop Ins	Revenue Assurance
1	0	0
2	0	0
3	0	0
4	0	0
5	0	0.61
6	0	4.61
7	38.64	41.61
8	38.64	46.61
9	38.64	48.61

chance of getting a price × yield of $147.90, for example. We can then determine whether and by how much insurance would pay off for each outcome. As shown in Table 18.5, multiperil insurance only pays off starting in outcome 8 and revenue insurance would payoff starting with outcome 5.

Finally, we can develop a payoff matrix as shown in Table 18.6. The revenue for each column is the sum of market returns from Table 18.4 and the insurance indemnity from Table 18.5, less the insurance premium. That is, there is 13% chance for outcome 1. You would have earned $147.90 if you had not purchased insurance. You would earn a bit less with MPCI or RA. However, if you had chosen MPCI, for example, you would nearly double your income against no insurance, if outcome 9 occurred—$76.71 compared to $44.00.

Table 18.6 provides a complete payoff matrix like that shown in Figure 11.8. You have everything you need to use Risk Ranker in Navigator SRM or SRM Lite.

TABLE 18.6
Payoff Matrix for Yield, Price, and Insurance Indemnities on Mountain View Farms

Outcome	Probability	No Insurance	Multiperil Crop Ins	Revenue Assurance
1	13%	147.90	141.97	141.31
2	13%	133.40	127.47	126.81
3	13%	127.60	121.67	121.01
4	13%	102.00	96.07	95.41
5	13%	92.00	86.07	86.02
6	13%	88.00	82.07	86.02
7	7%	51.00	83.71	86.02
8	7%	46.00	78.71	86.02
9	7%	44.00	76.71	86.02

Try filling in the SRM Lite payoff worksheet or graphic payoff worksheet in this chapter.

REFERENCES

Bastian, C., and P. Hewlett. 2004. *Safety-first: A RightRisk™ lesson guide*. Accessed on June 9, 2009 from http://agecon.uwyo.edu/rightrisk/school/RR-L-1%20January%2004.pdf.

Best Demo. 2009. *Demonstration of Simetar© simulation, 622 MS Simulation Course*, Agricultural & Food Policy Center, Texas A&M University, College Station, TX, http://www.afpc.tamu.edu/courses/622/.

Coble, K. H., T. O. Knight, G. F. Patrick, and A. E. Baquet. 1999. *Crop producer risk management survey: Preliminary summary of selected data: A report from the Understanding Farmer Risk Management Decision Making and Educational Needs Research Project*. Information Report 99-001, Mississippi State University.

Hoag, D., and J. Parsons. 2009. *Using Strategic Risk Management Lite (SRM Lite©): An example on Mountain View Farms*. RightRisk Education Team Lesson Guide LG_01_09, http://www.RightRisk.org, February.

Parsons, J., and D. Hoag. 2009. *Researching barley insurance options: A RightRisk Lesson Guide with Mountain View Farms*. RightRisk Education Team Lesson Guide LG_09_07, http://www.RightRisk.org, February.

Index

D

F

G

H

N

O

P

R

S

T

U

V

W